Thomas Hillier

Diseases of children; a clinical treatise based on lectures delivered at the Hospital for Sick Children, London

Thomas Hillier

Diseases of children; a clinical treatise based on lectures delivered at the Hospital for Sick Children, London

ISBN/EAN: 9783337229115

Printed in Europe, USA, Canada, Australia, Japan

Cover: Foto ©berggeist007 / pixelio.de

More available books at **www.hansebooks.com**

DISEASES OF CHILDREN

A CLINICAL TREATISE

BASED ON LECTURES DELIVERED

AT THE

HOSPITAL FOR SICK CHILDREN, LONDON

BY

THOMAS HILLIER, M.D., LOND.

FELLOW OF THE ROYAL COLLEGE OF PHYSICIANS, PHYSICIAN TO THE HOSPITAL FOR
SICK CHILDREN, AND TO UNIVERSITY COLLEGE HOSPITAL, LONDON

LONDON
JAMES WALTON
BOOKSELLER AND PUBLISHER TO UNIVERSITY COLLEGE
137, GOWER STREET
1868

TO

SIR WILLIAM JENNER, Bart., M.D. F.R.S.,

PHYSICIAN IN ORDINARY TO HER MAJESTY THE QUEEN, AND TO HIS ROYAL HIGHNESS THE
PRINCE OF WALES, PROFESSOR OF CLINICAL MEDICINE IN UNIVERSITY
COLLEGE, LONDON.

This Work is Dedicated

IN TOKEN OF ADMIRATION, GRATITUDE, AND ESTEEM.

PREFACE.

THE original intention of the Author was to publish the Clinical Lectures, as delivered by him at intervals at the Hospital for Sick Children, Great Ormond Street, London.

As he proceeded, however, he determined to abandon the lecture-style and to expand and modify the material, so as to include the experience derived from later cases of disease, and from more recent publications on the several subjects under consideration.

The book now assumes the form of a series of short monographs, on the more important diseases which have come under treatment at a Hospital into which are admitted only children, between the ages of two and twelve years. There is but little reference to the diseases of new-born children, or of infants under two years of age. Those diseases which fall exclusively under the care of the surgeon are also omitted.

Until recently the opportunities for the clinical study of infantile maladies in this country have been very few. The establishment of the Hospital for Sick Children has served to direct attention to the necessity of affording facilities for this study.

Several of the general hospitals in London have

recently opened children's wards, and hospitals for children have been established in some of the larger provincial towns of the kingdom.

It appeared to the Author that although several valuable manuals on the diseases of childhood were in existence in England, there was room for a book treating clinically on the same subject; and he hopes that this volume may prove a useful contribution to the study of children's diseases.

32, QUEEN ANNE STREET,
    CAVENDISH SQUARE.

CONTENTS.

	PAGE

INTRODUCTION. — Importance of studying children's diseases.—Their frequency and fatality.—Their peculiarities.—Clinical examination of children. — Great importance of the family history.—Choice of medicines 1

PNEUMONIA.—Mortality from this disease.—Increased mortality from bronchitis of late years.—Three diseases included under this name in registers of mortality—Pulmonary collapse.—Broncho-pneumonia.—Anatomy.—Symptoms. — Course and duration.—Diagnosis. — Prognosis. — Treatment.— Case of broncho-pneumonia 8

LOBAR PNEUMONIA.—Ages of patients.—Anatomy.—Course and symptoms. — Critical days. — *Cerebral* pneumonia. — Complications.—Diagnosis.—Prognosis.—Treatment.—Cases . . . 20

PLEURISY. — Its frequency. — Statistics. — Symptoms. — Physical signs—Diagnosis.—Case of congenital diaphragmatic hernia.—Cases of empyema.—Thoracentesis.—Pleurisy and pericarditis.—Circumscribed empyema.—Modes of termination of Pleurisy.—Treatment 45

RICKETS.—Definition.—Relations to other diseases.—Causes.—Is it ever congenital ?—Changes in the osseous system.—Modes of growth of bone.—Case.—Deformities of skeleton.—Other symptoms.—Treatment.—Statistics of disease at the Hospital for Sick Children 78

PAGE

TUBERCULOSIS.—Number of organs liable to tubercle.—Frequency of tubercle.—Tubercle of bronchial glands.—Cases.—Mesenteric glands.—Tabes mesenterica.—Lymphatic glands enlarged from other causes.—Omental tubercle.—Diagnosis.—Temperature—Treatment.—Cases of tubercular peritonitis.—Cancer of kidney . 103

DIPHTHERIA.—Cases of laryngeal diphtheria.—Relation of diphtheria to croup.—Epidemic croup is diphtheria.—Dr. Jenner's views.—Paralysis as a sequel to diphtheria not known till the present century.—Review of article in Reynolds's "System of Medicine."—Proportion of cases with laryngeal symptoms.—Varieties of diphtheria.—Nasal diphtheria.—Cutaneous diphtheria.—Description of the false membrane.—Seats of paralysis.—Theory of its production.—Extreme slowness of pulse.—Case of ischuria renalis after diphtheria.—Period of paralysis.—Albuminuria.—Causes of death 123
Diagnosis. — Treatment. — Local applications. — Emetics. — Mercury.—Chlorate of potash.—Iodide of potassium.—Tracheotomy.—After-treatment.—Prognosis 134

ACUTE HYDROCEPHALUS AND MENINGEAL TUBERCLE.
Nomenclature.—Anatomical characters.—Cases of hydrocephalus without tubercle.—Case of meningeal tubercle without premonitory symptoms.—Case of meningitis and hydrocephalus lasting three months without tubercle.—Difficulty of diagnosis.—Tubercular meningitis without tubercle in the lungs.—Whytt's stages of tubercular meningitis.—Three distinct sets of cases: those coming on in the course of phthisis, those without premonitory symptoms, and those with the usual prodromata 153
Symptoms, as affecting severally the organs of circulation, those of respiration, those of digestion, and those of the nervous system. — Prognosis. — Causes: age, tuberculous diathesis, undue intellectual exertion, external violence 163
Diagnosis from typhoid fever, from cerebral congestion during dentition, from idiopathic meningitis, from cerebro-spinal meningitis.—Case of cerebro-spinal meningitis with recovery.—Diagnosis from hydrocephaloid disease, from pneumonia . 167
Treatment of tubercular meningitis.—Short notes of 13 cases with brief remarks 171

CHRONIC HYDROCEPHALUS, TUBERCLE OF THE BRAIN, AND OTHER CEREBRAL AFFECTIONS.
Cerebral tubercle rare in adult, common in child.—Anatomical

CONTENTS.

characters.—Ages of patients.—Symptoms, usual course and termination 178
Caution required in prognosis and diagnosis.—Case diagnosed as cerebral tubercle recovering with permanent amaurosis.— Amaurosis a frequent result of tumours of encephalon.— Retrograde changes in tubercle deposited in the brain with arrest of the disease 181
Case of chronic hydrocephalus, probably tubercular, lasting six years 186
Tubercle in cerebellum, two cases reported 188
Diagnosis of cerebral tubercle: from other tumours, from large head in rickets, from hypertrophy of brain 190
Other causes of hydrocephalus : (1) Congenital malformation ; (2) Intra-arachnoid hæmorrhage ; (3) Inflammation of the lining membrane of ventricles ; (4) Dropsy of ventricles . . 191
Case of congenital hydrocephalus tapped 192
Diagnosis from epilepsy, spasmodic contractions, essential paralysis, and neuralgic headaches 193
Cases of intense headache proving fatal with no obvious lesions to account for death.—Case of hydrocephalus with emaciation. —Neuralgic headache cured by quinine 194
Renal and arterial degeneration with intense headaches, arachnoid hæmorrhage and effusion into ventricles.—Caries of sphenoidal bone, causing abscess within the skull pressing on the brain 200
Tabular abstract of cases of cerebral tubercle 204

PYÆMIA AND OTORRHŒA.

Surgical pyæmia and septicæmia.—The views of Mr. Lee and Sédillot.—Idiopathic pyæmia.—Slight and more serious forms. —Cases treated successfully with quinine 208
Pyæmia following suppuration of mucous membranes.—From disease of tympanum.—Treatment of pyæmia 214
Cases of disease of ear leading to cerebral abscess and to secondary deposits in lung.—Varieties of Otorrhœa.—Slight affections of the ear not to be neglected.—Symptoms of extension to brain and treatment 216

CHOREA.

Statistics as to age and sex.—Connection of chorea with rheumatism and heart disease.—Dr. Kirkes's theory.—Part of the nervous system affected 223

b

	PAGE

Causes of chorea : fear, season, climate.—Mortality.—Symptoms.
—Diagnosis.—Duration 229
Treatment.—Cases.—Periosteal swellings complicating chorea.—
Fatal cases 234

PARALYSIS.

Children subject to all kinds of paralysis.—Conditions causing
motor paralysis 245
The form of paralysis specially called "*infantile.*"—Two cases.
—Course of symptoms.—Age of Patients.—Parts affected . 250
Diagnosis 252
Cruveilhier's atrophy rare in children.—Two cases . . . 253
Pathological anatomy of infantile paralysis.—Prognosis . . 258
Treatment 263
Cases of paralysis of cerebral origin 266
Facial paralysis 268
Tabular abstract of cases of infantile atrophic paralysis . 270—276

ASCITES.

As a part of general dropsy.—As a result of peritonitis, tubercular or non-tubercular.—Obscure cases due to some obstruction of the portal circulation.—Cirrhosis of liver . . . 277
Diagnosis.—A case of Hydronephrosis simulating Ascites.—
Treatment.—Case of cirrhosis 283

SCARLATINA.

Great fatality of it.—Varying degrees of mortality in different
epidemics.—Period of incubation.—Properties of the virus.—
Difficulty in destroying the virus.—Recurrence in the same
person 285
Symptoms, mode of attack, rash, pulse, temperature, sore-throat,
tongue, desquamation, albuminuria 292
Scarlatina maligna, two forms of it 298
Cases.—Scarlatina without eruption 300
Diagnosis 301
Rubeola notha 303
Sequelæ of scarlet fever 303
Co-existence with other acute diseases.—Scarlatina with diphtheria.—Cases of 308
Treatment of scarlatina and sequelæ 314
Illustrative cases 320

CONTENTS. xi

PAGE

TYPHOID FEVER.

Infantile remittent, a term to be discarded.—Cases of . 325
Peculiar features of disease in children 328
Ulceration of intestines leading to exhaustion 331
Sex and age of patients.—Mortality.—Case with double pneumonia 333
Two very prolonged cases recovering 337
Treatment 340
Case with broncho-pneumonia 341
Typhus fever, usually mild, case of . . . 343

SKIN DISEASES.

Enumerated in order of frequency . . . 346
Pemphigus, treatment by arsenic 347
Purpura hæmorrhagica, treatment by turpentine . . . 350
Favus.—Tinea Tonsurans.—Tinea decalvans.—Molluscum.—
Pityriasis versicolor.—Scabies.—Pruriginous strophulus . . 352
Eczema.—Treatment.—Psoriasis.—Lupus-Psoriasis . . . 360
Syphilitic skin disease 366

EPILEPSY AND CONVULSIONS.

Epilepsy frequent in children.—Treatment by bromides.—Successful cases.—Eclampsia nutans 367
Convulsions.—Great frequency in children.—Causes.—"Inward Convulsions."—General convulsions.—Sequelæ of convulsions.
—Treatment 373
Laryngismus stridulus and spasm of respiratory muscles.—
Treatment.—Peculiar paroxysmal attacks 378

APPENDIX.

Case of broncho-pneumonia 383
Formulæ for medicines in children 385

CLINICAL TREATISE

ON THE

DISEASES OF CHILDREN.

INTRODUCTION.

THE diseases of children are for the most part diseases which are also met with in adults; but peculiar features are exhibited by many of them which render a special description necessary.

There are, however, a few diseases which are seen only in children, and there are many which are more frequent in childhood than at a later period of life.

The importance of carefully studying children's diseases may be enforced by a consideration of their great frequency, their enormous mortality, and by the influence which diseases in early life exert upon the development of the growing frame and all its organs.

Of 1000 children born, 150 die within twelve months; 113 during the next four years; giving 263 or more than a quarter within five years of birth. During the next five years 34 die; during the next five years 18 more die; so that at 15 years of age only 684 remain of the 1000 born. Of those who survive very many bear permanent marks of imperfect development, of defective nutrition, or of actual disease, due to maladies contracted in early life.

Many of the diseases of childhood, if not congenital, are due to inherited tendencies to disease. Many are caused by unfavourable conditions surrounding the child after birth. The human infant is, of all animals, one of the most helpless and

B

the most dependent on its mother for good nursing, warmth and pure air.

Instinct teaches the lower animals to provide for their young. Instinct as well as her moral sense would teach the human mother that she herself should nourish her child ; but this instinct she does not always follow when she can do so ; and circumstances sometimes prevent her nursing her own infant, when she would gladly do it. Ignorance, vice, and poverty, singly or in combination, frequently lead to children being ill-fed, ill-clothed, exposed to cold, supplied with impure air, without sufficient sun-light, and otherwise subjected to unwholesome influences.

The main peculiarities of children's diseases depend on a number of conditions, to some of which I will briefly refer. The processes of absorption, secretion, sanguification, circulation and respiration are carried on with great rapidity ; in fact, all nutritive processes are extremely active.

The spinal nervous system is in a condition of active growth, so that it plays a more important part in disease than at a later period of life ; convulsions, reflex phenomena, and spinal paralyses are more marked than at a later period of life.

The brain is rapidly growing and its higher functions are undergoing development ; diseases affecting the nutrition of this organ are peculiarly liable to cause mental deficiency. The mucous membrane of the alimentary canal is very sensitive, and liable to catarrh and congestion, from the use of crude or unsuitable articles of diet, from exposure of the patient to cold, or from slight disturbances of health. This circumstance must be borne in mind in the medicinal treatment of children. Irritating drugs must be avoided, and when diet and hygienic measures will suffice, drugs should be entirely omitted. The younger the child the more important is this caution.

Emaciation occurs with great rapidity, and recovery of flesh may be equally rapid. Want of food or a low temperature can be sustained but for a very short period.

Owing to the flexibility of the chest walls and the structure of the pulmonary tissue, collapse of lung is more common in infancy than at a later period; bronchial inflammation is therefore more dangerous.

The osseous system is undergoing rapid growth, which may be arrested or be abnormal; from this cause deformities of various kinds arise, as specially referred to in the chapter on Rickets.

A disease of nutrition, such as tuberculosis, often runs a very rapid course and involves a larger number of organs than it does in older subjects. Syphilis interferes with sanguification and general nutrition.

Febrile affections have a strong tendency to remission during some part of the day. From this circumstance the term "infantile remittent" has been used to include a large number of different diseases, accompanied with pyrexia.

Owing partly to the patient not being able to define his sensations, and partly to a closer sympathy through the nervous system of different organs with each other, it is sometimes more difficult in a child than in an adult to decide what organ is at fault when the disease is a local one. Pneumonia and pleurisy more frequently present this difficulty than any other diseases.

The clinical examination of children is an art which differs in several important particulars from that of adults. We have not the patient's account of his symptoms to guide us in the case of young children; we must trust entirely to our own observation and that of other intelligent persons who wait upon the child.

The timidity of the child at the approach of a stranger will sometimes increase the difficulty of our examination.

If we see the patient asleep, it is often a great advantage. We can then notice his position, the expression of his face, whether his eyes are closed, the frequency and rhythm of his respiration, and whether there is any movement of the alæ nasi with the respiratory process. Usually also the pulse can be counted and the temperature of the skin felt by the hand before the patient wakes.

If the child cannot be seen asleep, and he is likely to be

shy, we must not approach him abruptly nor look at him very intently, but talk to his mother or nurse, and by degrees endeavour to attract the child's attention to some toy, to a watch, a stethescope, or a book with pictures. All this time observations can be made as to his expression, his mode of respiration, the use of his limbs, the state of his eyes, and other matters.

Children are good physiognomists and know those who are fond of them, so that the physician who loves children will usually have much less trouble in examining a child than one who has no affection for them.

The pulse is sometimes one of the most difficult things to determine, owing to the restlessness of a young child. It should, if possible, be felt before anything is done likely to excite the patient, because its frequency after excitement will be much increased. The state of the abdomen may be felt under the clothes without laying the child on its back, which will often cause it to cry; and to ascertain whether there is any tenderness on pressure this may be important. Care must be taken that the hands are not cold when applied to the child's skin. It is impossible to examine a child satisfactorily unless its clothes be taken off; this is necessary in order to see the formation of its chest, the condition of the ribs and their cartilages, and the effect of respiration upon them, the state of its osseous system, of its nates and anus, and the general condition of its skin. The size of the liver and spleen can only be properly ascertained when the child is stripped and placed on its back. Auscultation of the chest should precede percussion, and this is practised most easily behind by the ear placed directly to the chest, although if the stethescope is lightly and properly used, few children will resist the employment of it. The most convenient position in which to examine the back of a young child is in the arms of its nurse, with the head over one of her shoulders. For the examination of the anterior aspect of the lungs and the heart, the stethescope is essential. Percussion should be very light and the finger used as a pleximeter; care must be taken in comparing the two sides to compare the notes obtained by percussion at the same periods of the respiratory process, or if this be impossible, to estimate the average of the sounds obtained at the height of inspiration, at the height of expiration, and between the two extremes, and compare the averages of the two sides.

The child must be symmetrically placed, the arms and shoulders on the two sides being in the same position, before any conclusion is drawn from the difference of a percussion note at corresponding points on opposite sides of the chest. If this precaution be not taken, one side will give a duller sound than the other one minute, and a clearer sound the next minute, owing to a change of posture. It must be remembered that during expiration, especially if forced, the base of the right lung is considerably duller than the base of the left, owing to the pressure of the liver when the diaphragm is at its highest point.

The examination of the mouth, the tongue, the gums, the teeth, and the throat, should be made last, because it is most likely to annoy the child and make it cry. The tongue may be seen sometimes by coaxing the child to protrude it; or, if too young or too wayward, it may be seen when crying. Sometimes the child will open its mouth when it is pressed gently upon the chin, or it may, in an extreme case, be necessary to pass the finger along the inside of the cheeks to the posterior part of the lower jaw, and then inserting it between the upper and lower jaw, the mouth is readily opened. To examine the throat a tolerably strong spoon, or still better a two-bladed tongue-depressor, should be slipped between the teeth over the top of the tongue to the neighbourhood of the papillæ circumvallatæ, and then pressed firmly down to expose the pillars of the fauces, the tonsils, and the back of the pharynx. This process is not a pleasant one, but gives no pain, and can be done with great rapidity. A good light is of course essential, and the back of the child's head should be well fixed behind. This may sound like a description of rather severe measures for a young child; but half measures give quite as much annoyance, and entirely fail in their object.

From what has been stated as to the rapid course of

children's diseases, it is obvious that it is necessary to visit children more frequently than adults, especially in acute diseases. It is also very important not to hurry one's visit, so that the patient may be seen, not only when agitated, but after he has become accustomed to the presence of a stranger.

The temperature is often a very important matter to be determined as an element in the diagnosis and prognosis of children's diseases. The hand cannot be trusted to give accurate information; the thermometer must be employed. It should not be used soon after the child's skin has been exposed to the air, as in dressing or washing; because a child's temperature will be reduced for some time after such exposure. The thermometer may usually be placed in the axilla, but considerable care is necessary to ensure its being in complete contact with the skin; and it will require to be left for five or ten minutes before it attains the maximum temperature. A more accurate method is to place the bulb of the thermometer in the rectum, and this can often be done in children without any resistance. The result thus obtained is more reliable, and the maximum is attained in a shorter period. It is usually half a degree or a degree higher than the result obtained in the axilla.

When engaged in the diagnosis of children's diseases, the hereditary predisposition is of great importance: it is desirable to know something of the father and mother's health and that of their nearest relatives, especially in reference to tubercular disease, scrofula, and syphilis. If this be not the first child it is important to know what is the state of health of the other children, or if they be dead what was the mode of their death. It is also necessary to know how the child was nourished during the first twelvemonths of his extra-uterine life, and whether his mother was in her usual health during his intra-uterine existence. All these, and other circumstances affecting the child's development, it is well to know, because they will often throw light on the nature of the disease under which the child is suffering. Inquiries of this nature should be made at an early part of the first visit, because they are less likely to excite curiosity or apprehension in the mind of the mother if made before the child has been examined, than if delayed till a later period of the visit. The time during which the child gets accustomed to the presence of the doctor may be thus profitably employed in the case of children under four or five years old. When children are above this age it is better to make inquiries of this nature out of the child's presence ; for it is of great importance in dealing with delicate children to discourage self-consciousness, and as far as possible to prevent their learning that the state of their health is a matter occasioning great anxiety.

In the choice of medicines for children, some peculiarities must be specially borne in mind. One of these is the great susceptibility of children to the action of opium in its various forms. Such preparations as syrup of poppies, which are very uncertain in their composition, should never be given to children. The action of opium upon a young patient must be carefully watched, and the doses must not be too rapidly repeated. Calomel is a useful aperient for children, either alone or in combination with jalapine. It has the advantage of being tasteless, and not bulky, whilst it is efficacious and not irritating.

As an alterative and as an antiphlogistic, mercurials have been too frequently employed. In congenital syphilis, however, grey powder is the most satisfactory remedy; and in membranous croup mercury is often of service. Emetics are of great value in infancy in catarrhal affections of the larynx and bronchi, also when the stomach has been imprudently loaded. Ipecacuanha is the emetic which I prefer in the great majority of cases. Antimony, given until it causes nausea, is useful in severe cases of croup. Depletion, even by leeches, is scarcely ever required by children in London practice at the present day.

Blisters are to be avoided in children, from the risk of their causing constitutional irritation and deep ulceration. If used at all, they should not be left on more than three or four hours at a time.

PNEUMONIA.

The mortality from this disease in children is very high. During the year 1861 there died in London, from pneumonia, 2660 children under five years of age, of whom 1424 were males. Nearly half of these died during the first year of life, and nearly one-third between the ages of one and two years. The mortality from pneumonia at all ages was at the rate of 1·48 per thousand males, and 1·06 per thousand females; under five years the rate of mortality was, amongst males, 7·87 per 1000 living at that age, and amongst females 6·8 per thousand. From these figures it appears that the death-rate from pneumonia was more than five times greater amongst children under five years of age than in the entire population. Between five and ten years of age the deaths from pneumonia were only 80 in number, which gives a death-rate of only 0·26 per 1000 living at that age; from 10 to 15 years only 26 deaths were registered from pneumonia. Of 389 cases in the Children's Hospital, registered as "Inflammation of Lungs and Air Passages," there were 104 deaths, which gives 26·7 per cent. as the rate of mortality. This includes both forms of pneumonia and bronchitis; it must be remembered that mild cases of bronchitis are seldom admitted as in-patients.

During the last ten or twelve years there has been in England a considerable increase in the mortality from bronchitis at all ages;* this increase has been most marked amongst infants and old people. In the seven years, 1848-54 the rate of mortality per 1000 from bronchitis was of males

* Registrar-General's 25th Annual Report, p. 187.

under five years of age 2·65, and of females 2·26; whereas, in the five years, 1858-62, the death-rate was 4·41 and 3·75 amongst males and females respectively. The death-rate from pneumonia at this age in the same period has remained almost stationary, whilst from phthisis there has been a slight diminution in the rate of mortality.

It is not at all evident on what circumstances the increased fatality of bronchitis depends. Meteorological conditions do not explain it. The mean temperature had not varied much, and the lowest temperature in 1858-62 was 47·0°, and in 1848-54, 47·7°.

The cases registered as deaths from pneumonia in infants belong to three distinct categories:—1. Some of them are ordinary *lobar pneumonia*, which, though common in the adult, is rare in the infant, and especially rare as a cause of death. 2. Others are *lobular pneumonia*, a disease rare in the adult, very common and very fatal in the child. It is usually preceded and accompanied by capillary bronchitis. It has been called *broncho-pneumonia*, and by Trousseau, *peripneumonic catarrh*. This disease has been confounded with another affection, which is also very fatal in infancy, and frequently follows and is accompanied by bronchitis. This constitutes the third class of cases, *pulmonary collapse*. It resembles the unexpanded fœtal lung, the condition called *atelektasis*. At one time all cases of collapse were looked upon as pneumonia; but of late years, since the time of Legendre and Bailly, there has been a disposition in some writers to go to the opposite extreme, and to regard all cases of lobular pneumonia as collapse. The two morbid conditions are quite distinct; they may, however, be combined in the same patient; they are both often preceded by bronchitis, and the distinction between them, though generally easy, is at times very difficult. A portion of collapsed lung may become the seat of pneumonia, and the transition is not always well marked.

The anatomical characters of collapsed lung may be briefly stated as follows:—It is tough, does not tear readily, is flesh-like to the eye and hand (hence the term "carnified" applied to it); it varies in colour from grey to dull brown, violet, or purple; does not crepitate; contains little fluid or froth; sinks in water; occupies less space than healthy lung tissue around, and is consequently, when near the surface, depressed below the surrounding parts of lung. Its section is smooth, by insufflation it may, with greater or less ease, be distended, when it assumes the aspect of healthy lung tissue. Except by insufflation it is in some cases impossible to distinguish collapsed lung from that which is the seat of lobular pneumonia. This morbid change may be confined to one lobule, or it may involve nearly the whole lung. A favourite seat of collapse is the anterior margin of the lower lobe or the middle lobe of the right lung, and the posterior margin of both lower lobes. This condition often exists in weak children whose lungs have never been well expanded; it is common in rickets, in hooping cough, in croup, in diphtheria, in measles, in pleurisy with effusion, and in many other affections of infancy.

The condition of pulmonary collapse may be combined to a greater or less degree with congestion. In this case the tissue is more friable, breaking down more readily on pressure; the section is violet in colour, and there escapes from it on pressure bloody fluid mixed with purulent mucus. There is no plastic effusion in these cases, and no exudation granulations to be seen under the microscope.

The congestion may advance until it merges into inflammation, and it is not always easy to say when true pneumonia has commenced.

Pulmonary collapse in bronchitis is mainly due to plugs of tenacious mucus, formed in the bronchi, and drawn by each inspiration into tubes of smaller diameter; the entrance of air is thus obstructed, whilst the escape of a small quantity of air is permitted at each expiration, the plug acting like a ball-valve in the mouth of a syringe.

Lobular pneumonia occurs in a *disseminated* form, or it is *generalized;* in the latter case it produces a condition similar to that due to lobar pneumonia. When disseminated there are found scattered through the lung, here and there, patches varying in size from a hemp-seed to an almond, of a dull red colour, sometimes mottled with yellow. They are circumscribed (though not so sharply as collapsed patches), solid, and on the surface prominent; on section they are not granular. On the surface of the lung they present nodular elevations, and sometimes much resemble tubercles. Insufflation of the lung has no effect on the affected portions. They are less tough than healthy lung, a sanious, frothy fluid may be pressed from them, and they sink in water even before pressure. If the disease has gone on to the third stage, these patches are of a yellowish-grey colour, are still more friable, and a purulent fluid escapes on pressure. Sometimes there is a single hepatized nodule very distinctly isolated from the surrounding tissue by a more or less marked fibrous capsule; this nodule may have passed into the third stage, and an abscess be formed, having but one cavity or multilocular. Besides these are to be seen numerous smaller purulent collections about the size of hemp seeds; they have smooth walls and communicate directly with small bronchi. The smaller bronchi leading to the hepatized portions of lung are usually filled with tenacious mucus, their walls are more or less injected and thickened, and their calibre is enlarged. The dilatation is usually cylindrical or globular, rarely, if ever fusiform, as it is sometimes seen in adults. When lobular pneumonia is generalized it closely resembles lobar, but it may be distinguished by this circumstance, that in lobar pneumonia the inflammatory process advances uninterruptedly from one centre, usually from the base, upwards; the lower portion of lung is usually in a more advanced stage than the upper; whilst in lobular pneumonia there are a number of independent centres, exhibiting very various degrees of progress, so that nodules in the third, or purulent stage, will be seen distributed promiscuously amongst others in the second stage or that of red hepatization, and others in the first stage or that of simple engorgement. Subpleural ecchymoses are often seen near the base of the lungs; they are of irregular form, and vary in size from that of a pin's head to a six-penny piece. They have the appearance of lying in the pleura, but on close examination they are found to involve the outer layer of pulmonary tissue, slightly elevating the pleura.

This disease is almost peculiar to childhood; it is most frequent between the ages of 12 months and 6 years. In infants of less than 12 months lobular pneumonia is rare, collapse and lobar pneumonia are more usual conditions. Between 5 and 16 years of age about one-quarter of the cases of lobular pneumonia are met with. Lobular pneumonia is very often secondary to hooping cough or measles;* when it occurs primarily or only secondary to bronchitis it is usually in debilitated subjects. It seems to be induced by the atmosphere of badly ventilated hospitals. The fact, that capillary bronchitis in the child often terminates in pneumonia, whilst in the adult it does so very rarely, is probably in great measure due to anatomical differences between the child's lung and that of the adult. In the child the separate lobules are more distinctly bounded by cellular tissue. This would have the effect of limiting the inflammation to a lobule or small cluster of lobules. Collapse of lung is very frequently the prelude to lobular pneumonia. The collapsed tissue becomes firmer and the seat of large exudation cells with single nuclei. It then becomes of a darkish mahogany colour, its consistence is friable and soft, and a small quantity of adhesive, bloody fluid escapes on pressure. The centre of these nodules loses its colour, whilst the periphery, still of a dark reddish brown, is less distinctly marked off from the healthy tissue; other nodules become pale throughout and very soft. The puriform fluid which escapes on pressure consists of large cells with nuclei undergoing fatty degeneration and smaller granular cells with multiple nuclei (pus cells). There is no fibrinous exudation as in lobar pneumonia. Ziemssen says, that in chronic cases another form of pneumonic consolidation is sometimes seen, which is more uniform and firmer, and constantly takes its origin in collapsed portions. If this

* Of 98 cases observed by Ziemssen, 43 succeeded measles, 23 hooping-cough, and 32 bronchitis and chronic bronchial catarrh.

change involves a whole lobe it maintains the normal volume, feels uniformly firm, and exhibits, with considerable injection of the pleura, a more or less thick exudation on the surface. On section is seen a pale reddish or bluish surface, which is not granular, but smooth, dry, and faintly glistening. Here and there the parenchyma is pale yellowish, or as if strewn with a fine whitish-yellow sand. No fluid or air escapes on pressure.

On microscopical examination there is found a moderate growth of cell elements, and a considerable swelling of the interstitial tissue, which is at the same time finely granular. In very old cases there is found an extensive growth of interstitial areolar tissue with dilatation of bronchi: a kind of cirrhosis. This has been described by Bartels, but has not been seen by Ziemssen.

The pneumonic nodules are sometimes changed into cheese-like masses, exhibiting under the microscope an abundance of larger and smaller clusters of fat globules, together with free fat and free nuclei. This cannot be distinguished from yellow tubercle.

Symptoms, Course, and Duration.—Lobular pneumonia runs a more or less chronic course, varying in duration from six or eight days to a month or two. The more acute cases occur after capillary bronchitis, and especially that form which is secondary to measles. (See case of Emily Thacker, Appendix.) The more chronic cases usually supervene on bronchial catarrh and hooping cough. The acute cases very closely resemble lobar pneumonia in their clinical aspects. The onset is sometimes well marked by a slight rigor and a sudden elevation of temperature; the pulse and the respiration are quickened. There is movement of the nostrils in respiration and energetic action of the muscles, usually employed only in deep inspiration; for example, the sterno-cleido-mastoids, the scaleni, and the pectorals. Cough, which had previously existed, often becomes altered in character; if there had previously been paroxysms these cease, and give place to frequent short attacks of cough, which are accompanied with painful contraction of the countenance or a piteous groan or cry. The child becomes very restless, especially towards night—very young children are occasionally drowsy—and the mucous membranes livid. These symptoms are usually due to the occurrence of collapse of lung tissue. The *physical signs* at first are those of capillary bronchitis. If collapse takes place there may be, on light percussion, slight dulness in the interscapular regions, or on the inner margins of the lungs which cover the heart; vocal fremitus is either normal or reduced in amount. At the end of two days signs of consolidation are possibly to be detected. There is increased dulness on percussion, more distinct vocal fremitus and bronchophony with bronchial breathing, and fine crepitation of a rather metallic character.

These signs usually begin at the lower part of one lung, and are then soon met with in some part of the other lung.

Lobular pneumonia runs its most rapid course in children under 18 months of age, in whom it sometimes proves fatal in the course of six or seven days, mainly by the supervention of collapse. Except in such cases, death seldom takes place under ten or twelve days. After great restlessness and signs of increasing dyspnœa, stupor sets in, and the respiration becomes more superficial. It subsequently becomes very hurried; involuntary movements, grinding of the teeth, and convulsions close the scene.

Generalized lobular pneumonia is sometimes with great difficulty to be distinguished from lobar pneumonia. The main points of distinction are the more frequent existence of fine râles over the chest, and the fact that both lungs are attacked more frequently in the former than in the latter. Cases of this kind occur most frequently after measles. The temperature ranges from 102° to 104° Fahr. In lobar pneumonia there is usually, after about five or ten days, in cases of recovery, a *sudden* fall of temperature, accompanied with pallor, faintness, or sweats; in lobular pneumonia the temperature reaches the normal standard after three or four days of *gradual* sinking. It then attains for a day or more a degree even lower than is usual in health; this is a phenomenon observed in both forms of pneumonia.

A more chronic course is usually taken by lobular pneumonia following bronchial catarrh or hooping cough. The fever never attains so high a stage; consolidation takes place very slowly. This form may last as long as six weeks or two months. The temperature seldom rises above 103°. After hooping-cough the child becomes much emaciated and exhausted; there is thirst and loss of appetite, constipation or obstinate diarrhœa. There is either restlessness and want of sleep or great listlessness and dislike to being disturbed.

Lividity increases, the face appears bloated, the skin is pale and of a dirty tint, and the cuticle separates in fine branny scales. Small boils, blebs with reddish serum, or abscesses often appear on the nates, the back, or the face. Scratching makes these places worse, and leads to more or less extensive ulceration.

In the course of lobular pneumonia which runs a chronic or subacute course, remissions often occur of nearly all the symptoms, giving rise to hopes of recovery which are frequently disappointed.

This is a much more fatal disease than lobar pneumonia. M. Barrier gives as the result of his experience in Paris 48 deaths out of 61 cases. Ziemssen gives 37 deaths out of 98 cases of broncho-pneumonia. He arranges his cases as follows :—

43 after measles . . . . 11 deaths 32 recoveries
32 after bronchitis and chronic catarrh 14 ,, 18 ,,
23 after hooping cough . . . 12 ,, 11 ,,

From these numbers it appears that after measles less than one-third died, whilst after hooping cough more than half of the cases died. The most fatal cases were those that occurred during the first year of life.

In the Hospital for Sick Children, London, on looking through a record of 127 post mortem examinations, I find that 21 presented more or less lobular pneumonia; in many it was primary and the sole cause of death; in other cases it was secondary to other diseases. The ages of the 21 cases were as follows :—two under 12 months, four between 1 and 2 years, nine between 2 and 3 years, one between 3 and 4 years, four between 4 and 5 years, and one 7 years old.

Death usually occurs by slow asphyxia and exhaustion. That broncho-pneumonia sometimes passes into tuberculization there can be little doubt.

Diagnosis.—From simple capillary bronchitis it is often impossible to distinguish the earlier stages of broncho-pneumonia, if there has been no sudden exacerbation of symptoms. In bronchitis alone the temperature seldom exceeds 101° or 102°, and there are no signs of consolidation of lung.

From *collapse* again the diagnosis is often most difficult, the more so because the two conditions often co-exist in the same subject. Collapse will cause no elevation of temperature, and does not often give rise to so much dulness on percussion, nor to sub-crepitant râles; as a general rule, too, there is no bronchophony.

From *lobar pneumonia* the distinction is very difficult, when broncho-pneumonia has become generalized. The points usually to be relied on are its secondary character, its involving both lungs, and the abundance of subcrepitant rhonchi. Lobar pneumonia may however be preceded by diffuse capillary bronchitis, though this is rare. The course of the two diseases is usually different. The difference as regards temperature has been referred to above. Another point on which Ziemssen has insisted is, that broncho-pneumonia does not often involve the whole of the lower lobes, so that the dulness caused by it does not extend forward into the axilla, but is limited to the dorsal aspect. In this respect I have not been able to confirm his observations.

Pleurisy can be always readily distinguished from broncho-pneumonia, unless the pleuritic exudation is scanty, not extending into the axillæ, and attended with bronchophony. In this, case the vocal fremitus must be looked to, being decreased in pleurisy, and increased, or at any rate not less than normal, in broncho-pneumonia.

Acute Tubercular Infiltration, if involving the lower lobes only or chiefly, gives phenomena in almost every respect similar to those presented by lobular pneumonia. The history of the case and the symptoms of tubercle in other organs are the chief guides to diagnosis.

Prognosis.—This is influenced, to some extent, by the disease to which it is secondary. It is more favourable after measles than after hooping cough. It is most serious under the age of 12 months. In weakly children, especially if rickety, the prognosis is most unfavourable. The acute cases, which approach lobar pneumonia in their characters, are more favourable than the subacute cases in which the fever does not run so high. The extent of pulmonary tissue which is attacked by capillary bronchitis is also an important element in the prognosis. The pulse, when small and soft, is an indication of the gravest import; it is an expression of extensive obstruction to the circulation through the lungs, in consequence of which but little blood is brought to the left side of the heart; moreover, in judging of this phenomenon, the paralyzing influence which carbonized blood has on the muscular force of the heart must be taken into account. Equally or even more unfavourable symptoms are, diminished sensibility of the bronchial mucous membrane, and a disabling of the muscles of respiration, indicated by cessation of cough and permanent rattling in the trachea.

Convulsions in this diease are almost uniformly followed by death.

Treatment.—Seeing that lobular pneumonia usually follows some other malady, especially bronchitis, it is of the utmost importance to take precautionary measures to ward it off in all cases of measles, hooping cough, and bronchitis. It is of primary importance to maintain an equable temperature; at the same time the rooms must be well ventilated. The child should be wrapped in flannel and kept in bed; the temperature of the room should be kept between 60° and 65°. When catarrh is extending into the finer bronchi the use of an ipecacuanha emetic is desirable; this has the effect of rendering the bronchial secretion less viscid and promotes its expulsion, thus checking the tendency to collapse and lobular pneumonia. In rickety or very weakly children a little carbonate of ammonia may be combined with the emetic, or its depressing influence may be counteracted by the administration of wine. As an antiphlogistic, blood-letting, either general or local, is totally inadmissible. In very rare cases is antimony allowable, and this only in very small doses in robust children quite at the outset of the febrile symptoms. Debility is the great thing to be dreaded in this disease. The local application of cold has been much lauded by Bartel; it has also been tried by Ziemssen; his experience leads him to the following results:—It has no influence on the progress of catarrhal pneumonia. It has, however, the effect of reducing fever, dyspnœa, and restlessness, more than any other agent. This effect is only accomplished after its application for a period of several hours, and is not permanent. In later stages, when the pulse is small and frequent, the protracted application of cold is hazardous from the risk of syncope.

Mustard plasters are of use; blisters should never be resorted to. The administration of ipecacuanha emetics from time to time after pneumonia has set in is indicated. After a while, if the case is progressing unfavourably, they cease to act, and must be discontinued. Resort must then be had to stimulant expectorants, such as senega, squills, and muriate or carbonate of ammonia. These remedies will require to be used from the first in the case of rickety and other weakly children. The inhalation of steam, with the addition of small quantities of oil of turpentine, is said to exert a beneficial influence. The diet of the patient must be especially regarded; it must be light and nutritious; strong beef-tea, milk, and raw eggs, in small quantities, frequently administered. Wine is required in cases of great weakness, and may be combined with egg.

In chronic cases the use of iron may be advisable. During convalescence great care is required; a change of air to a milder place, if possible, should be given ; and in cases disposed to tubercle cod liver oil is to be recommended.

Bronchitis. Lobular Pneumonia after Measles. Recovery.

Charles Webb, aged 5 years, had measles on the 11th October ; the rash was very fully developed. After this he continued very ill with dyspnœa, cough, and fever. On the 22nd October he came under my care. He was very hot and slightly delirious. The alæ nasi acted freely in respiration. Pulse 180. Respiration 74. Temperature of axilla 102·5°.

Oct. 23rd. Countenance had a dusky tint. Very restless. Pulse 144. Respiration 64. Temperature 102°. Cough very frequent. He constantly passed his hand across his face and forehead. Tongue had a brownish red streak down the centre and was dry. He lay constantly on his back, with legs drawn up. He was too ill to be examined as to physical signs in thorax. No appetite. He was ordered on admission, Ammoniæ carbonatis, gr. ij ; syrupi, ʒj ; decocti senegæ, ʒiij. Tertiâ quâque horâ. Brandy 6 ounces in the 24 hours.

24th. Countenance less dusky. Passed a quieter night. Pulse 180. Respirations 56 easier. Temperature 101·5°. There was much recession of the lower half of chest walls in front during inspiration. Over front of right lung inferiorly the percussion note was dull. Fine moist rhonchi heard over both lungs posteriorly.

25th. Seemed better. Pulse 172 small. Respirations 55. Temperature 101°. Duskiness of countenance had disappeared. Bowels comfortably open.

26th. Improving. Rhonchi heard in chest were not so fine. There was less recession of chest walls during respiration. Skin is hot but moist. Tongue not so dry and cleaner. Temperature 100·5°.

27th. Not so well. More dyspnœa. Troublesome cough. No appetite. Wears himself out with irritability of temper. A great deal of fine and coarse bubbling heard over the whole posterior aspect of chest. Respiration 52. Pulse 168 small.

29th. Not so well. Drowsiness. Bowels relaxed. Profuse sweating. Physical signs much the same, except that there was bronchial breathing under the left clavicle. Patient could not lie down on account of dyspnœa.

30th. Better. Less dyspnœa. Pulse 144. Respiration 43. Child seemed weak. Still very little appetite.

Nov. 2nd. Much better. Pulse 132 with more power. Respiration 30. Appetite improving. Less rhonchi and of a coarser character.

7th. Has been steadily improving since last note. Rhonchi much less. No signs of consolidation.

16th. Still improving. Pulse 115. Respiration 20.

20th. Discharged well.

Lobar Pneumonia.—This is an affection which is common to children and adults equally. It attacks strong children more frequently than those that are weakly. One attack renders a child more liable to a second. The period of dentition is favourable to the occurrence of pneumonia. It is very often primary, and not uncommonly secondary to measles, typhoid fever, scarlet fever, and other acute specific diseases. Of 186 cases observed by Ziemssen, 12 occurred in the 1st year of life :—

In the second year	25	cases.
„ third year	15	„
„ fourth year	27	„
„ fifth year	16	„
„ sixth year	22	„
„ seventh year	11	„
„ eighth year	12	„
„ ninth year	11	„
„ tenth year	12	„
„ twelfth year	8	„
„ thirteenth year	1	„
„ fourteenth year	0	„
„ fifteenth year	2	„
„ sixteenth year	3	„

Of the whole number 117, considerably more than half, were under six years of age. Boys are more liable to the disease in the proportion of 2 or 3 to 1. The exciting cause is in a small proportion of cases only, distinctly ascertained to be exposure to cold or wet. In the majority of cases no obvious exciting cause can be made out.

In comparison with adults, the pneumonia of children presents a few peculiarities in anatomical characters. On section or fracture the granular appearance is less marked, or may be quite absent. This has been supposed to depend on the smaller size of the alveoli in children. Rilliet and Barthez consider that it is due to exudation taking place into the parenchyma between the alveoli, whilst Ziemssen ascribes the absence or indistinctness of granulation to the exudation containing but little fibrin.

It is not uncommon to see the hepatized lung marked with depression by the ribs. It is very seldom that pneumonia of children terminates in abscess or gangrene. Whether it undergoes degeneration into tubercle, as some believe, I am not prepared to say. It is, as in the adult, commonly enough combined with pleurisy. Together with pneumonia of the upper lobe, fluid exudation has been occasionally found in the lower part of the pleura of the same side.

Hyperæmia of the brain is often found; meningitis seldom exists as a complication.

The right lung is more frequently attacked than the left in the proportion of 139 to 115, from the statistics of Rilliet, Barthez, and Ziemssen combined. Of 311 cases, in 21 only were both lungs involved.

Usual Course and Symptoms.—It begins much the same as in adults; the attack is usually sudden, and sometimes accompanied with rigors. In very young children, instead of rigors there are slight chilliness, pallor of the skin, with a bluish discoloration of the nails, and sometimes convulsions. There are frequently also vomiting, violent headache, thirst, dyspnœa and pain in the side, or epigastrium, with a short dry cough. In a few hours febrile heat succeeds and sometimes sweating. The little patient lies usually on his back, with flushed cheeks, and nostrils dilating with each inspiration; respiration is quick and shallow. There is not the powerful action of the inspiratory muscles of the neck which is seen in bronchopneumonia, nor that restlessness and anxious expression which are characteristic of this disease. Sometimes, however, there is acute pain in the side at the onset, which seldom lasts more than about 24 hours. Fever runs high, the temperature reaching from 103° to 105°, the pulse is frequent, full, and hard; there is no appetite, the bowels are confined, urine is scanty and high-coloured. Herpes labialis is often seen about the fourth or fifth day; it is not either a good or bad omen. Physical signs appear on the second day, or even later; in pneumonia of the upper lobe they are scarcely perceptible till the fourth day. They are of the same kind in children as in adults. About the fifth or seventh day, or occasionally not until the eleventh or thirteenth day, a crisis takes place, often marked by profuse perspiration and considerable depression. The patient often falls into a quiet sleep, disturbed by an occasional cough. About this time the temperature falls within twenty-four hours from 104° to a degree lower than in health; the pulse also becomes slower as well as the respiration. The rapid fall of temperature is a very striking and characteristic phenomenon. In cases of greater severity and extent, the crisis is not so abrupt. The temperature does not fall all at once, but makes several rapid falls, separated by intervals of rest in the course of thirty-six or forty-eight hours. The cough is still present, and accompanied with moist sounds in the bronchi and trachea; physical signs of consolidation remain, and gradually disappear in the course of several days or a week. Recovery from pneumonia is more rapid in children than in adults. In unfavourable cases there may be a slight remission of temperature, but it is incomplete, and fever returns the next day with its former activity. Death usually occurs in fatal cases during the second week; it may be preceded by œdema of the portions of lung not hepatised, or by pleurisy or pericarditis. In other cases, life is prolonged to the third or fourth week; fever continues high, the signs of consolidation remain unchanged. Emaciation sets in, the cough is very troublesome, and death ensues from exhaustion, sometimes ushered in by convulsions.

The urine, as in the adult, is deficient in chlorides during the process of hepatization, and it again contains them when resolution sets in.

Critical Days.—Ziemssen has noted specially the days on which defervescence or convalescence has commenced, marked by a sudden fall of temperature. He says, that of 107 cases, in 95 the crisis was on uneven days, and in 12 only on even days. There were 9 on the third day, 3 on the fourth day, 31 on the fifth day, 5 on the sixth day, 35 on the seventh day, 4 on the eighth day, 9 on the ninth day, 8 on the eleventh day, and 3 on the thirteenth day. These results accord with what were obtained in adults by Traube, but do not correspond with those of Wunderlich, who found the crisis in 46 cases on the uneven days, and in 29 cases on the even days. Ziemssen suggests that the discrepancy may arise from a difference in the methods of defining the crisis. He himself considers the crisis to occur, not on the day on which the greatest depression of temperature occurs, but on the day on which the first considerable fall takes place, accompanied by other phenomena, such as sweating, sleep, increased bronchial secretion. The temperature often falls still lower on a subsequent day.

When several lobes are involved, either of one lung or of both, the course is usually protracted, and the crisis takes place on the ninth, eleventh, or thirteenth day. In cases of pneumonia of the upper and middle lobe of right lung, the crisis usually occurs on the seventh day.

Occasionally the fever does not run an uninterrupted course; but, after a cessation of fever, lasting from 24 to 28 hours, there is a return of fever which is coincident with the inflammation of another portion of lung.

There is a class of cases of pneumonia in which cerebral symptoms prevail to such an extent as to mask the pulmonary symptoms, and often to mislead the practitioner. They have been described by Rilliet and Barthez as cerebral pneumonia, and divided into two categories, the meningeal, and the eclamptic or convulsive. These terms are not, in my opinion, very good, because they imply too much, but they serve to indicate peculiarities in the symptoms.

In the former variety headache and vomiting are constant prominent symptoms, and frequently delirium is present; then drowsiness or even coma succeeds.

In the latter variety, convulsions occur and sometimes recur for several days, and these are accompanied or followed by a tendency to coma.

As compared with meningitis, these cases of pneumonia are usually characterised by the absence of sighing, of the sudden piercing cry, and of the frequent changes of colour; it is also unusual to see such deep coma as occasionally accompanies meningitis. The temperature is much higher in pneumonia than in meningitis, and is maintained at a high degree until a sudden fall occurs at the crisis.

The case of W. H. S. (page 38) is a good specimen of the convulsive variety. The distinction between meningeal and convulsive pneumonia appears to be impossible and inconvenient; the symptoms characteristic of the two forms, as described by Rilliet and Barthez, are often intermixed to a greater or less degree.

It is much more common to meet with cerebral symptoms, when pneumonia attacks the upper lobe of the lung, than when it involves the lower lobe only or primarily. Why this should be so I am not able to say.

Lobar pneumonia is more frequently primary than is lobular pneumonia. It is, however, often secondary to measles or scarlatina, and more rarely to other acute specific diseases. The prognosis in these cases is more unfavourable than in the primary disease.

Complications.—Pleurisy is the most frequent complication in the child as in the adult ; it is nearly constantly present when superficial portions of lung are inflamed.

Pericarditis occurs but seldom, and meningitis still more rarely as a complication of pneumonia. Bronchitis is occasionally met with to a slight degree, but very seldom to any great extent, lobar pneumonia in this respect differing entirely from lobular pneumonia. Noma is seldom or never seen after pneumonia in England, but in the Parisian hospitals it is occasionally met with.

Symptoms.—*Fever.*—The temperature is by far the best measure of fever, it follows the same laws in the child as in the adult. The constantly high temperature from the beginning of the illness to the commencement of the crisis, the slight oscillations during the 24 hours, and the remarkably sudden fall at the end of a certain number of days, are its distinctive character both in children and adults. The highest temperature which I have noticed has been 105·2° in the axilla; it was seen on the tenth day of a case of pleuro-pneumonia involving the whole of the right lung, and proving fatal on the eleventh day. Ziemssen has observed a temperature of 106·7° in the rectum, and 105·8° in the axilla of children who have subsequently recovered. The lowest temperature observed by this author was 97·2°, which corresponds to about 96·3° in the axilla. A depression of temperature to about 2° below the normal standard occurs about 30 hours after the commencement of defervescence or cooling.

The range of temperature noted during the course of the disease in 24 hours prior to the sudden fall at the crisis seldom exceeds 1° to 1·5°; the lower temperature usually occurs in the early morning and the higher one in the afternoon. In a few cases the highest temperature has been noted near mid-day. A fall sometimes takes place on the third day, but much more frequently on the fifth or seventh, the depression does not usually reach its full extent on one day, but is completed on the next; the temperature then falling below what is normal to the individual, and this to a greater extent in cases where the previous elevation has been very great. The temperature will sometimes fall within 36 or 48 hours from 105° to 97°. The low temperature is sometimes maintained, and at other times slight exacerbations occur for several consecutive evenings and then the normal degree is attained. If an elevation of temperature exceeding 100° occurs after the lowest depression, there is reason to fear that a relapse has taken place, and inflammation is involving fresh pulmonary tissue.

Pulse.—This, as a general rule, varies somewhat in the same ratio as the temperature. It is not, however, nearly so much to be relied on; it is not so constant, its fall takes place later than the temperature, and it is liable to disturbance from excitement or other accidental causes in children. The pulse in very young children is often excessively frequent; as many as 200 in the minute under 12 months of age, and from 150 to 160 in children from 18 months to 5 years of age.

After the commencement of convalescence the pulse may become much elevated for a time without any corresponding elevation of temperature. As to quality it is at first full and hard, subsequently small and soft. It is often distinctly dicrotous at the time of the crisis.

Cough.—This is a symptom seldom entirely wanting; it is usually short and suppressed, more rarely paroxysmal. In cases where cerebral symptoms predominate it is often so slight as to escape notice. Suffocative paroxysms are but rarely seen in this form of the disease. The strongest paroxysms of coughing often take place at the time of the critical phenomena. This depends probably on a more abundant secretion taking place in the mucous membrane and the walls of the alveoli.

The respiration is always quickened, and that out of proportion to the pulse, so that instead of being about one-quarter as frequent as the pulse, it is one-third or one-half as frequent.* It is usually regular in rhythm; at times it is sighing, and in pneumonia of the apex irregular. Expiration is sometimes accompanied with a moan. There is often a pause at the end of inspiration, but none between expiration and inspiration. A symptom, which is generally observed in the earlier days of the disease, is movement of the alæ nasi so as to expand the nares —this is either synchronous with inspiration or immediately precedes it.

Sputa are seldom seen in young children. Under the age of 3 or 4 years they are of great rarity; above this age they are occasionally ejected, especially when the cough gives rise to choking or vomiting.

Pain in the chest is very generally present, but is often not ascertained on account of the age of the patient and his inability to define his sensations. In older children the pain may be referred to the hypochondrium of the affected side, the lower part of the axillary region near the junction of the fourth, fifth, and sixth ribs to the sternum, or in the epigastrium. Children not unfrequently complain of "pain in the

---

* The frequency of the respiratory movements is much affected by the extent of pulmonary tissue inflamed; the co-existence of bronchitis also adds much to the frequency of it.

stomach or belly" in pneumonia; this probably arises from a misapplication of the words.

Skin and Mucous Membranes.—A general redness of the skin, which has been mistaken for scarlatina, is occasionally seen. It is more of a rose colour than the rash of scarlatina, and is seen in isolated and well-defined patches. It occurs at the outset of the disease, within the first 24 hours, and disappears in the course of a few hours. It was seen in the case of W. H. S. (page 38), and was mistaken for scarlatina. It is a symptom of rare occurrence.

Redness of the cheeks is a common symptom; it is sometimes limited to one cheek, but not necessarily the one corresponding to the side on which the lung is affected, or the side on which the patient usually lies.

Herpes of the lips or nose is often met with about the third day of the disease, it is not necessarily of good omen, though it frequently accompanies mild cases.

Perspiration is generally met with at the commencement of the crisis, and precedes the depression of temperature. It usually lasts, with intermissions, from 12 to 24 hours, but sometimes it recurs at intervals for 36 or 48 hours, or even longer. In cases ending fatally it is not seen till the seventh or eighth day, and usually lasts till death.

Jaundice is a symptom in pneumonia of the right lower lobe, but rarely met with.

The mucous membranes are usually injected, especially the conjunctivæ, the nares, and the mouth, during the height of the fever. The nares are dry until resolution commences, when coryza often shows itself.

Digestive Functions.—Vomiting is met with in half the cases of pneumonia at the outset. It is usually violent; the vomita consist of food, clear fluid, and bile. Sometimes the vomiting lasts 24 hours or even 2 or 3 days, as in the case of Holman (page 34).

Diarrhœa occasionally accompanies the sickness, at other times the bowels are confined, not usually obstinately costive.

There is complete loss of appetite and much thirst during

the febrile stage. Appetite often returns immediately after the crisis.

There is great muscular prostration caused by pneumonia, so that patients usually keep their beds from the first. Loss of flesh and anæmic pallor are also seen during convalescence.

Urine is usually reduced in quantity. The chlorides are either absent or diminished in quantity. Albumen is but rarely present in the urine in primary pneumonia.

Nervous System.—The cerebral symptoms which often accompany pneumonia have been already referred to. Headache is more frequently present than not ; convulsions, drowsiness, and delirium are only occasionally seen.

Physical Signs.—Inspection reveals but slight difference in the expansion of the two sides of the chest ; exaggerated respiratory movements of the cervical muscles are not commonly observed, as in the case of broncho-pneumonia. On measurement there is sometimes found a slight increase of circumference on the affected side, when the entire lower lobe of the left lung is hepatized or the middle and lower lobes of the right lung.

On palpation increased vocal fremitus is to be felt during crying, or singing, or loud speaking. This sign cannot always be obtained in young children, as they will not always cry, sing, or talk loudly when one wishes them to do so.

Percussion signs are much the same in the child as in the adult ; there is always increased resistance over the affected part, the pitch is raised, the duration of the note is reduced, and the quality is either wooden, amphoric, or tympanitic. The upper part of the lung sometimes gives a tympanitic resonance when the lower half of the lung is hepatized. In rare cases a distinct crack-pot sound has been elicited on percussion under similar circumstances.

On auscultation the fine crepitation, usually heard in adults during the advance of the disease, is sometimes heard in children, but it is not unfrequently missing. Instead of it a faint vesicular respiratory murmur, or an indistinct respiratory murmur with faint bronchial expiration, is heard at the earlier stage of the disease. On the second or third day true bronchial or tubular respiration is to be heard. When the pneumonia begins centrally, as it often does in the upper lobes, bronchial breathing is sometimes not to be heard until the fourth or fifth day of the disease. In these cases bronchial breathing is heard first in the supra-spinous region, and soon afterwards under the clavicle. Bronchophony is a sign usually met with, and when present is of the same value in the child as in the adult.

When pleuritic effusion complicates pneumonia, the percussion note is absolutely dull, the sense of resistance is increased, the depressions of the intercostal spaces obliterated, the semi-circumference of the chest is increased, and the vocal fremitus diminished. Bronchial breathing is sometimes exaggerated, and occasionally a cavernous sound is heard. It might be expected that the respiratory murmurs would be reduced in intensity, but this is seldom or never the case; the layer of fluid effused is not sufficiently thick to produce this result.

Crepitation is more frequently heard during resolution than at an earlier period. It is quite possible, in the child as in the adult, to be misled by a crepitant sound heard on full inspiration, in a lung which is not habitually expanded by tranquil respiration; this sound will not be heard after a few full inspiratory movements.

Diagnosis.—The absence of sputa, the frequent absence of rigors, the occasional occurrence of head symptoms, the slightness of pain in the side and of the cough, concur to make the diagnosis of pneumonia in the child, in some cases, a matter of considerable difficulty.

From bronchitis it is distinguished by greater heat of skin, less dyspnœa, and by the physical signs of pulmonary consolidation.

From *atelektasis* and *pulmonary* collapse it is distinguished by the elevation of temperature and the flushed face, in contradistinction to the low temperature and the pale or cyanotic face seen in atelektasis. In the latter disease the respiratory murmur is weak and not bronchial, and the voice is not bronchophonic.

Tubercular infiltration of the lung resembles pneumonic consolidation in physical signs. The course of the disease is less rapid, but it may come under notice in such a way that the earlier stages of it cannot be determined. The antecedents of the child and its hereditary predispositions must be taken into consideration. The temperature in tubercle will be found elevated, but not to the same degree as in pneumonia; there will usually be some periods in each 24 hours when very great remissions occur, whilst this is not observed to any marked extent in pneumonia, until a great fall occurs near the crisis, and then oscillations last only for a day or two.

From *pleurisy* the diagnosis will be made partly by the temperature, which is more elevated, but chiefly by the physical signs; in pleurisy the dulness begins below and extends upwards, the upper limit being nearly straight; there is bronchial or diffused blowing respiration and diminution of vocal fremitus, with sometimes a diminution of vocal resonance. It is the exception to hear friction sounds; when there is much effusion the intercostal depressions are effaced, the side is enlarged, and the heart, liver, or spleen is displaced.

From *meningitis* the case must be distinguished by a physical examination of the chest. There may be very little to excite suspicion of pulmonary mischief; usually, however there is some movement of the alæ nasi, some quickening of the respiration, and an occasional cough; in all cases it is wise to examine the chest. The temperature in tubercular meningitis is much lower than in pneumonia, and is usually subject to remissions unlike those of pneumonia.

Prognosis.—Primary lobar pneumonia of children is a disease in which the mortality is very low if the patients are judiciously treated. Ziemssen gives, as the result of 201 cases treated by him, 7 deaths from the primary disease, 2 deaths and 2 cases of chronic illness from secondary diseases, either tuberculosis or induration and bronchiectasis. Rilliet and Barthez lost about one-eighth of their patients; Lesczinsky lost 25 out of 245 patients. Ziemssen's practice has been in a small town on the sea coast amongst a very healthy population, and this may explain the more favourable character of his results. The circumstances which most seriously modify the prognosis are the previous health of the patient, the elevation of temperature, and the extent of pulmonary substance involved. The occurrence of pericarditis or meningitis, which is very rare as a complication, would add much to the seriousness of the prognosis.

It is always a serious omen when the seventh day is passed without any decided fall of temperature; still more so, if by the end of the ninth day no such depression is observed.

Treatment.—Blood-letting, which was formerly considered a *sine quâ non* in the treatment of pneumonia, is now for the most part entirely discarded. I have never had occasion to resort to it. Cases might, however, arise in which it would be right to recommend it. If the disease had only commenced 24 or 36 hours previously, if very much pulmonary tissue were inflamed, the pulse full and bounding, the dyspnœa and pain very great, the temperature 105° or more, and the previous health of the patient had been good, it would be wise to let blood from the arm to the extent of a few ounces. Such a case would be likely to occur in a very strong child who had been exposed to intense cold for several hours, or had fallen into water and kept on its wet clothes for a long time.

Usually, however, the best treatment is to keep the patient in bed in a room about 60°, well ventilated, without a draught, to give a simple saline mixture, containing citrate of potash or nitre, a milk diet during the height of the fever, and, when the temperature falls, some good beef-tea. Pneumonia must be regarded as the local manifestation of a general disease in the great majority of cases. The tendency of the disease in children is towards recovery; the great point is to do nothing to interfere with a rapid convalescence. Antimony is seldom necessary or desirable; if given at all it should be confined to those cases in which the pulse is full and strong, the temperature very high, and the skin and mucous membranes very dry and injected;·and it should be given only for a short time at an early stage of the disease.

The local application of cold over the affected part, by means of ice or cold wrappings, has been recommended by F. Weber, of Kiel,* and Niemeyer,† and tried by Ziemssen in many cases. According to the latter writer, it often has the effect of relieving pain, but it does not do anything towards reducing the fever or shortening the duration of the disease.

Counter-irritation is an expedient not much to be relied on in pneumonia. Where there is severe pain in the side a mustard plaster is often of service. Blisters are seldom or never to be recommended; certainly not in the acute stage. If resolution goes on very slowly and there is persistent pleuritic pain, an occasional flying blister will be of service.

Digitalis is strongly recommended as an antiphlogistic in pneumonia; its effect on the circulation requires to be carefully watched. I have not myself had any experience in its use in the treatment of this disease. I believe calomel is a drug which is not to be recommended except as an occasional aperient.

If the pneumonia is complicated with bronchitis, and the bronchi contain much mucus, an emetic or stimulant expectorant, such as senega, carbonate of ammonia, or benzoic acid, is indicated.

When fever has subsided, if the cough is very troublesome, anodynes may be used, such as hydrocyanic acid, henbane, or morphia. If there is much accumulation in the bronchi, opium is contra-indicated, lest the sensibility of the mucous membrane should be too much reduced, and the bronchi remain clogged for want of expulsive cough.

For diet, during the height of the fever, fluid should be freely administered — barley-water, toast-and-water, or soda-water; milk may be given with a little arrow-root, corn-flour, or sago; mutton or chicken-broth and beef-tea are also allowable if the patient is naturally weak, and in all cases, when the

---

* " Beiträge zur Anatomie der Neugebornen." Kiel, 1852, vol. ii. p. 63.
† " Lehrbuch der Speciellen Pathologie und Therapie." Vol. i. p. 147.

fever is subsiding, egg and milk or lightly boiled egg may also be given under the same conditions. Wine is not often required, unless the patient be in a very cachectic state, and does not take nourishment well. It may be often given on the subsidence of fever with advantage.

During convalescence the use of iron in a mild form is of service, steel wine (Ph. L.), or ferrum redactum ; quinine is also useful at this stage. A change of air is often of great benefit. Care must be taken against exposure to cold, but it is desirable that the patient should go out of doors well protected by warm clothing if the weather will at all admit of it.

CASE I.—*Right Pleuro-pneumonia, fatal on 11th day; not seen by me until the 9th day. Caused by exposure to wet.*

Amelia F., æt. 8, a healthy child, was exposed to the wet for some time on 12th June. On the following day, at 11 a.m., she had a fit of shivering, with pain in her right side. She was treated by a medical man, but continued to get worse, so that on the 21st June she was brought to the Hospital for Sick Children. I found her lying on her left side. Face pale, skin hot ; she dreads any movement of her body. Much expansion of nares with respiration. Tongue furred on dorsum, red at edges and tip. Pulse 126 soft. Respiration 75. Temperature 104·6°. Patient is drowsy.

Thorax.—There is dulness over nearly the whole *right* lung, and tenderness on pressure. The base of the lung both anteriorly and posteriorly is less dull than above the middle. There is a coarse friction sound audible over the upper part of right lung, behind as low as the 7th rib, and in front as low as the 3rd rib. From the 3rd to 6th rib anteriorly there is a fine bubbling sound, and here the percussion sound is rather tympanitic. On the *left* side percussion is normal, and respiration is exaggerated. ℞. Pot. iodid. gr. j ; vin. ipecac. ♏v ; æther. chlorici, ♏v ; mist. camph. ad ʒss. 4tis horis. Wine 2 oz. Catapl. Sinapis to anterior aspect of right lung.

June 22nd. Patient looks better. Pulse 118. Respiration 64 easier. The percussion dulness now extends to base of right lung. Respiration over the whole lung is bronchial, with friction in some parts and coarse crepitation in others. Left lung as before, except that there are some moist rhonchi near the angle of scapula. Heart's sounds indistinct. Apex beats 1¼ inch below and ½ inch outside nipple. Cough troublesome, but not so much pain as yesterday. Urine scanty, contains a quantity of albumen. Drowsiness continues. Tongue brownish fur, red at tip and edges. In the evening temperature 105·2°. Four leeches were applied to the right side ; these gave marked relief for a time.

23rd. Delirious through the night. Bowels open three times in the night. Tongue more furred, of a creamy colour. More movement of alæ nasi. Pulse 136 regular, small. Respiration 62. Temperature 104·8°. Only the front of the chest examined in consequence of weakness of patient. The respiration sounds on right are faint, except near the nipple, where they are bronchial and accompanied with bubbling sound. ℞. Ammoniæ sesquicarb. gr. iss.; liq. ammon. acet. ʒj; syrupi, ʒj; aquæ, ad. ʒss. 3a. qq. hora. During the day the dyspnœa became more urgent, and patient was weaker. Dry cupping was administered without relief. Death occurred at 7 p.m.

Autopsy.—Body well nourished. On cutting through costal cartilages on right side much fluid escapes. Right lung extends over middle line from 1st interspace to 5th rib; adhesions of medium firmness over middle of upper lobe, and a few posteriorly over both upper and lower lobes. Pleural surface of posterior part of this lung injected, and on diaphragm, covered with granular lymph. Apex of this lung is externally puckered, and of a pale colour. On section four-fifths of the upper lobe is hepatised; the lobular structure is well mapped out. The middle lobe is œdematous and pale; it is bound to the lower lobe by vascular adhesions. The greater part of the lower lobe is solidified, breaking down under the finger, not exhibiting the lobular structure nearly so well as the upper. Bronchial glands at root of lung engorged with bloody serum. The pleura contained about 12 or 14 oz. of clear fluid. Right lung weighs 24½ oz. Left lung weighs only 6¼ oz. A little thin frothy mucus escapes from the bronchial tubes on section. In the lower lobe, just beneath the pleura, is a small nodule of the size of a pea, of cartilaginous consistence on section, and externally of an opaque white colour. In the centre it contains some yellow gritty matter enclosed in a cyst. Other organs appear healthy, except the kidneys, of which the right one is rather congested, whilst the left is pale, flabby, and the cortical substance is unduly opaque.

Remarks.—This case presents an illustration of intense pleuro-pneumonia from exposure to wet.

The treatment during the first week of the disease is not known. At the time when she came under my care, all treatment appeared futile. The leeches, which appeared to give temporary relief, had probably better not have been applied at so advanced a period of the disease.

CASE II.—*Pneumonia of Right Apex. Rapid convalescence from 6th day.*

J. H., a healthy boy, aged 9 years, was in his usual health until the morning of the 25th April, when he returned from play at 10 o'clock complaining of frontal headache; he was chilly and vomited. His bowels were costive, and the motions offensive. His breathing was hurried.

April 26*th*. He became hot, and towards evening delirious. The delirium returned towards evening, until the 30th. His mother states that he has had a cough with some expectoration for some time past. It became worse on the 25th. The sickness continued after food or drink until the 29th. He complained of pain under the middle of the sternum from the first day of illness; it was increased by a deep breath. He was admitted into the Children's Hospital under my care on the 30th April. He was weak, but able to walk into the ward. On each angle of his mouth, and on the upper lip, were small patches of herpes of two days' standing. Pulse 120 weak. Respiration 34. No action of the nares. Tongue red, moderately moist, with a slight whitish fur. His cough was frequent and troublesome. Temperature in axilla 100°. There is evidence of consolidation over the upper part of right lung, with fine dry crepitation at the end of inspiration. Coarse rhonchi under left clavicle.

May 1*st*. He passed a tolerable night, but wandered occasionally. He seemed much better. Temperature 99·75°. Cough easier. No pain in chest.

2*nd*. Percussion under right clavicle, very notably improved, and the respiration much less bronchial. Pulse 116. Respiration 30. Temperature 99°. An abundance of chlorides in the urine. On admission he was ordered acetate of ammonia and citrate of potash, with 2 oz. of wine; beef tea, and milk diet. He made a rapid recovery. On the 5th May he was anxious to get up. He seemed quite well.

CASE III.—*Pneumonia (right upper lobe). Delirium and pain in head for several days. Sudden fall of temperature during* 8*th day.*

Francis Hawk, æt. 8, said to be always strong and healthy. No hereditary tendency to phthisis. On 21st November, having been in usual health all day, without any obvious cause, refused to take his tea, vomited, and brought up the remains of his dinner. He asked to go to bed, saying that his head ached in front. Whilst in bed he became very hot and thirsty, and continued to retch every quarter of an hour, bringing up phlegm.

Nov. 22*nd* (2nd day). The retching continued through the night. This morning took a "fever powder." Bowels acted three or four times loosely.

23*rd* (3rd day). Retching all day. Much thirst. Bowels very loose, motions of a dark brown colour.

24*th*. (4th day). Still continuing to retch, bringing up phlegm. Towards evening he had an attack of shivering. At 8 p.m. was seen for the first time by a doctor; he was then delirious, trying to get out of bed. The doctor said there was "congestion of lungs and a fever," and ordered a mustard plaster to back of the chest.

25th (5th day). Continued restless and delirious.

26th (6th day). Still wandering. Seen by another doctor, who said there was congestion of lungs with typhoid fever, and advised his being brought to the hospital.

27th (7th day). Notes on admission.—Was very delirious last night, crying out, " Oh, my poor head ! " Mother thinks he has suffered from headache for several days during delirium. Bowels have continued loose. Has been hot and thirsty. Skin now very dry. Has had a slight cough since 21st, but not troublesome till yesterday. Has complained since 21st of abdominal tenderness ; yesterday was afraid to be touched near the epigastrium. Mother states that breathing has been difficult for several days and she has noticed his nostrils working. Pulse 132. Respiration 56. Tongue white. Under right clavicle, and at upper part of right lung behind, percussion not deficient in resonance. Respiration somewhat tubular in front, and blowing with a slight tubular quality behind. Some fine rhonchus heard behind on coughing.

27th, 9 p.m. Pulse 136. Respiration 56. Temperature 103°. Slight action of nares.

28th, 10 a.m. Was very delirious all night, tried to get out of bed ; frequently called for water. Has passed three stools in last 24 hours. Urine 34 oz. Has taken 1¼ pint of beef-tea and about as much milk. Pulse 132 moderately full, slightly dicrotous. Respiration 60. Very slight action of nares. Mouth partly open. Tongue dark red at tip and edges, elsewhere furred. A slight dusky red colour in cheeks. Temperature 104·2°. Skin dry. No enlargement of spleen or liver. Abdomen not distended or retracted.

Chest.—Some pain near right nipple, increased by pressure. The physical signs presented are those of consolidation of the upper half of right lung, back and front, with a distinct friction sound near right nipple. No sign of mischief in the left lung.—3 *p.m*. Pulse 132. Respiration 64. Temperature 103·2°.—9·30 *p.m*. Pulse 108. Respiration 52. Temperature 98·6°.

29th. Looks much better. Says that he feels better. No delirium since yesterday afternoon. Physical signs as before. Pulse 108. Respiration 36. Temperature 97·8°. Scarcely a trace of chlorides in urine.

30th. Tongue clean, moist. Pulse 112. Respiration 40. Temperature 98°. Urine 1018, acid, contains more chlorides. Bowels open.

Dec. 1. Less dulness at upper part of right lung. At 3rd interspace in front a percussion note almost of amphoric quality is heard. Respiration is less tubular and more blowing than it was ; a medium-sized rhonchus is heard under the right clavicle on coughing. Patient seems much better. Urine turbid, acid, 1016, contains abundance of chlorides, no albumen. Pulse 108. Respiration 36. Temperature 98·2°. From this time he continued rapidly to improve, the temperature never exceeding 98·3°. By the 8th December the hepatised lung had undergone resolution, the physical signs having resumed almost their normal character.

Remarks.—This case presents several features of interest. The continuance of delirium from the fourth to the eighth day, the occurrence of rigors on the fourth day, the persistence of diarrhœa for many days, and sickness for two or three, are not usual symptoms. The *sudden* fall of temperature between 3 and 9 o'clock on the 28th of November, at the end of the eighth day of illness, was very marked.

CASE IV.—*Pneumonia (left lung, lower lobe). Slight Endocarditis. Convalescent from 7th day.*

Fanny Keefe, æt. 6. Attack sudden; shivering followed by heat, anorexia, pain in head and left shoulder. Delirium during 2nd and 3rd days. Admitted on 4th day. Countenance dusky, drowsy. Pulse 144. Respiration 32. Temperature 103°. Signs of consolidation of lower part of left lung. On 6th day, pale, skin moist, temperature 101°. Pulse 144 irregular. Still has pain in left side. On 7th day temperature fell to 98°. Convalescence not rapid. Systolic murmur at left apex, came on during convalescence.

Treatment.—Bicarb. of potash and nitrate of potash, with a purge. On the 6th day wine.

CASE V.—*Pneumonia (right lung, upper lobe), ushering in Hooping-Cough. At the onset rigors, drowsiness, headache, and vomiting.*

C. M., æt. 7, a girl of apparently good constitution, was quite well until the morning of the 27th February, when she complained of feeling chilly, was heavy, and inclined to sleep. She remained so throughout the day, and in the evening complained of pain in her head and stomach. During the night her feet were cold, and her body seemed alternately hot and cold. She was very thirsty, and vomited.

Feb. 28*th.* She took no breakfast, and vomited at noon. She had no cough, her breath was not observed to be hurried; her skin was very hot, and her thirst considerable. Bowels active.

March 1*st.* Bowels acted from castor oil. Seemed quite as ill as on the previous day. Her head ached, she was sleepy and flushed; but had no cough.

2*nd* (4th day). A short hacking cough was first noticed; this increased through the night, and became very troublesome the next day.

3*rd* (5th day). No expectoration. Rather less fever. Bowels now open twice a-day; motions semi-relaxed, dark and slimy. A patch of herpes appeared on the lip. She continued in much the same state until her admission on 6th March. Her skin was then rather hot and dry, her face not flushed. Her tongue was red with elevated papillæ at the tip and sides; a dryish white fur on the dorsum. Fauces red and angry looking, not ulcerated. Pulse 128, of moderate volume, sharp and feeble. Respiration 60 shallow, with moderate action of alæ nasi. Patient has a loose cough. Right lung: *anteriorly* from beneath clavicle to level of nipple there was almost absolute percussion dulness; below this, for 3 or 4 inches, marked amphoric resonance; from clavicle to nipple breathing is tubular and voice-sounds bronchophonic. For 2 inches below nipple there is a crackling, somewhat creaking sound like friction, with inspiration and expiration. *Posteriorly* there are also signs of consolidation over the upper half of right lung. Over the left lung there are no abnormal physical signs.

Urine, high coloured, free from albumen, with a very scanty amount of chlorides.

Treatment.—Liq. ammon. acet. ʒss.; pot. iodidi, gr. ij; misturæ ipecacuanhæ opiatæ, ʒiij. *M.* 4tis horis. Emplastr. lyttæ, 3 × 4, infra clavic. dextram.

7th. Passed a restless night. Coughs occasionally without expectoration. Pulse 124. Respiration 48, with less action of nares. Urine contains more chlorides.

8th. Slept well. Appetite improving. Cough troublesome, with an occasional hoop. Pulse 124. Respiration 36, without movement of nares. Physical signs in thorax much the same.

11th. There is still consolidation of upper lobe of right lung. Respiration still blowing in character before and behind, within former limits, but less tubular. There is still a creaking sound at the level of nipple. Urine contains an abundance of chlorides. Pulse 120 small and feeble. Respiration easy. She has hooping-cough, and is therefore discharged. No further note was taken of her condition.

CASE VI.—*Pneumonia (right upper lobe). Mistaken at first for Scarlatina, and then for Hydrocephalus.*

W. H. S., a moderately healthy boy, aged 14 months, was taken from London to Plumstead on the 25th June. On the 28th June he was suddenly seized with chills and convulsions, which lasted with intermissions for several hours. The next day a red rash appeared on the skin, which was supposed by the medical attendant to be scarlatina; this disappeared in a few hours. The face was flushed, the skin hot, the anterior fontanelle pulsated strongly, the bowels were relaxed, breathing was hurried, there was occasional shrieking in a shrill tone, and he had a slight hard cough. A leech was applied to the head, 3 gr. of grey powder were given, to be repeated every night; and the parents were told that the child had water on the brain with bronchitis, and that there was scarcely any hope of his recovery. I was called to see him on the 3rd July. I found him quite sensible, having passed a restless night. The pulse was quick, and in proportion to the respiration as 3 to 1. The nares expanded scarcely at all in respiration. The skin was not very hot. On examining the chest I found dulness under the right clavicle and in the right supra-spinous fossa, with tubular breathing and no crepitation. Elsewhere the physical signs in the thorax were normal. I recommended hydr. c. cretâ, gr. j ; pulv. cretæ, gr. iij. Ter die. And a croton oil liniment. In a few days the dulness disappeared at the upper part of the right lung, and the child recovered.

Remarks.—The case is an instructive one, showing the importance of carefully examining the chest in a child seized with sudden cerebral symptoms and fever.

A red rash, simulating scarlatina, has been occasionally observed by others, as mentioned above. The grey powder was, I now believe, objectionable because unnecessary. In similar cases I now give a very simple saline. When I saw the patient the disease had probably almost run its course, and resolution would have taken place immediately without drugs.

CASE VII.—*Pneumonia (left lung, upper and lower lobes). Recovery. Resolution on 9th day.*

M. A. H., a not very strong girl, æt. 7 years. On the first day of illness had pain in head and vomiting. The next day pain in her left side, with heat of skin. Her cough, which had been present for some time, became worse. She continued much in the same state during the next six days. She was not admitted to the hospital until the eighth day. She was then very feverish, flushed ; eyes suffused ; lips dry ; a small patch of herpes under nostrils. Pulse 166. Respiration 64. Temperature 102·4°. A short suppressed cough. Tongue red at edges, dry white fur elsewhere.

Thorax.—Physical signs. *Right side :* Percussion everywhere extrasonant. *Left side :* Anteriorly percussion note is tubular ; posteriorly up to spine of scapula it is rather dull ; there is still a certain amount of resonance ; above the spine of scapula it is tubular. There is but little movement of the left side in respiration.

Auscultation.—*Left side :* In front there is scarcely any breath sound audible ; behind, except at the very apex, respiration is strongly tubular ; in the scapular region there is a little coarse crepitation. Bronchophony perfect. Bowels not open. One dose of calomel. R. Ammon. carb. gr. ij.; liq. ammon. acet. ♏xv. ; mist. ammon. citr. ad ℥ ss. 3a. qq.

hora. At 5 p.m. the temperature had fallen to 99·3°; and at 9 p.m. to 98°.

(10th day). She looks very much better, sits up in bed almost as if well. Temperature, 9 a.m., 101·2°; 5 p.m., 98·2°. Pulse 136 regular. Respiration 48. Tongue moist, less furred. Dulness over left back less than yesterday; tubular respiration continues.

(14th day). Looks well. Pulse 100 regular. Respiration 28. Skin natural. Tongue clean. Bowels open. Dulness over left back is but slight. Respiration rather coarse, but not otherwise abnormal.

CASE VIII.—*Double Pneumonia (right, upper, and middle lobe, left central part). Pleurisy right side. Delirium and Deafness. Convalescence began on 7th day.*

Alice Butler, æt. 8. Not a strong child, was under treatment for rickets when 4 years old, and last year had a troublesome cough. Was in her usual health on 25th October, when she was struck on her right side by another girl, and very much hurt. She complained in the evening of some pain in that side. Passed a good night.

Sunday, Oct. 26th.—Went to church as usual, but on returning home she complained of feverish headache, of pain in right shoulder and side, and shivered considerably. She took no dinner or tea, and went to bed at 8 o'clock.

27th. In bed all day with headache and much fever. Vomited once after taking a powder. No cough was observed.

28th. Too weak to stand.

29th. Notes on admission to Hospital for Sick Children.—Lying on back. No dyspnœa, no action of nares. Skin hot and dry. Occasional cough. Complains of pain under right armpit; pressure near right nipple and under armpit causes pain. She is very weak.

Chest.—Percussion note over middle of right lung posteriorly is higher pitched and more resistant than left. Elsewhere resonance normal. Respiration over same part has a slightly blowing quality; no crepitation. A suspicion of grazing friction sound in right infra-mammary region. Skin hot and dry. Slight flush on left cheek. No headache. Tongue slightly brown. Bowels costive. Last night took calomel, gr. iss.; rhei. gr. viij. Now to have mist. ammon. citratis, ʒij; ammon. carbon. gr. ij. 4tis. horis. Pulse 138. Respiration 36. Temperature 103·4°. Last night patient wandered much, and has continued to do so to-day.

30th. Much delirium last night. Bowels open four times since last note. Stools dark in colour, loose. Left cheek constantly much flushed. Face has a dull, anxious look. Tongue dark brown, rather dry. Physical signs much as yesterday. No eruption. Skin dry. Still complains of pain in right mammary region, also a little near left nipple. Pulse 130. Respiration 28. Temperature 103·8°.—3 *p.m.* Nares act. Dulness at angle of *left* scapula, with blowing respiration. Right infra-clavicular region higher pitched percussion note than left, and respiration diffused blowing.

Diagnosis.—Double Pneumonia. Urine contains a very small quantity of chlorides.

Oct. 31, 10.30 *a.m.* (5th day, end). Was very delirious up to 10 o'clock last night, after which passed a quieter night than she has had before. Bowels open twice. Still has a flush on left cheek. Lips dry. Tongue thickly furred, except at top ; less brown than yesterday. Scarcely any action of nares. Percussion over lower two-thirds of right lung, behind less resonant ; respiration slightly diffused blowing, no rhonchus or friction. (Fluid in pleura probably.) On left side signs as yesterday. Right supra-spinous fossa percussion notably higher pitched and more resistant than left; respiration bronchial, with tubular quality ; well marked bronchophony. Under right clavicle, as low as second rib, percussion high pitched and resistant ; respiration here weak and harsh. Some bronchitic rhonchi at left infra-clavicular region. Urine acid, pale sherry, 1025 sp. gr. Chlorides present in small quantity. Pulse 128. Respiration 38. Temperature 102°.

Nov. 1*st.* (6th day, end). Passed a good night. Takes her food (beef-tea, and milk, and wine) well. Coughs but seldom. Pulse 120 weak. Respiration 38.

2*nd,* 11 *a.m.* (7th day, end). Slept well. Very little cough. Looks better. Skin feels cool. Tongue less brown and moister. Percussion note certainly improved at right back ; a weak diffused blowing respiration is heard over whole right back. Right inter-scapular supra-spinous and infra-clavicular regions dull ; respiration bronchial. Percussion under left scapula now almost normal. Some bronchitic rhonchi over left back. Since yesterday patient has become somewhat deaf. Chlorides now abundant in urine. Pulse 100. Respiration 34. Temperature 99°.

3*rd.*—Much as yesterday. Pulse 118. Respiration 30.

5*th.* (10th day, end). Lies in bed on back, drowsy. Slept well last night. Is decidedly rather deaf. Has much improved in appearance. Dulness over upper part of right lung much less. Respiratory sounds weak over the back, and slightly bronchial in front of right lung. Pulse 108. Respiration 26. Skin cool.

7*th.* (12th day). Percussion note is now but slightly duller right than left. A rub of friction is heard at inferior angle of scapula. Deafness less. Pulse 100. Respiration 24.

12*th.* Much improved. Sits up in bed, reads and plays with her toys. Physical signs of consolidation in right lung not quite gone. Pulse 98. Respiration 24.

17*th.* Is as well now as before the late illness.

Dec. 1*st.* Discharged. Not strong. Right clavicle more resistant and duller on percussion than left ; infra-clavicular region duller right than left. Expiration somewhat prolonged and blowing in this region ; elsewhere physical signs normal.

Remarks.—In this case inflammation commenced centrally in the right lung, and subsequently attacked the left lung. The prostration was out of proportion to the extent of the lung involved. There was delirium, which is not unusual; and deafness, which is very rare in primary pneumonia, though a common symptom in continued fever. I find no mention of deafness as a symptom of pneumonia in writers on this disease.

The diagnosis on admission was not easy; the general aspect of the child was that of typhoid fever; there were delirium and deafness with but little dyspnœa, and the physical signs in the chest might easily have been ascribed to old tubercular consolidation; the ratio of pulse to respiration was not much perverted, 138 : 36. There was, however, more restlessness than is usual in typhoid, and some tenderness on percussing the right side of the chest; the attack was sudden and attended with rigors, which are not usual in continued fevers. The blow on the chest appears to have been the cause of the attack. It is to be regretted that the temperature was not more frequently observed. In pneumonia the temperature is maintained for five, seven, or nine days, and then falls abruptly and maintains afterwards a low level, unless a relapse occurs with fresh inflammation; in typhoid, on the contrary, there is throughout a greater tendency to a morning remission of fever, and for several days before the normal temperature is maintained there is a high temperature in the evening, with a normal temperature earlier in the day.

CASE IX.—*Pneumonia (right upper lobe). Headache, delirium, constipation, and vomiting, prominent symptoms. Fourteen days of languor, nausea, and headache followed by rigors. Fall of temperature on 8th day. Recovery.*

Elizabeth Fitzgerald, æt. 7 years, had usually enjoyed good health, was stout and strong. Her father died of consumption, and his family were said to be all consumptive. She is the fourth child of her mother, who is healthy—one child younger than Elizabeth is said to have had disease of the brain.

Until the 9th December she appeared quite well; about this date began to complain of headache, felt sick and lost her appetite; she was restless at night. She continued to go to school for a week after this; but was obliged to stay at home after the 17th December, having become more languid, headache continuing as before. Her bowels throughout were regular. On December 23rd, she had a violent fit of shivering, vomited frequently during the day, especially after taking anything. This sickness continued almost constantly for three days, when she was seen as an out-patient, and ordered sodæ sesquicarbonatis, gr. v., with misturæ gentianæ c. sennâ, ℥ss. nocte maneque.

Dec. 27th and 28th. Sickness was abating. On admission to the hospital, on 30th December, it was stated that she had been coughing more or less for the past week; had been hot and thirsty for four or five days; had been delirious for five nights; had complained much of headache, chiefly at the back. She had been losing flesh; had occasionally become flushed, especially on left cheek; had frequently started in her sleep; on two or three days she had ground her teeth much.

On 31st December, the following notes of her condition were taken: Lying on right side, with a dull weary expression of face; pupils equal, of medium size; no action of alæ nasi. Lips dry; tongue red, slightly furred, with prominent, rather red papillæ. Skin of face rather opaque, dirty look, with some colour; skin of body generally has a slight tawny tinge. Pulse 124 weak. Respiration 32. Has an occasional short cough. Temperature 99°. Sternum from 3rd rib to ensiform cartilage somewhat depressed. Expansile movements of chest rather deficient on both sides. Percussion note right infra-clavicular region decidedly higher pitched, and more resistant than left. Respiration weak, diffused blowing, with some fine and coarse crepitant rhonchi. Percussion note also dull over posterior aspect of upper third of right lung. Respiration sounds as in front, with some bronchitic râles over the whole back on deep inspiration and coughing. No enlargement of liver or spleen. Abdomen rather flat, not tender. No eruption on skin. Bowels very confined. Some delirium and restlessness during the night. Is now drowsy.

Jan. 1st. Patient slept better, was not delirious. Skin cool. Pulse 120. Respiration 28. Percussion signs same as yesterday. The rhonchi rather coarser in character.

3rd. Pulse 108. Respiration 26. Patient looks better.

6th. Improving. Upper part of right lung more resonant. Cough has quite ceased.

10th. Quite convalescent; allowed to get up. Dulness and other signs in right lung have almost disappeared.

Remarks.—The history of this case led one to suspect cerebral mischief rather than pulmonary. The frequent vomiting, the constant headache, the delirium with constipation and loss of flesh were very much like the symptoms of tubercular meningitis. There was no mention of dyspnœa or pain in the chest, and but slight reference made to cough as a prominent symptom. The occurrence of rigors, the heat of skin, thirst, and cough, made one think of pneumonia as a possible cause of the symptoms.

Without a careful examination of the chest this diagnosis could not have been made. If we had been content with examining the bases of the lungs merely, no light would have been thrown upon the case. When the child came under observation, the fever had almost subsided, and redux crepitation was already audible.

Another disease which was simulated in this case, especially remembering the hereditary tendency, was acute tuberculosis with infiltration of the apex of the right lung; the acute course of the case, with the cessation of fever after the eighth day, excluded this hypothesis. During the fortnight of premonitory symptoms a diagnosis would probably have been impossible; the symptoms appear to have been those mainly of gastro-hepatic disturbance. It is quite possible that had she been properly treated at this time the attack of pneumonia might have been averted.

At the time the patient was admitted no treatment was needed beyond rest in bed and nutritious light diet. After a few days quinine was given.

Cases of pneumonia of the upper lobe of one lung occurring in delicate children with an hereditary tendency to tubercle are often difficult of diagnosis. The signs of consolidation in the lung may be ascribed to tubercular deposit, if the patient's lungs have not been examined at some recent period previous to the attack; the sudden accession of fever, with great heat of skin, some dyspnœa, and a deficiency of chlorides in the urine, would prove the pneumonic character of the attack.

PLEURISY.

THIS is a more common disease in childhood than is generally supposed. Its presence is often overlooked, the symptoms not being well marked, and the chest not being examined. When an examination is made it is often mistaken for pneumonia. I have frequently seen cases of pleurisy diagnosed as "Infantile Remittent," or when chronic, as "Phthisis;" or "Disease of the Liver," which is often a cloak for ignorance.

Statistical returns fail to give a correct impression of its frequency. Of 1756 deaths from pleurisy in England during 1856 and 1857, according to the Registrar-General, there were 185 of children under 5 years of age, or 10·5 per cent.; between 5 and 15 years there were 101 deaths, or 5·7 per cent. of all the deaths from this disease.

The deaths from pleurisy at all ages were only 2·1 per 1000 of the deaths from all causes; under 5 years of age, the deaths from pleurisy were only ·05 per 1000 of the deaths from all causes; between 5 and 15 years of age, the deaths from pleurisy were 1·8 per 1000 of the deaths from all causes.

In London, during 1842 and 1845, of 202 fatal cases of pleurisy at all ages, only 14, or 6·3 per cent. were under 5 years of age; in 1856-57, of 282 deaths from pleurisy, 38, or 13·4 per cent. were under 5 years of age; and in 1863, of 155 at all ages, 16, or 10·3 per cent. were under 5 years. The deaths from pleurisy appear to have been more numerous in 1863 than 20 years before that date, and a larger proportion of these deaths occurred in young children at the later date than at the earlier one. The difference is probably due to more exactness in diagnosis than to any real difference in the frequency or fatality of the disease.

In the Hospital for Sick Children, of 4100 patients admitted between 1860 and 1866 there were only 123 cases of pleurisy, of which 21 were fatal, the majority of which, if not all, were cases of empyema, and many of them secondary to other diseases. Idiopathic pleurisy is much rarer than pleurisy secondary to tubercle, to scarlatina, and to Bright's disease.

I do not propose to enter at length into a general description of pleurisy, but to dwell mainly on the points in which this disease, when it attacks children, differs from pleurisy when occurring in adults.

Pain in the Side.—When this symptom is present it is an important one. It is more frequently absent in children than in adults, and when it exists in infants it is often not ascertained by the physician or nurse. There is nearly always uneasiness or discomfort of some sort at the earlier periods of the disease. There is a general dislike to being disturbed, and sometimes an appearance of general soreness of the surface. Pain is often referred to the abdomen, especially the hypochondriac regions; this is increased by pressure from interference with the descent of the diaphragm. Sometimes there is pain in the shoulder and arm of the affected side.

There is usually, as in the adult, *cough*, which is short, small, and stifled, causing pain; occasionally this symptom is almost absent throughout. A certain amount of *dyspnœa* generally exists at the outset, and later if there is much effusion into the pleura. As a general rule, the younger the child the more marked is this symptom.

The *temperature* in pleurisy may reach as high as 103° F.; there is usually a remission during some part of the 24 hours. The very high temperature is maintained only for a few days, it then *gradually* falls (not suddenly as in pneumonia), and it does not attain the normal level until absorption has taken place, varying from 99·5° to 101° during that stage. In case of empyema there is an increase of temperature towards evening even as high as 103°, with a morning remission to about 100° or 99°.

The *pulse* in pleurisy is very quick at the commencement; in cases of great effusion, by compression of lung the right ventricle is unable to drive the blood into the left side, consequently the arterial system is not distended, and the pulse is small and weak. This condition will be aggravated by pericardial effusion, or by abundant effusion in the left pleura, displacing the heart, and thus impeding its action. Frequency of pulse usually continues as a symptom after a considerable remission of temperature has taken place, it is more than in the convalescence from pneumonia, quickened by bodily movement or mental emotion.

The *urine* is scanty in amount, of high specific gravity, and dark in colour during the height of the fever ; as the pleuritic fluid is absorbed it becomes more abundant, of lighter colour, and less dense. Albumen and casts of tubes have been observed in a few cases of idiopathic pleurisy ; they have disappeared as the fever subsided. Urine becomes much more abundant immediately after the evacuation of a quantity of fluid by paracentesis. (Traube, "Ueber den Zusammenhang von Herz und Nierenkrankheiten.")

There is considerable loss of strength even in mild cases, together with emaciation and pallor. Recovery, however, takes place rapidly.

In chronic pleurisy emaciation sometimes occurs to a very marked degree ; this appears to depend on the loss of appetite and the waste due to the purulent nature of the exudation.

Vomiting, usually of a bile-coloured fluid, occurs in about half the cases as an early symptom; it is rarely seen in adults. Constipation is usual at first, alternating with diarrhœa if the case become chronic.

Headache is usually present when the patient is old enough to explain his sensations. *Convulsions* are rare. I have seen them once in a girl aged 7 years ; they lasted 4 hours, in Case III. (Isabella Irons).

Delirium is not often present, but sometimes exists, as in Jessie Trimm (Case II.).

Drowsiness is occasionally observed. It was a very marked symptom in two cases recorded in my note-books.

Physical Signs.—On inspection, in addition to the feebler expansion of the affected side, there is often seen at the outset a drawing-in of this side, with a voluntary curving of the spinal column in the upper dorsal region towards the affected side. An appearance is presented similar to that subsequently produced by absorption and contraction. As exudation increases, and pain diminishes, this contraction gives place to dilatation and obliteration of the intercostal spaces.

Ziemssen mentions a phenomenon which he has once observed, pulsation of the pulmonary artery, both visible and palpable. It occurred in a boy, five years old, with very extensive contraction of the chest from pleurisy. The cause assigned was the pressure of the lung from behind, by which the artery was pushed directly on to the thoracic wall. I have not met with any cases of pulsatile empyema simulating aneurism, such as are described by Dr. Walshe in adults.

Displacement of the Heart to the right, or the left, is always to be felt in cases of great effusion. It is remarkable that Barthez and Rilliet make no mention of this sign. It occurs at any age; it was well marked in a child aged only five months, on whom paracentesis was performed. The liver and spleen are also pushed downwards in cases of abundant effusion.

Vocal Fremitus is diminished, but this sign cannot usually be much relied on, especially in young children, because the vibration of their vocal cords are not often transmitted to the chest walls even in health, and because of the difficulty of inducing them to exert their voices powerfully at the right moment, it sometimes may be effected by making them cry. If present on the healthy side, it is reduced or nil on the diseased side.

Percussion.—In case of slight exudation there is slight dulness on superficial percussion; this is often not readily detected owing to the restlessness of the patient. When the fluid is abundant, dulness becomes absolute. At the upper anterior part of the chest and in the interscapular region, however, where compressed lung comes near the surface a tympanitic note is usually heard. The displacement of the heart, liver, or spleen may be determined by percussion and palpation. Sometimes, but not often, the limits of dulness may be changed by changing the posture of the child. When this can be clearly made out, it is a very valuable diagnostic mark.

Auscultation.—It is very seldom in the child that there is complete absence of respiratory murmur, even over a large amount of fluid.

Bronchial Respiration of a diffused blowing character is a more important sign of pleurisy in children than in adults. It is met with very early in the disease (Barthez and Rilliet say as early as the first day), and it may persist in fatal cases to the last, notwithstanding very large accumulations of fluid. It is not so tubular or so metallic as the respiration often heard in pneumonic consolidation. It is frequently mistaken for the sign of pneumonia, and has often led to errors in diagnosis.

The bronchial respiration may persist after a large amount of fluid has been effused and the chest walls are much distended, so long as there is dyspnœa and the patient is able to draw powerful inspirations. As the fever subsides and the breathing becomes more tranquil the respiratory murmur may become weak or nil. Together with bronchial respiration there is also often some degree of increased vocal resonance, a fact which renders the diagnosis of pleurisy in children much more difficult than in adults. It appears to depend on the smaller dimensions of the child's chest, and the consequent comparatively thinner layer of fluid. There is, however, sometimes as in the adult a diminution of vocal resonance.

When there is pulmonary consolidation in addition to pleuritic effusion, there is usually an exaggeration of the bronchial character of the respiration, or even a true cavernous character is heard, and if this is combined with rhonchi a cavity

E

in the lung may be suspected. The voice is still more resonant, and the dulness is even greater on percussion. The diagnosis from physical signs in these cases is sometimes most perplexing.

Ægophony is occasionally, but not often, detected at the upper limit of dulness. The child's voice is, in health, somewhat ægophonic, so that this sign is not one that can be much relied on.

Friction Sounds are very seldom heard except during the stage of absorption. This is owing partly to the character of the exudation, and partly to the shallowness of the inspiratory movements.

Metallic Tinkling is a very rare phenomenon in the child. Ziemssen says he has never heard it in children. When present it is unmistakeable evidence of air and fluid in the pleura. It occurred in Trimm's case, and lasted for 3 or 4 days; about the same time occurred suddenly a copious expectoration of purulent fluid. There is no doubt that a bronchopleural fistula was here established. (See page 55.)

Pleuritic fluid in children, as a general rule, becomes purulent much earlier than in the adult. It is sometimes purulent from the first. The matter often points externally, and this may be either in any space from the second to the sixth space, or even lower. There is more tendency in the child for the pericardium to become involved by extension from the pleura.

Diagnosis.—The chief diseases with which acute pleurisy may be confounded are pneumonia, hydrothorax, typhus or typhoid fever, acute hydrocephalus, pericarditis, and some abdominal affections. The onset may be as sudden as in pneumonia, but rigors or convulsions are more rare; the elevation of temperature is not so great, not often exceeding in the axilla 102° Fahr., it falls more gradually, not suddenly to the normal level. Dyspnœa is marked only at the outset, and when there is very great effusion. In regard to physical signs, there is impairment of resonance both in pleurisy and pneumonia, in the child the dulness of pleurisy in the earlier stages is less marked than in the adult; in pneumonia there is sometimes heard at an early period fine dry crepitation, but in pleurisy a distinct friction sound is seldom heard. The respiration is of a diffused blowing character in pleurisy and more tubular and harder in pneumonia. In pneumonia, there is sometimes detected increased vocal resonance and fremitus, whilst in pleurisy vocal fremitus is diminished, and vocal resonance is sometimes increased, or at others diminished.

The ratio of the frequency of the pulse to that of the respiration is higher in pleurisy than in pneumonia. In the latter the respiration may be half as frequent as the pulse, in the former it is seldom more than one-third as frequent. The voice does not aid the diagnosis in the child as it does in the adult. If vocal fremitus is increased we may conclude that there is consolidation of lung; if decidedly diminished we may diagnose pleuritic effusion, but very often we fail to get fremitus either on the healthy or diseased side. If vocal resonance is much increased consolidation is probably the cause, if slightly increased there may be pleuritic effusion; there is decided deficiency of vocal resonance only when there is very considerable effusion in the pleura. When ægophony is well marked pleurisy may be diagnosed. Percussion in different postures may aid the diagnosis in pleurisy. The absence of sputa in the child renders the diagnosis more difficult. When there is distension of the affected side, with obliteration of the intercostal spaces, and displacement of the heart, the mediastinum, and the liver or spleen, the diagnosis is easy.

When broncho-pneumonia or collapse of lung is limited to one side, the diagnosis of these affections from pleurisy is not always easy. Vocal fremitus is important when the voice of the child is strong enough to produce it. From temporary obstruction of a bronchial tube, however, the vocal fremitus may be reduced instead of intensified in collapse or bronchial pneumonia. A single examination is therefore not always sufficient to render a correct diagnosis possible. The history of the case will throw light on it; suddenness of attack and great pain would point to pleurisy.

Hydrothorax is usually double, but not always. After scarlatina the diagnosis is sometimes very difficult, especially seeing that both hydrothorax and pleurisy are common at this time. There is but little cough, pain in the side, or dyspnœa at the commencement of hydrothorax. The fluid is more easily moveable, and the limits of percussion dulness are more readily affected by position in hydrothorax. In pleurisy, too, the dulness is usually, but not always, higher behind than in front when the patient sits upright; whereas in hydrothorax the upper margin of dulness is almost horizontal, or describes a parabolic course, as pointed out by Damoiseau (Archives Générales, 1843). Occasionally, in chronic pleurisy, dulness may extend higher before than behind, effusion being limited.

The early symptoms in pleurisy sometimes lead to a suspicion of *cerebral* mischief. There may be vomiting, severe headache, and screaming, with irregular respiration. Hurried breathing is an occasional symptom of cerebral disturbance. The cough may be very slight or quite absent. *A careful examination of the chest in all doubtful cases of acute disease in children is very important;* without it, frequent errors in diagnosis will be made. There would probably be found in a case of pleurisy, slightly impaired resonance over the lower part of one lung posteriorly, and weak rather bronchial respiration at the same part. The onset is almost too acute for tubercular meningitis, and the attack probably has not been preceded by loss of flesh, capricious appetite, and constipation. There is occasionally observed in pleurisy of infants a symptom which is also not uncommon in cerebral disease, namely, a drawing down of the head towards the shoulder of the affected side.

The statement of children that they have pain in the belly with some tenderness on pressure in the hypochondrium, together with vomiting, may lead to the suspicion of abdominal rather than thoracic mischief. Careful examination, both of the chest and abdomen, will decide which part is at fault.

Rheumatic Fever in children often occurs with but little implication of the joints. There is a great tendency to endo- and peri-carditis, and sometimes to pleurisy.

Pericarditis of rheumatic origin, or from other causes, may simulate pleurisy; in both there are dyspnœa, cough, fever, and pain in side. A careful auscultation of the heart will generally detect friction, and in a day or two there will be elevation of the apex beat, an extension upwards of the area of cardiac dulness, which takes a triangular form with the base downwards.

I once met with a case of congenital deficiency in the right wing of the diaphragm, leading to hernia, with escape of a quantity of large and small intestine into the right pleura. The physical signs very much resembled those of pleurisy with effusion. There was dulness on percussion, with a tympanitic character, under the right clavicle, bulging of the chest walls, displacement of the heart to the left, dyspnœa, and a dry cough with pain. The case is reported in the Transactions of the Pathological Society, vol. xii., p. 115.

The chief points of distinction from pleurisy which were overlooked were the great variability of the symptoms and physical signs dependent on the condition of the intestines, the paroxysmal character of the dyspnœa, the absence of fever, and the fact that the liver could not be felt below the false ribs, although the right side of thorax was so much distended.

The case is a very interesting one, both on its own account and from the resemblance it bore to effusion of fluid in the pleura. I will therefore give a short sketch of it.

CASE I.

A girl, aged 6 months, was brought to me at the beginning of 1861 with the following history : She was born in August, apparently in good health ; nothing abnormal was observed in her except that her sternum was unusually arched from above downwards. Soon after birth she had a dry irritable cough which caused some lividity of the countenance and fingers ; this symptom was also observed when she cried at all violently. In November she was suddenly seized with very violent screaming, which caused great lividity and clenching of the hands ; this paroxysm lasted nearly two hours ; a doctor saw her, said it was colic, and ordered a warm bath ; after this there was some persistent dyspnœa, and the cough was more troublesome. On the 27th November she was seen by a hospital physician, who wrote on her paper—"Pleurisy. Right side of chest quite dull. Heart reaches to outer side of nipple." Iodide of potassium was ordered to be given internally, and iodine ointment locally. On December 4th she was again seen by the physician, who noted "Right side of chest becoming resonant." The same treatment was continued. On 11th December her medicine was changed to ipecacuanha and carbonate of ammonia. On the 15th January I first saw her, and noted "Lividity on crying. Right side dull. Heart beats outside nipple." I resumed the use of iodide of potassium. On February 4th I made this note : "In about the same state as for last six weeks, alternately better and worse for a day or two at a time ; constantly has a harsh and somewhat ringing cough ; at times suffers much pain. Is at the breast, which she takes well, but is often sick soon after being suckled. She is generally cold ; has been hotter the last week or two ; perspires a good deal at times. The sternum is arched forwards. During inspiration it is raised a good deal, the soft parts are drawn in, and there is recession at the epigastrium. The intercostal spaces are less obvious on the right than the left side. Percussion note at right infraclavicular region is rather tubular, and at lower part of the chest dull. Respiration at the upper part of front of right chest is very weak, and at lower part quite inaudible. Heart's sounds are not heard in the usual situation, nor until about half an inch outside the left nipple. Posterior aspect of chest.—The right side is generally dull ; when lying on her face there is a little resonance in the infra-spinous region, but when sitting up there is universal dulness. Respiratory murmur is very weak at upper part of right back, deficient lower down, and absolutely *nil* at base." The child gradually became worse, losing flesh, and the paroxysms of dyspnœa becoming more frequent.

On the 19th March I saw her, and noted—"Not so well. Left side of chest now dull. Respiration inaudible." On 21st March she died.

The physical signs in this case very closely resembled what are caused by pleurisy with effusion, and this was the diagnosis made, although it was felt there was some degree of mystery about the case; the paroxysms of dyspnœa occurring so frequently, the early history of the case, the arching of the sternum, and the slight amount of bulging of the intercostal spaces were not explained by this diagnosis.

After death there was found to be no pleurisy, but a quantity of small and large intestine had found its way, through a congenital deficiency in the right wing of the diaphragm, into the right pleura ; the aperture was nearly round, and about an inch in diameter.

CASE OF THORACENTESIS.

CASE II.—*Acute pleurisy of right side. Paracentesis after two months. Evacuation of 24 oz. of purulent fluid. Establishment of broncho-pleural fistula. Closure of wound. Recovery, with restoration of form of chest. After nine months cerebral tubercle leading to amaurosis.*

Jessie Trimm, a girl aged 7 years, of sallow complexion, the third of six children, two of whom were born dead, and one had a bad knee; the rest living and well. Her father and mother were living and healthy. On the 10th June she woke up suddenly with great pain in the right shoulder and under her arm, increased by breathing or coughing. This continued for several days together, with heat of skin, thirst, and loss of appetite. For the first week she was *delirious* both night and day. On the 24th June she was admitted to the hospital with a troublesome cough, and pain in her right side. She was pale, not oppressed or livid. Pulse 128 weak. Skin not hot. Some sudamina on chest.

Physical Signs.—Right side of chest not expanded well on inspiration, which is jerking.

Front.—Heart's apex not displaced. Percussion sounds good over left lung and sternum, and immediately to the right of sternum. In the right acromial angle it becomes dull and is absolutely so in the anterior part of right axilla. There is exaggerated respiration sound over left lung; over right lung the sound is weak, and accompanied by sonorous rhonchus.

Back.—Absolute dulness below the spine of right scapula, except close to the spine, where there is still some resonance. Respiration normal over left lung with some sonorous rhonchi. Over right lung weak especially below, blowing and rather metallic between scapulæ. Hydr. c. cretâ, pulv. ipecac. co. āā. gr. ij, nocte maneque. Ung. potassii iodidi lateri dextro applicetur.

June 26th. Not so well. Much pain in right side, restless nights. Right side of chest looks decidedly larger than left; right costal angle much more obtuse than left. Heart's apex now beats more to the left, immediately below nipple. The dulness extends farther upwards and forwards in front. Vocal fremitus is abolished over the right lung posteriorly; also vocal resonance. Jugular veins distended. Liver reaches five fingers' breadths below margins of ribs. Pulse 132. Respiration 36. Catapl. Sinapis. ℞. Potassii iodidi, gr. ij ; mist. potássæ citratis, ʒss. Ter die.

29th. Not any better. Sleeps well. Still has great pain on right side ; occasional paroxysms of dyspnœa. Pulse 124. Respiration 26 irregular in depth and rhythm.

Front of Chest.—Absolute dulness to upper border of 2nd right rib ; above that amphoric in character and a *cracked-pot sound* easily produced. Dulness now extends to the left of sternum, one finger's breadth. Under the left clavicle the percussion note is not so good as it was. Below the 2nd right rib the respiratory murmur is very weak, and is gradually lost lower down. Heart's impulse felt a little outside nipple. Right side below nipple measures 12 inches; left 11¼ inches.

July 2nd. Very pale, rather oppressed and drowsy. Lies either on back or right side, Pulse 132 weak. Respiration 30 regular. Dulness does not now reach quite so far to left, only to the left margin of sternum. Right side over nipple 11¾ ; left side 11 inches. Heart's impulse a little inside line of left nipple.

6th. Much the same. Pulse 132 weak. Right side 11¾ ; left side 10½ inches over nipple. Liver dulness reaches as low as umbilicus.

14th. Child has wasted much. Pulse 144. Respiration 36. Bowels regular. Spine curved with convexity to right. Physical signs much the same.

Aug. 13th. No improvement has taken place. It was very evident that a large accumulation of fluid existed in the left pleura which caused obliteration of the intercostal spaces, displacement of the heart to the left, depression of liver, absolutely dull percussion note over the whole of right lung, except close to the spine behind and to the clavicle in front, weakness or absence of respiration sounds, and absence of vocal fremitus and resonance over the right side of chest. An exploratory needle was passed into the 8th interspace in the right axilla, and pus escaped. An opening was now made with a knife, and 24 oz. of pus were let out; the last 6 oz. were sanious in character. Much relief was afforded by the operation; the patient's appetite and spirits were much improved. Pus continued to flow all night, but ceased next day.

Sept. 1st (19 days after the operation). Right shoulder depressed. Spine strongly curved with concavity to right as low as 10th dorsal vertebra. Right angle of scapula tilted outwards. Right side measures 11¼ inches; left 11½. Right front of chest much flattened ; no expansion on inspiration. Slight pulmonary resonance on percussion over whole of right front. Respiratory murmur blowing under right clavicle, scarcely audible about nipple and below. Respiration harsh, expiration prolonged under left clavicle.

Back.—Percussion note tolerably good on right side close to spine, but becomes less resonant on passing outwards. Respiration blowing in interscapular region, becoming very weak on passing outwards. Heart's impulse felt immediately below nipple, lower margin of liver dulness on level of umbilicus. Occasionally cries with pain in the side. The opening in the side is probed daily, and from 1 to 2 oz. of pus flow from it. The discharge is rather on the decrease.

26th. For three or four days past she has once a day brought up, during a sharp attack of coughing, 5 or 6 oz. of muco-purulent fluid. For the same period distinct *metallic tinkling* has been heard on coughing, over the greater part of right lung. The cough was on several occasions brought on by introducing a probe into the wound in the side. There was no doubt a communication established between the pleura and the bronchi.

Oct. 12th. Patient still very pale, but much improved. Coughs but very little. *Wound quite healed. Front of Chest.*—Much depression of right side, especially below the clavicle. No expansion on this side. Right nipple lower than left ; the sternum is pushed forwards, and the costal angle is very acute, with some flattening over lower ribs on the right side. Respiration blowing over whole of the right side, loudest near sternum and under the clavicle.

Back.—Close to spine on right side, percussion note good, passing outwards it becomes absolutely dull. Respiratory murmur very weak, audible only near spine ; it is of a blowing character, especially the expiration. No metallic tinkling. Spine much curved with concavity to right ; right shoulder much depressed. Heart beats at epigastrium. Left side over nipple measures 10¼ ; right 10¾ ; 3 inches above nipple left 10¼, right the same ; left 2¼ inches below nipple 10½, right side the same. She was now sent to the country. In May of the following year she was brought back to the hospital suffering from cerebral symptoms, probably due to tubercle in the cerebrum. There was now very slight flattening in the right infraclavicular region. No curvature of spine. Respiration sounds healthy. She recovered from the cerebral attack, but with permanent loss of sight.

Remarks.—This case is a remarkable one in several respects. The extension of dulness into the left infra-clavicular region with bronchial breathing from compressed lung on the sound side ; the occurrence of metallic tinkling and the expectoration of puriform fluid, which probably came from the pleura, and the completeness of the recovery after the admission of air freely into the pleura, are points of great interest. The quantity of fluid evacuated by paracentesis (24 ounces) was very large for a child seven years old.

CASE III.—*Empyema of the right side, following what was called "Remittent Fever." Recovery with permanent fistula.*

Isabella Irons, æt. 7, the eldest child of healthy parents. About two months before admission she had "remittent fever." This was accompanied with diarrhœa, and lasted a month. It was ushered in by convulsions, which lasted four hours, and severe pain under the right arm, much increased by breathing or moving the arm. This pain lasted five weeks. She had no cough. The right side of chest was observed to become larger, and about five weeks before admission a swelling was seen under the right nipple. This became red and was lanced ; a quantity of pus (about a quart) was discharged, and there had continued a free discharge until the time of her admission to the hospital. Her appetite has much improved, and she was regaining the flesh she had lost in the earlier part of her illness.

July 9th. State on admission.—Very thin. Sleeps well, perspiring during the night. No cough. Bowels costive. Appetite moderately good. Tongue pale, moist. Pulse 116 weak, regular. Lower part of sternum and ensiform cartilage much depressed. Two inches below and a little outside right nipple is a small fistulous opening, from which a constant discharge takes place. The front of right chest much contracted, very flat under the clavicle. Spine much bent with concavity to the right. Lower ribs very oblique, almost touch each other. Right side over nipple measures 10 inches. Left side 10¼ inches. Percussion over the whole of right front is absolutely dull; below left clavicle tolerably clear. Respiration over left lung in front, harsh and exaggerated. Over right side, close to sternum and under clavicle, very high pitched and blowing; the same character noted over the whole side, but very weak below. Heart's apex felt an inch outside, and half an inch below left nipple. The impulse is seen in 2nd, 3rd, and 4th interspaces.

Back of Thorax.—No expansion on right side. Percussion absolutely dull over whole right side, except close to spine. On left side extraresonant. Respiration harsh and blowing over right side, very weak below. Harsh and loud on left side. Vocal fremitus felt over the whole of left side; not felt at the right base, but slightly increased above. It was evident that the right pleura had been the seat of purulent effusion, the right side of the chest was now undergoing contraction. Pleurisy had evidently existed from the commencement of the so-called remittent fever, and was probably the primary disease, which was not recognised by the doctor in attendance upon her. She was treated with tonics internally and iodine injections. Her general condition much improved. The discharge from the side continued to be very abundant, and when the injections were omitted became very foetid. The right side measured near nipple 9⅜ inches, and the left 10 inches. On admission the right was half an inch larger than the left. The angle of the right scapula was much tilted outwards. Spine was more curved, with the concavity to the right. The heart's apex still beat outside left nipple, not having receded with the contraction of the side. In this condition the patient was discharged after three months' stay in the hospital.

CASE IV.—*Empyema of left side of six months' standing if not much longer. Paracentesis. Permanent Fistula with contraction of side and curvature of spine.*

Henry Brown, æt. 6 years, a healthy boy until he was 3½ years old. He then had scarlatina; a few months after this his mother noticed a swelling of a bluish colour about the size of a large plum under his left arm. His health continued pretty good. About eighteen months later, 14th May, he was seized with chilliness, and pain on the left side, which was much increased by coughing and breathing. He was feverish, lost his appetite, and was confined to bed ; he had a slight cough. The fever lasted a month. On the 10th of June his chest was punctured on the left side, and "three pints of matter" are said to have been removed. This opening continued open until 12th September, when he came under my care in the hospital. He has had a frequent hacking cough, and for two months past rather copious expectoration. When admitted he was very pale and thin. His whole body strongly bent to the left ; his left shoulder much depressed, and the lower angle of this scapula much tilted out. Very great curvature of spine, reaching from the last cervical to 10th dorsal vertebra, with concavity to the left. Left infraclavicular region much flattened ; no expansion or elevation of left side on inspiration. Left side measures 10⅝, right 10¾ inches. Absolute dulness over the whole of right lung in front and over most of sternal region. Respiration on left side close to sternum, blowing and expiration metallic ; over infraclavicular and mammary regions, blowing but not metallic ; below nipple respiration is lost. At the back some pulmonary resonance close to spine on left side, very dull in other parts ; respiration only heard near the spine on the side. The right side of chest appears to be healthy. Heart's impulse seen in 2nd, 3rd, and 4th *right* interspaces close to sternum. There is a fistulous opening, from which abundant discharge proceeds between the 9th and 10th ribs in the left infra-axillary region. He was treated with tonics, cod-liver oil, and nutritious diet. Iodine injections were also employed, but very little could be made to pass into the opening ; no effect was produced by them. He continued under notice until May with very little change in his symptoms, the opening continued to discharge matter freely, which was at times fœtid and at others inoffensive. He was sent to Margate, where he remained for several months. His general health was improved, but his chest remained in the same condition.

Remarks.—In this case the operation was performed after the disease had existed for so long a period that the lung had probably lost its power of expansion and remained permanently collapsed. The cavity existing in the pleura could not be filled up by the lung and the chest walls coming into apposition, although great contraction of the thoracic cavity did take place by approximation of the ribs and curvature of spine, with depression of the shoulder.

The swelling under his left arm, which appeared after scarlatina, was probably a localised empyema, and at the later period there was diffuse and severer pleurisy set up. It might have been allowable to make a counter-opening and attempt to pass a narrow drainage tube from one opening to the other, with the view of preventing undue accumulations of matter and keeping it free from putridity, as recommended by Chassaignac.*

This plan I have tried in two instances, but have been obliged to close the second opening, owing to the amount of irritation set up by it.

CASE V.—*"Low fever"* at age of two years, followed by Empyema on left side. Paracentesis after three years. Pyrexia next day. Death in 18 days.

William Dodkins, aged five years. One of three children, healthy. Father and mother living. No consumption known in family. When he was two years old he had a "low fever," which lasted three or four weeks. After this his belly began to swell, and his cough set in. His belly is getting bigger and his cough worse. He is wasting much, though his appetite is moderately good. Bowels regular; no vomiting. Cough disturbs his rest at night.

April 14*th*. Admitted to the hospital. Moderately well nourished. Has a tolerably good colour in cheeks. No lividity. His fingers are most unusually clubbed. Pulse 100. Respiration 24. Skin cool. Appetite good. No obvious dyspnœa.

Chest.—The left half is obviously larger than the right. Right side (1¼ inch below nipple) measures 11⅜ inches; left side, 12¼. Impulse of heart is strongest at the costal margin of the right half of epigastrium; it is also felt distinctly in 5th right interspace, half an inch to left of right nipple. The veins over thorax are much distended, especially on the left side. Veins of neck are full. There is considerable bulging of the region usually occupied by heart. Left intercostal spaces filled out. Left half of chest is both elevated and a little expanded during inspiration.

Percussion.—This is absolutely dull over the whole left side of thorax, except at apex in front where there is a little resonance, close to the spine behind; and at base of thorax, about two inches below nipple, where the stomach note is heard. Dulness does not extend to right of sternum.

Auscultation.—At apex in front, the respiration is blowing, and tolerably loud; it is also heard near spine; elsewhere it is weak; at the base and in axilla almost inaudible.

Right side.—Percussion extra-resonant and respiration puerile. Liver dulness extends from 1½ inch below right nipple, 5 inches downwards.

---

* Two cases are reported by Dr. Goodfellow in the Medico-Chirurgical Society's Transactions, vol. 42, in which this plan of treatment was followed with much benefit.

CASE OF CHRONIC EMPYEMA.

Intestines appear to be filled with scybala, which were voided after the use of a simple enema.

21st. A trocar was passed into left pleura, through the 6th interspace in the line of the anterior fold of axilla. It was done in a warm bath ; a large quantity of pus escaped. One hour after the operation, patient expressed himself as much relieved. The wound was closed.

22nd, 3 *p.m.* Slept well till half-past 7. He then complained of headache, and was sick. Vomiting repeated five or six times until 11 o'clock. Bowels open twice, previously costive. He now looks pale and heavy. Lips a little bluish. Skin hot. Temperature of axilla 101·2° Fahr. Pulse 168 tolerably full. Respiration 40. Less fulness of veins of chest walls.

Percussion.—Left side, tubular resonance everywhere in front, least under clavicle ; behind, tubular down to angle of scapula, dull below. Respiration scarcely audible in front ; behind, heard faintly everywhere, most distinctly near spine. Tongue moist, thin white fur. No rash ; no sore throat, but fauces are red and swollen. (There was a case of scarlet fever in the ward a few days ago.)

23rd. Skin cooler. Tongue moist, red, rather smooth, with enlarged papillæ. Pulse 144. Respiration 31. No discharge from opening. Tubular percussion note reaches now to the base of left lung behind. Respiration is louder over the lower part of left lung.

25th. Is getting thinner. Looks heavy and flushed. Sweats a good deal in his sleep. Bowels open twice in twenty-four hours, semi-solid, light coloured. A good deal of nausea. Tongue moist, red, enlarged papillæ, not unlike the tongue of the 5th or 6th day of scarlet fever. Pulse 124 sharp. Respiration 34. Coughs a good deal, swallows sputa ; but his breath, when coughing, is fœtid. Left side measures 11¼ inches ; right, 11¼ inches. The veins on left side are more distended again. A distinct succussion splash can be easily produced in left pleura. The neighbourhood of wound is red and swollen. Wound discharges thin serum.

30th. He is getting worse ; much thinner ; lips dry and scaly. Small ulcers on tongue. Pulse 138 weak. Respiration 38. Left side now measures 12¼ inches. The wound still looks unhealthy ; is now opened up, and three pints of the most abominably fœtid dark-coloured fluid let out. The cavity is washed out with dilute permanganate of potash.

May 2nd. Weaker. Ulcers on tongue continue ; the rest of tongue glazed and smooth. Appetite gone. Pulse 128. Respiration 40. Thorax injected daily with Condy's fluid. No albumen in his urine.

5th. Much weaker. Pulse 148 very sharp. Veins of neck full. Tongue much ulcerated. Discharge very profuse and fœtid.

9th. Dying of exhaustion.

Autopsy.—Weight 19½ lbs.

Chest.—Fistulous opening on upper border of 6th rib. The upper border of rib is carious. In left side of thorax is a large cavity containing a little fœtid pus. The walls of it are ragged; there are irregular ridges of thick white substance, and depressions of a dirty purple colour. It looks as if a thick false membrane were undergoing ulceration.

Left Lung reduced to a very small size, close up to spine, scarcely at all affected by powerful inflation of bronchus. There are two small apertures on the surface of upper lobe which permit the escape of air.

Right Lung did not collapse on opening thorax. Weight 5 oz. 4 dr. No tubercle. A little central pneumonia of lower lobe. In pleural sac of this side was about 1½ oz. of turbid serum; the pleura itself injected and covered by a soft puriform exudation.

Heart, 2 oz. 10 dr. Right side not specially engorged.

Liver, 21½ oz. Left lobe adherent to diaphragm; texture coarse, yet structure distinct. *Spleen*, 3 oz.; a few old adhesions.

Kidneys.—Cortex pale and opaque; together weigh, 5 oz.

Lymphatic Glands.—Free from tubercle.

Remarks.—In this patient empyema followed what was designated *low fever* at the age of two years. It was most likely simple pleurisy with ill-defined symptoms. The chest was never examined. The small amount of inconvenience arising from a large accumulation of pus in the pleura is remarkable. The great clubbing of the fingers was very striking.

I believe that it would have been better not to have emptied the chest by paracentesis. In an empyema of three years' standing it might have been expected that the lung was collapsed and atrophied past recovery, and as the patient was not suffering greatly from dyspnœa or irritative fever, it would have been probably wiser to leave it quite alone.

The experience afforded by this case would lead me in another similar case to leave nature undisturbed. The admission of air to the pleura led to ulceration of the thick false membrane, and to fœtor of the pus. The two small openings of the lung were most likely the means by which air got into the pleura. The operation was performed under water and the wound closed at once, so that probably no air was admitted by the wound. When the matter was evacuated there was of course during each inspiration a great reduction of atmospheric pressure in the pleural cavity, and a tendency to distension of the collapsed lung; in this way probably the orifices were produced. The establishment of a communication between the bronchi and the pleura after paracentesis is not at all uncommon. It is often supposed that this is due to the softening of tubercle, but a more common cause is the rupture from distension on the removal of atmospheric pressure, as I have just explained. It must be remembered that collapsed lung is not at all disposed to become the seat of tubercular deposition. The case of Jessie Trimm is an illustration of the rupture of lung tissue into the pleura after paracentesis, whether from softened tubercle or not is uncertain. (See page 55.)

CASE VI.—*Idiopathic Pleurisy, left side. Onset not attended with pain in side or sickness, but with drowsiness and chilliness. Recovery at the end of a month complete.*

G. R., æt. 9, a healthy boy. Was in his usual health until 4th December, when he suddenly felt, whilst at school, drowsy and heavy. In the afternoon of the same day he felt chilly. The next day he did not seem well, but only remained in bed to breakfast. A cough was noticed about the same time, but it was unaccompanied with pain. He continued to get up for several hours each day until 13th December, when he was admitted to the hospital.

State on admission.—Moderately well nourished, but pale. Pulse 120 of moderate volume and strength. Respiration 26 in the minute, with slight action of nares.

Chest.—Very little movement of left side on inspiration; to the eye the left side appears larger than the right. Intercostal spaces less apparent on the left than on the right side. Right side (1 inch above upper part of ensiform cartilage) measures $11\frac{2}{8}$, the left $12\frac{3}{8}$ inches.

Percussion.—Almost absolute dulness over nearly all the left lung except close to spine in interscapular region. Right lung normal.

Auscultation.—Feeble distant breath-sound from left apex to 2 inches below nipple, from which point breathing is bronchial as well as distant. Behind—feeble breath sound from apex to lower half of scapular region, below which it is bronchial, except close to spine, where it is nearly normal. Heart's impulse most distinctly felt midway between right nipple and ensiform cartilage. Empl. lyttæ 4 × 3, infra scapulam sinistram. ℞. Hydrargyri chloridi, gr. iss.; pulv. ipecac. co. gr. ij. Ter die.

December 16th. Slightly improved. Pulse 120. Respiration 28 regular. Physical signs as before, except that the percussion note under left clavicle is not absolutely dull; and over cartilages of 2nd, 3rd, and 4th ribs, the resonance is somewhat amphoric. The blister did not rise well; another to be applied in the axilla.

18th. Tongue rather dry, clean at tip, coated on dorsum with thin white fur. Pulse 108. Respiration 28. Amphoric percussion note not heard over upper left ribs; there is feeble vesicular breathing to be heard for 1½ inch below clavicle; from this point to base distant bronchial breathing, with confused vocal resonance. Behind—physical signs as before. Vocal fremitus absent over left lung. Ung. Hydrarg. to be applied to abdomen on a flannel bandage.

20th. Improving. Pulse 116 full, sharp. Respiration 24. Physical signs much as before, except that from three inches below left nipple to base creaking friction sound is heard for the first time.

23rd. Improved. Gums slightly swollen, red at edges. Pulse 100 of moderate volume. Respiration 28. Respiratory movements almost equal on the two sides. The percussion note in front now very nearly normal over left lung; behind dulness exists only from angle of scapula downwards. Friction sound is now audible over lower part of left lung, back and front. Heart's impulse has regained its normal situation. Omit the mercurial belt.

30th. Pulse 104. Respiration 24. Friction sounds still heard. R. Syrupi ferri iodidi, ʒss.; ol. morrhuæ, ʒj. Ter die.

Jan. 13th. Much improved. The right side measures 12⅔, left side 12⅛ inches, at an inch above nipple. There is a little more expansion of left side in inspiration than of right. Other signs normal, except that there is a slight impairment of resonance over the extreme base of left lung.

Remarks.—In this case a slight mercurial action was set up. The improvement was coincident in time with the evidence that the gums were slightly affected by mercury.

Longer experience has induced me to discontinue the use of mercury as an antiphlogistic in pleurisy, and to trust to simple saline treatment, with the employment of local depletion by leeches on the first or second day, if there is much pain and fever and the patient is not previously cachectic or reduced by antecedent disease.

CASE VII.—*Left Pleurisy. Pericarditis. Chronic fistulous openings. Waxy degenerations of liver, spleen, and kidneys.*

The following case of pleurisy complicated with pericarditis, presents features of interest. It is not clear what was the primary disease, whether she had scarlatina, pericarditis, pleurisy, or caries of the sternum in the commencement. I am inclined to think that she had scarlatina; and that, in the next place, disease of bone set in, that this caused pericarditis,

CHRONIC PLEURISY AND PERICARDITIS. 65

and subsequently pleurisy. Then occurred, as a result of long standing suppuration and bone disease, waxy degeneration of the liver, spleen, and kidneys.

Emily J. Rance, one of five children; two, besides herself, are living and healthy, two died in infancy of convulsions. Father, aged 56, is subject to bronchitis. Mother is living, aged 35. Neither of them is rheumatic or consumptive.

Emily was five years old when brought to the hospital. She had usually enjoyed good health ; six weeks previously had been taken suddenly ill after dinner, vomiting, and complaining of pain in her left back. She became blue in the face ; a doctor saw her, and ordered castor oil and a warm bath. Before the warm bath, at 9 p.m., the mother observed a general redness of her skin, not in spots. It was quite gone the next morning. During the night she was very hot and thirsty, wandered a little, and had pain in her forehead. She did not complain of pain in her throat then or subsequently ; she was kept in bed two or three days only ; she was, however, far from well, had a slight cough, which became worse at the end of a month, and had frequently been delirious at night. For a few days before admission her feet and left arm were slightly œdematous. She complained of no pain in her limbs or joints, nor was there any redness or swelling observed in them.

Nov. 12th, 1862.—*State on admission.*—She appears a tolerably well-built child ; face full, eyes blue, superficial veins of temple very distinct. A thin sanious discharge from nares. Lips dry and cracked. Pulse 120 weak. Respiration 40 irregular. Some œdema of legs and left arm. Cough troublesome, dry, more or less paroxysmal ; no whoop. Right supra-clavicular space is full, bulges during coughing ; the jugular veins are full and pulsate; they become much distended during coughing. Well marked distinct bulging in the præcordial region. It commences almost imperceptibly about an inch to the right of sternum, gradually increasing towards the left till within $\frac{3}{4}$ of an inch of the left nipple where the maximum of elevation is reached ; it ceases rather abruptly at the nipple ; upwards it extends to the lower border of the 2nd rib ; downwards to the level of the ensiform notch. Distance of right nipple, from median line, $2\frac{2}{8}$ in. ; of left nipple, $2\frac{3}{4}$ in. On applying stethescope over the bulged part, it is observed to pit considerably. Percussion from 1st left rib to base of lung in front gives a dull note. Absolute præcordial percussion dulness extends to $\frac{3}{4}$ of an inch to the right of middle line ; but there is deep-seated dulness until the right nipple line is reached. On *right side*, percussion note is high pitched under the clavicle, moderately good in other parts. The point at which heart's apex beats cannot be made out ; there is a slight impulse at the ensiform cartilage, a slightly diffused weak impulse can be felt over the præ-

F

cordial bulging. Heart's sounds distinctly heard, without murmur. No respiration heard over left lung in front. Patient complains when the left side of chest is examined; there is a good deal of tenderness over the præcordial region. Right side measures on level of nipple, 10¾ in.; the left, 10⅞ in. Respiratory murmur under right clavicle exaggerated, and expiration rather long, separated from inspiration by an interval; vocal resonance increased.

Back.—Right side quite normal.

Left.—Percussion note tolerably good, except over scapula, where the note is high pitched and resistant. Respiration rather weak over whole left side, with a subcrepitant râle at base. Left axillary and infra-axillary region dull when lying on back, but decidedly less so when she lies on her right side.

Urine was free from albumen.

Remarks.—Here were signs of pericardial effusion and effusion in the pleura, presenting several abnormal features. The bulging of præcordial region was too great in proportion to the displacement of the heart and the distinctness of its sounds; the effusion in the pleura was not, as it usually is, most marked in the back, but almost exclusively in front. There appeared to be considerable obstruction to the venous circulation, indicated by the condition of the jugulars and the œdema of the legs. It was not easy to conjecture what had been the order of supervention of phenomena, because six weeks had elapsed before any physical examination of the chest was made. The diagnosis made was pericarditis with circumscribed pleurisy.

Nov. 24th. Slowly improving. Præcordial bulging distinctly diminished. Left side now measures 10⅝ in., instead of 10⅞ in. Heart's sounds are heard more distinctly than they were; at the base, and towards left nipple, an occasional distinct friction sound is heard. The præcordial region has been painted with tincture of iodine since admission; she took for several days, 1 grain of calomel and 1 grain of Dover's powder, three times daily; since then she has taken this mixture:—R. Potassii iodidi, gr. iss; pot. chloratis, gr. v; syrupi, ʒj; aquæ, ad ℥ss. Ter die.

27th. Better. Distinct pleural friction of a grazing character can be heard both with inspiration and expiration near left nipple. No pericardial friction.

Dec. 1st. To-day pericardial friction is distinctly heard just to the right of sternum. Heart's sounds are loudest, and impulse most dis-

CHRONIC PLEURISY AND PERICARDITIS. 67

tinctly felt midway between ensiform cartilage and right nipple. Percussion good in left axillary and infra-axillary regions; over left lung behind not quite so resonant as right, *especially the upper half*.

3rd. A slight swelling and redness have appeared two inches below and one inch outside left nipple, about the size of a small walnut; it is tender and semi-fluctuating. On the 5th December it was punctured, and about a teaspoonful of pus and blood escaped.

18th. The punctured spot in side continues to discharge a little. There is still friction audible much as on the 1st instant.

26th. The friction sound is gone. Discharge continues. Præcordial bulging less; dulness much the same.

Jan. 16th. Patient is sallow and considerably emaciated; takes food well.

24th. A distinct systolic murmur heard at 2nd right interspace, near sternum, rather of a cooing character. She has been on a liberal diet with some wine, for several weeks past.

Feb. 10th. More emaciated; weighs only 23 lbs. From 2 to 3 oz. of pus escape daily from wound in side.

21st. A tendency to diarrhœa. The amount of discharge has increased; there is about 4 oz. daily. Patient is weak and restless. Pulse 128 feeble. ℞. Pulv. ipecac. co., gr. j, hac nocte. ℞. Mist. aromaticæ, ℨij; sp. ammon. co. ♏v. 6tis. horis.

March 10th. Patient continues much the same. Sternum at lower part and ensiform cartilage project forwards. The bulging of præcordial region has given place to a slight flattening of the left mammary region. ℞. Tinct. opii, ♏iij; aquæ, ℨij, p. r. n. for diarrhœa. Brandy, 3 oz. daily.

April 15th. Occasional attacks of diarrhœa. The discharge from side is even more copious.

May 15. Diarrhœa has ceased. Patient has a dusky tint of skin, much emaciated. Prominence of lower end of sternum increased. ℞. Ol. morrhuæ, ℨj. Bis die. ℞. Mist. quinæ, ℨij. Bis die.

18th. For a week past patient has complained of pains and tenderness under left nipple, where there is a red tender spot.

June 1st. Three days ago the tender spot of skin gave way, and about a pint of pus escaped; the opening is ¾ inch below nipple, and about 1 inch above the former opening.

11th. Is improving in appearance and spirits. Has gained flesh; now weighs 25 lbs. Slight curvature of spine, with convexity to right; left shoulder depressed. Neither the spleen nor liver can be felt.

After this she went on, at one time worse, at another better; often having attacks of diarrhœa; gradually losing ground. On the 26th November the following note was taken :—Pulse 116. Respiration 32. Is thinner than in June last. Anterior aspect of thorax much the same; the two fistulous openings discharge freely. She complains of pain under sternum near the 3rd rib. Heart's impulse indistinctly felt to the right of sternum under nipple ; and here the sounds are most distinctly heard. No murmur with heart's sounds. There is a reddish tense swelling over lower end of sternum, about 1½ inch long by ¾ of an inch wide.

She gradually became weaker, worn out by suppuration and diarrhœa. A few days before death, on 1st February, 1864, she was generally anasarcous.

Autopsy.—Heart universally adherent to the pericardium. The fistulous openings below left nipple lead into the pleura. Slight caries of sternum, and thickening of the cartilages of the ribs over the heart. Left lung collapsed and almost universally adherent. Right lung emphysematous, adherent at apex and in one or two other spots, especially near the pericardium. Cretified tubercle at right apex, and at the root of left lung.

Spleen.—Large, moderately firm ; touched with tincture of iodine, a dark brown colour is brought out in parts over the surface.

Liver large and firm. The section is seen to be infiltrated with a semi-transparent material, which is coloured reddish-brown by tincture of iodine.

Kidneys small, very pale and firm. Seen to be coloured brown in spots by diluted tincture of iodine.

Heart very flabby. Valves healthy.

Circumscribed accumulations of pus in the pleura are more common in children than would be supposed from the space devoted to the subject in systematic books on the subject. No mention, indeed, is usually made of this affection. Several such cases have come under my notice. The site of it may be in any part of the pleural cavity, in front of, behind, or below the lung.

CASE VIII.

In the following case the diagnosis was not made during life. There were circumscribed abscesses in the lower part of each pleura, besides small purulent deposits in the middle and lower lobe of right lung. The origin of the disease was, no doubt, in the temporal bone ; this gave rise to coagulation of blood in the lateral sinus and internal jugular vein, the clot became disintegrated and passing into the circulation gave rise to secondary deposits.

A delicate boy, aged 5 years and 9 months, had suffered from otorrhœa with fœtid discharge, ever since an attack of scarlet fever when he was

CIRCUMSCRIBED EMPYEMA.

4 years old. He was also subject to an eczematous rash on the scalp. On the 22nd August this rash suddenly disappeared, and the ear ceased to discharge. The next day he complained of headache and pain in ear and was very languid. He continued to go to school until the 27th August. He then had a severe attack of shivering, and uncontrollable vomiting. He was very hot and feverish.

Aug. 29th. He was admitted to hospital. He was suffering from severe paroxysmal headache, with alternate flushing and chilliness. A free discharge of fœtid pus from right ear. Slight retraction of abdomen. Bowels open twice in 24 hours. Tongue coated with a thick greyish fur in the centre. Pulse 136 weak, regular. Respiration 26. Nothing very notable in physical signs. He was ordered 15 gr. of quinine in the 24 hours for his headaches.

Sept. 2nd. Less headache. Left pupil decidedly larger than right. Has pain around navel. There is a want of resonance at left base of thorax, at posterior part of right apex and anterior part of left apex. Auscultation not abnormal. Vomiting frequent. Pulse 148. Respiration 50.

3rd. Pulse 120. Respiration 50. Skin hot and dry. Tongue cleaner. Has much pain in abdomen near navel, which is retracted.

5th. Has a typhoid aspect. Tongue cleaner, very red at tip and edges. Bowels open five times in 24 hours; some stools dark, some light yellow. Pulse 164 hard. Respirations 58. Is always hawking up phlegm. Respiratory murmur very weak over left back; a little dry crackling at right apex behind. A trace of albumen in urine.

6th. Much the same. More albumen in urine.

7th. Looks worse; complexion pale and muddy. Pulse 164 sharp, dicrotous. Respiration 56. No typhoid spots. Bowels open 7 times in 24 hours. Stools loose, light yellow. Has for several days complained of pain in left leg; nothing to be seen. He continued gradually to get worse without any notable change of symptoms. He was constantly hacking and coughing up small pellets of muco-pus. His urine was sometimes albuminous, and sometimes without albumen. Bowels generally loose; motions generally yellow, sometimes granular, like the stools of typhoid fever. Occasionally he vomited. His pulse continued very rapid and weak, and respirations from 50 to 55 in the minute. There was not much change in the physical signs in the chest, beyond a slight increase of percussion dulness and occasional rhonchi dry and moist, heard in different parts. He died on the 17th September very quietly, after having been insensible for about 2 hours.

Remarks.—The diagnosis was very obscure; it lay between typhoid fever, tubercle, and purulent infection from disease of the ear. The diarrhœa was somewhat like typhoid, the headache was almost too intense and paroxysmal for this disease;

the shivering about the fifth day, the severity and continuance of the vomiting, the flatness of abdomen, and the fluctuations in the frequency of the pulse were also unlike typhoid fever.

The state of the lungs and the frequency of the respiration were consistent with acute tuberculosis, the headache and vomiting might have been caused by tubercle in the brain, whilst the diarrhœa might have been due to tubercular ulceration of the intestines.

The occurrence of a severe rigor with pain below the ear, and a temporary cessation of discharge from the ear, pointed rather to systemic infection; the frequent occurrence of ultimate flushing and chills also suggested the same thing.

CASE IX.

In another child, aged only 5 months, who had a cough which seemed to give pain for three weeks; signs of effusion were discovered in the left pleura, absolute dulness over the whole lung, great weakness of respiratory murmur, and displacement of heart to the right of sternum, and of spleen downwards. The pulse was 168, very weak. Respiration 72 in the minute. A few hours later the child seemed moribund, and paracentesis was performed in the 6th interspace in middle line of axilla; between 3 and 4 oz. of pus were drawn off. The child rallied, became quite lively, and was able again to take the breast, which he had not done for 36 hours. He sank, however, the next day; his breathing again becoming laboured, owing to tympanitis and collapse of the right lung. On post-mortem examination, in addition to general collapse of both lungs, recent adhesions were found in the upper part of the left pleura, and in the lower part about 3 oz. of pus not fœtid. Besides this was found an encysted collection of concrete pus enclosed in a thin firm membrane, between the two lobes of the lung at the posterior part of the fissure. The lung was deeply indented by the cyst. It would appear that this collection was of some duration, and was quite distinct from that contained in the general cavity of the pleura. There was no disease in the vicinity to explain its existence.

Modes of Termination.—Pleurisy may terminate in recovery either complete or incomplete. In slight cases, with serous effusion, complete recovery without leaving appreciable traces of the disease is not uncommon. Instead of this complete restoration of the affected side, there is sometimes permanent elevation of the diaphragm and the liver or spleen, with incomplete expansion of the lower part of the lung, and sometimes the heart is not restored to its place.

When the exudation has been more abundant and of longer standing, the contraction of the side owing to permanent condensation and atrophy of lung is more marked, the ribs approach nearer to each other, the angle which they form with the spinal column is more acute, they are twisted on their axes with an inclination outwards, and the false ribs approach nearer to the crest of the ilium. There is often marked curvature of the spine, with the convexity almost invariably towards the sound side, and a compensating curve in the lumbar region. This curving occurs to a greater extent in children than in adults, in consequence of the laxity of their ligaments and intervertebral fibro-cartilages.

The shoulder is lowered, and the lower angle of the scapula is tilted outwards. There is often, in addition to the general contraction of the side, a local flattening and depression at the upper anterior part of the chest. The lower end of the sternum is turned towards the sound side.

The expansion of the lung after long-standing exudation in pleurisy is impeded, not only by false membrane enveloping it, but by atrophy of the lung itself. It is not known how long a lung may be compressed before this atrophy takes place.

Pus in the pleural cavity, if left to itself, may perforate the external wall of the chest, in almost any situation from the first to the eighth or ninth intercostal space. Dr. Walshe says that a spontaneous opening from empyema has occurred in the adult above the clavicle. Of 16 cases that have come under my own care, two opened in the 5th interspace a little outside the nipple, one on the right and one on the left side; one pointed and was opened on the left side in the 9th interspace in the axillary line, another in the 9th left space; and in a fifth two openings occurred in the 7th left interspace, one in the line of nipple, and one an inch and a half farther back; in the sixth an opening was established two inches below and an inch outside the left nipple, a second opening near it appeared at a later period. In none of these cases did spontaneous perforation occur higher than the 5th interspace.

Pus left in the pleura does not always take this course, but may be enclosed in a dense pyogenic membrane, the fluid portion becoming absorbed. This would very likely have occurred in the case of Wm. Dodkins (Case V.) if left alone.

Mr. Hilton, in his work on "Rest and Pain" (p. 375), has pointed out that the fluid contents of abscesses may be absorbed, leaving but very little solid residue, the pus cells being probably first broken up. Communications have sometimes been formed with distant parts by straight or tortuous sinuses, as with the intestine, or with a lumbar abscess opening in the groin. (Krause, Das Empyem. Luschka, Brit. and For. Med. Chir. Review, vol. xiv. p. 527.) It occasionally perforates the lung, and a quantity of pus escapes into the bronchi, as in the case of Jessie Trimm. Two cases of this kind are recorded by Barthez and Rilliet, and several by Heyfelder. This occurrence has, in many instances, been followed by recovery, with or without cough and paroxysms of fœtid expectoration.

Amongst the sequelæ of pleurisy may be enumerated tubercle, chronic bronchitis, dilatation of bronchi, emphysema, Bright's disease, amyloid degeneration of viscera and lymphatic glands. Tubercle is not so common as might be expected. Chronic bronchitis, with or without dilatation of bronchi, is more common, and is often confounded with phthisis. There may result from this, hypertrophy and dilatation of the right ventricle of the heart, with intensity of the second sound of the heart over the pulmonary cartilage. Bronchitis may exist in both lungs, and dilatation of the bronchi chiefly in the affected lung; this may lead to rupture and the establishment of a broncho-pleural fistula. *Emphysema* of the sound side occurs occasionally to a slight extent. *Bright's Disease* is not a result of pleurisy, but it may accompany it, as it often does when pleurisy is a sequela of scarlet fever. Amyloid degeneration may follow empyema as it follows any other long-standing purulent discharge. It may be suspected from a shotty induration of the lymphatic glands, enlargement of the liver and spleen, and emaciation, with occasional diarrhœa.

Treatment of Pleurisy.—This disease does not often, in the present day, require active antiphlogistic treatment. Cases occur occasionally in which, *at the very outset,* local or even general depletion is useful. These cases are very rare in town practice. They are cases of primary pleurisy in strong, well-nourished children, where there is much pain, dyspnœa, and high fever. After the first or second day blood-letting is inadmissible. The application of cold to the side is strongly recommended by Ziemssen.* A large soft towel, wetted with cold water, is squeezed as firmly as possible, and then applied to the affected side; over it a second dry towel or oiled silk is placed. The application is renewed every five or ten minutes for an hour or two, until the pain is relieved. It is then discontinued, but may be repeated if the pain returns.

Formerly I gave mercury to all cases of primary pleurisy, but this practice I have discontinued, except in the form of an aperient. Instead of it salines, such as acetate of ammonia, nitrate of potash or soda, the citrate of potash, and nitrous ether are given. The iodide of potassium is also given as a valuable diuretic and to promote absorption of the exuded fluid. If there be great pain and irritable cough, Dover's powder is of great benefit; the bowels being regulated by calomel or hydrargyrum cum cretâ with jalap or scammony. It is a good plan to limit the movements of the inflamed part by applying long strips of diachylon plaster from the spine to the middle line in front over the whole of the affected lung.

When the fever has subsided, the use of iodide of iron with iodide of potassium, nitrate of potash, and other diuretics, is indicated. A nutritious diet of milk, fresh underdone meat, and small quantities of wine should be given. Cod-liver oil is of great service in chronic cases where the stomach and bowels tolerate it. Change of air to a warm, dry climate is to

---

* "Pleuritis und Pneumonie im Kindesalter." Berlin, 1862.

be recommended; when this cannot be secured, any change of air is often useful.

In regard to *paracentesis*, it is not easy to lay down rules to indicate when it should be performed. It is more successful in children than in adults. When there is great distension of the side, causing much dyspnœa, and if, after a fair use of internal remedies, exudation seems inclined to increase, delay is pernicious and reduces the patient's chances of recovery. The exudation of pleurisy more speedily becomes purulent in children than in adults; in secondary pleurisy it is commonly purulent from the first. This is an additional reason for not unnecessarily postponing the operation. The longer the operation is delayed the less probability is there of the lung being capable of expansion. On the other hand, in a case of many months' duration, if the patient is not suffering from dyspnœa or hectic, it will be wiser to leave the side unopened, although it is much distended. The operation in Dodkin's case (page 60), whilst relieving his respiration, set up irritative fever, and the pyogenic membrane lining the pleura was broken down and the pus became decomposed.

In fixing the site for the operation, the first point is to determine where the lung is not adherent to the side; the absence of breath-sound and of vocal resonance and fremitus will indicate non-adhesion. The fifth or sixth space in front of the insertion of the serratus magnus is generally to be chosen; in children one space may be preferred to another on account of its greater width. If the operation fail to obtain fluid, it may be that there is a firm, thick false membrane under the incision; in this case, if the diagnosis of fluid is certain, a second opening may be made in another spot, or the original one may be enlarged.

Of 33 cases of paracentesis collected by Ziemssen, 8 died and 25 recovered more or less completely. The ages were as follows:—Two were under 12 months, of which one recovered; from 1 to 2 years one case, which recovered (Steinbeck, 1843); from 2 to 3 years, one case fatal; from 3 to 4 years, three cases; from 4 to 5, four cases; from 5 to 6, none; from 6 to 7, five cases ; from 7 to 8, four cases ; from 8 to 9, one case ; between 7 and 9 years, four other cases ; from 11 to 12, one case ; from 12 to 13, one case ; from 13 to 14, one ; at 14 years, one case. In four instances the age is not stated.

Dr. H. Guinier has also collected 31 cases, and published the results in the "Journal für Kinderkrankheiten," 1865, p. 308. His cases, no doubt, include some of those mentioned by Ziemssen. He records one case in which he operated successfully on a boy of 12 months, and speaks of it as the only recorded case so young. He has failed to note a similar case reported by Fereday ("Provincial Medical and Surgical Journal," Sept. 1849) of a child only 10 months old. The largest number of operations have been performed between the ages of 6 and 9 years, about half of the whole number. Boys have been more frequently operated on than girls: Guinier says four times as frequently ; in Ziemssen's cases, the sex is only mentioned in 15, of which 8 were males and 7 females. Guinier says that the left side was six times as frequently involved as the right ; in Ziemssen's cases, on the other hand, 7 were right, 7 left, 1 both sides, and in 18 the side affected is not mentioned.

Of 12 cases under my care, 5 recovered completely and 2 with permanent fistulous openings, and 5 died. Besides these, 5 pointed and were lanced or opened spontaneously, of which 4 recovered, 2 with permanent fistulæ, the other died 14 months afterwards with necrosis of sternum and waxy degeneration of viscera. Of the fatal cases, one was a child only 5 months old, who died of collapse of lungs ; one was only 12 months old ; and another was a child 5 years old, in whom empyema was of long standing, probably three years (see Case V.).

Where there is a fistulous opening into the pleura, great deformity often ensues by contraction of the affected side. The discharge may be very profuse so as seriously to exhaust the patient's strength, or it may be scanty. It is often very fœtid, in consequence of its being long exposed to air before its escape. To promote closure of the opening, everything must be done to improve the patient's health by pure, bracing air, nutritious food, tonics, and cod-liver oil. Injections of various kinds have been recommended. M. Boinet (in the "Archives de Médecine," May, 1853) relates cases in which iodine injections have been useful. The strength of the solutions was from 10 to 50 parts of tincture of iodine, and from 1 to 4 of iodide of potassium to 100 parts of water. The injections may be repeated once, twice, or thrice a week, according to circumstances. I have tried this plan in several instances with the result of correcting the fœtor of the discharge, but no other result.

Another plan is that of Chassaignac, recommended by Dr. Goodfellow (Medical and Chirurgical Transactions, vol. xlii. p. 231); it consists in passing a long probe from the opening in front to the posterior wall of the thorax, and making a counter-opening down upon the end of this probe. A small india-rubber tube, with small holes cut in it at intervals, is then passed from one opening to the other and the ends tied together; this acts as a drainage tube, preventing the accumulation of matter and its consequent decomposition and offensive odour. This plan I have tried once, and I have seen it adopted by others. In these cases it became necessary to allow the posterior openings to heal on account of troublesome ulceration which occurred around them.

One great obstacle to the healing of the pleural fistula is the constant movement upon each other of the costal and pulmonary pleura lined by pyogenic membrane. The condensed and atrophied condition of the lung also increases the difficulty of bringing the sides of the abscess together; in the child, however, the thoracic walls are so compressible that this obstacle is less than in the adult. To ensure rest of the sides of the abscess and approximate them as closely as possible, it would be well to apply strips of diachylon around the affected side as firmly as possible. To prevent great curvature of the spine the use of a Tavener's belt, which has a band round the body and a crutch under the arm, is of service.

In performing the operation of paracentesis, I once took the precaution of putting the patient into a warm bath, and opening the pleura under water to prevent the entrance of air. I have also evacuated the contents of the chest by means of a long narrow elastic tube, whose free end opened under water for the same purpose. Without the entrance of air it is difficult to draw off nearly all the contents of the empyematous sac; to accomplish this the employment of an exhausting syringe as recommended by Dr. Bowditch * is to be recommended. Dr. Bowditch operates behind, between the ninth and tenth, or tenth and eleventh ribs, in a line from the lower angle of the scapula. In 26 cases where serum was evacuated, 21 got well; in 24 where pus was drawn off at first, 8 got well, 9 were relieved, and 7 died.

It is desirable, in any case, to close the opening after the operation, and to open it again from time to time if necessary.

Postscript.—Professor Lister, of Glasgow, has recently shown that the danger of opening chronic abscesses may be much diminished by not allowing any air to enter their cavity until it has been filtered through carbolic acid. He describes in the "Lancet" for July, 1867, his method of dealing with such abscesses. The plan appears to me to be worth adopting in the treatment of empyema requiring to be tapped. I have tried it in one case of left empyema, in a girl aged 4 years, with very good effect; there was a fistulous opening for 7 weeks after the operation, but the matter was never in the slightest degree fœtid, the discharge ceased, and the child made a complete recovery. I intend to try the plan again on the next opportunity.

---

* "American Journal of Medical Sciences," vol. xx. p. 325. Oct. 1850. And " Gardiner's Clinical Medicine," pp. 380, 718.

RICKETS.

THERE is no disease of children which it is more important to recognise in its various aspects than rickets. Its frequency would not be suspected from a study of statistical tables. It is a very frequent cause of death; yet so seldom is it assigned as the cause that the Registrar-General has not found it necessary to devote a column of his tables of mortality to this disease. The secondary diseases are recognised, such as bronchitis, collapse of lung, atrophy, measles, hooping-cough, or convulsions; but the primary disease, which renders these secondary diseases fatal, is ignored.

It is very important to be fully alive to the earliest indications of this disease of nutrition; because it is an important guide to treatment, and it can very generally be readily cured in its early stages.

If a child cuts its teeth late, if it does not walk so early as other children, if the fontanelles are late in closing, the probability is that it is the subject of rickets.

What, then, is intended by this word?

Definition.—A general disease of nutrition chiefly affecting infants, characterised at first by unhealthy alvine secretions, pains in the limbs, perspiration about the head, and subsequently by great muscular weakness and retarded ossification and dentition, softness of bones, with abnormal growth of cartilage, causing various deformities in the head, trunk, and limbs. In the spleen, lymphatic glands, and liver there is degeneration with enlargement, sometimes also in the cerebrum.

It is a disease of nutrition quite distinct from struma, tuberculosis, and syphilis.

Its relation to tuberculosis requires a few words. Both diseases are characterised by abnormal nutrition of different parts of the body. Defective hygienic conditions will produce tubercle in one child and rickets in another. Rickets seldom commences after the age of four or five years. Tuberculisation begins at any period of childhood, or later in life; it often follows rickets. The two conditions are not mutually exclusive, as maintained by Trousseau; they may co-exist in the same subject; the two processes do not often go on actively at the same time. Rickets and syphilis, also, are quite different diseases; syphilis is generally most marked in the first-born children, and each child born in succession bears less marks of the disease. Rickets is usually more and more marked in every successive child born. Syphilitic children who survive very commonly present signs of rickets.

Is the disease known as "mollities ossium" the same as rickets attacking adults? It is almost universally admitted to be totally distinct. The anatomical characters are quite different; both diseases are, it is true, characterised by a want of lime salts in the bones. In mollities there is absorption of the earthy part of completely formed bone; the bone becomes more and more porous and brittle, whilst the cancelli become filled with a jelly-like very vascular medulla. In rickets there is abnormal growth and extensive preparation for the development of bone, with an arrest of progress in ossification. In some cases there is absorption of already-formed bone even in rickets; but this does not occur to nearly so great an extent as in mollities. The progress of mollities is usually towards death; whereas in rickets, with proper treatment, recovery is the rule. When death occurs it is usually from secondary diseases, such as bronchitis, hooping-cough, or diarrhœa.

Trousseau maintains that mollities ossium is nothing more than the rickets of adults, and reports two cases in which recovery took place from the administration of cod-liver oil.

A case is reported by C. O. Weber, of Bonn, in which a patient, aged 22, who died from mollities, with 67 fractures and numerous deformities, had also signs of rickets in the enlarged epiphyses and the swollen cartilaginous ends of the ribs. The skull and clavicle were free from deformity. The parents assigned the first symptoms in this patient to the age of seven years. There is no reason why patients who have rickets in infancy may not suffer from mollities ossium in adult life, and yet the diseases be in no way related to each other.

Persons who have been rickety in childhood are often very strong in adult life, and live to a good old age. Amongst the large number of adults who were rickety as infants, the proportion of cases in which mollities ossium appears is excessively small, probably not greater than the proportion of such cases amongst the community at large.

According to the careful observations of Ritter von Rittershain,* it appears that chronic tuberculosis of the father is a predisposing cause of rickets. Of fourteen fathers with rickety children, whose health was carefully ascertained, seven were the subjects of tubercular disease. *Syphilis* of the parents also slightly inclines the child to rickets.

Is rickets ever hereditary? It is generally believed not to be so. Herring affirms that it is so in the highest degree. The observations of Ritter von Rittershain also indicate that it is so. Of 71 mothers of rickety children, 27 themselves bore the traces of having had rickets. Thirteen of these 27 mothers had their first-born children rickety. Of 10 cases of rickets in which symptoms appeared under the age of 3 months, the mothers were rickety in 5. Of 13 cases in which all the children of one couple having several children were ascertained to be rickety, there were traces of rickets in 8 of the mothers, and in a ninth case both the parents were rickety. Of 15 cases which were nourished exclusively at the breast for six months or more, the mothers were rickety in 7.

The results which I have obtained by analysing cases at the Children's Hospital are not so conclusive as these; but there is no doubt that many women have suffered from rickets in

---

* "Die Pathologie und Therapie der Rachitis." Berlin, 1863.

infancy, who do not, in adult life, present such evidence of it as is easily to be discovered in a hasty visit to the out-patients' room of a hospital.

Ritter's results certainly appear to prove that rickets in the mother is a powerful predisposing cause of rickets in the child.

It is also certain that when the mother is in delicate health, in a state in which anæmia and want of power form the prominent features, without being the subject of actual disease, her children are apt to be affected with rickets to a most decided degree, even though the father is in robust health, and when the hygienic conditions surrounding the children are favourable. It is also generally the case that if a woman bear one rickety child, her subsequent offspring will be also rickety. This is especially the case amongst the poor, and is explained by Dr. Jenner thus:—"Among the poor the parents are generally worse fed, worse clothed, and worse lodged, the larger the number of their children. And among the rich and poor alike the larger the number of children, the more has the mother's constitutional strength been taxed, and the more likely is she to have lost in general power."—(Medical Times and Gazette, March, 1860.)

Occasionally it is observed that a mother who is weak when pregnant and suckling, bears a child which becomes rickety, and then recovers her strength and subsequently bears healthy children.

Unsuitable food is, without doubt, one of the most common causes of rickets. The practice of feeding children under six months of age with all kinds of unsuitable things, especially farinaceous substances imperfectly cooked, often lays the foundation for this disease. These things are sometimes given as a supplement to breast-milk, sometimes with cow's milk, and in many cases with scarcely any milk at all. When the mother is very weakly, the infants may be exclusively nourished at the breast and still become rickety. Friedleben analyzed the milk of mothers under these conditions, and found it to contain too much water (90 per cent. instead of 86 or 87), too little casein, sugar and butter, and too little earthy salts. Impure air and want of light also conspire to aggravate the disease.

Measles and hooping cough sometimes give an impetus to the progress, and sometimes act as exciting causes of rickets, though more rarely than is the case with tuberculosis.

The following numbers are obtained from the analysis of 116 cases of rickets amongst the patients at this hospital :—Of 116 children suffering from rickets, the mothers' ages at their births were as follows :—Under 20 years there were 4 mothers ; between 20 and 25 years, 33 ; between 25 and 30 years, 24 ; between 30 and 35 years, 24 ; between 35 and 40 years, 11 ; and above 40 years of age, 3. The age of the mother does not seem to have any connection with the production of rickets.

Of 117 rickety children, 32 were first-born children ; 18 were second children ; 24 were third ; 17, fourth ; 11, fifth ; 7, sixth ; 5 were seventh in the family ; 2 were eighth ; and 1 was a tenth child of its mother.

The mothers' conditions of health were thus given :—Good in 64, weak in 31, syphilitic in 2, tubercular in 19, rickety in 1. These numbers fail to give a correct impression of the effect of the mothers' health in reference to rickets. It is very rare to see a rickety child which has been nursed by a healthy mother with an ample supply of milk.

In 103 cases in which inquiries were made as to the fathers' health, 53 were said to be very healthy, 20 moderately strong, 13 weakly, 16 tubercular, and 1 syphilitic. There was no case in which the father was said to be rickety ; but as they were many of them not seen, this information cannot be relied on.

Of 113 cases, none of the fathers were under 20 years of age, and only four were over 50 years.

Of 116 cases in which inquiries were made as to the feeding of the child during the first six months of life, 10 were brought up by hand from birth, 4 were only suckled for 14 days, 13 others were weaned before the age of 6 months. In 60 cases there was from birth a mixture of natural and artificial feeding; of these, 9 were weaned under 6 months, 20 were weaned between 6 and 12 months, 25 between 12 and 18 months, and 6 were partially suckled after the age of 18 months.

Injudicious feeding is, as already stated, one of the most frequent causes of rickets. Of cases where children were fed exclusively from the breast, 7 were weaned between 12 and 18 months, and 3 over the age of 18 months.

The ages at which the children came under my notice were: under 3 months, 1 ; between 3 and 6 months, 6 ; between 6 and 12 months, 27 ; between 12 and 18 months, 22 ; between 18 months and 2 years, 18 ; between 2 and 3 years, 25 ; between 3 and 4 years, 15 ; and above 4 years, 3. Many of these cases had been under treatment for a considerable time before I saw them, and in nearly all the cases symptoms of disease no doubt existed for a considerable period before they were treated at all. Taking this into consideration I believe that we may conclude that rickets begins in a majority of cases under the age of 12 months, at the period when growth is most active. Of 521 cases observed by Ritter, 266 were under 12 months of age, and of these 91 were under 6 months.

Bednar ("Krankheiten der Neugebornen und Säuglinge," Theil iv., s. 36) speaks of congenital rickets as a common occurrence. It is not clear whether he speaks from his own experience or not.

Virchow speaks of a preparation in the Warzburg Museum which has been usually considered congenital rickets. The thorax is said to be characteristic ; the thigh is much bent and the tissue is harder than natural, so that the medullary cavity is almost obliterated. Ritter von Rittershain has given in his book the drawing of the skeleton of a newly born child supposed to be rickety. The original is in the museum of the Franz-Joseph Hospital for Children in Prague. There is much curving of the thighs and legs, also of the pelvis and spine. The epiphyses of the ulna, the phalanges of the fingers, and the anterior extremities of the ribs are enlarged. The child was also the subject of hydrocephalus and of sclero-derma. The skull was large, and its ossification not arrested.

It would seem that this must be admitted as a case of congenital rickets. Virchow has seen the ends of the ribs enlarged by the end of the 2nd month of life; Ritter has seen them so within 3 weeks of birth.

Chemical Changes.—Rickety bones are characterised by a deficiency of earthy salts; instead of 63 per cent., as in health, the newly-formed bone in extreme rickets yields not more than 21 per cent.

There is also in rickety subjects an increase of carbonic acid in the newly-formed bone, and a reduced specific gravity. The animal organic matter of rickety bones is said by Marchand, Lehmann, and Simon, not to yield gelatine on boiling; whilst on the other hand, Friedleben and Schlossbergen say that rickety bones yield perfect gelatine.

Anomalies in the growth of Bone in Rickets.—The main peculiarity of rickety bone, as just mentioned, is the want of earthy salts. This deficiency is even more marked in the layers which form the transition from cartilage to bone. The yellowish white layer in which the cartilage cells are gathered into oblong groups, with their long diameters placed transversely, is from four to ten times thicker than in healthy bone. The epiphyses are also enlarged laterally, as well as extended in length.

Another important difference is a want of regularity in the progress of the line of ossification. The original cartilage is only partially absorbed, and medullary spaces are formed by the absorption not of calcified substance, but frequently of cartilage not yet calcified. The medullary spaces in rickety bones are much narrower than in perfectly formed bones. The walls of these spaces do not undergo absorption with the same rapidity and freedom as in health.

Instead of a rapid collapse of the cartilaginous tissue, there occurs, according to Müller, a slow transformation from homogeneous cartilage into a pale, soft, uniformly-striped mass, whilst the cells, increasing in number, come to resemble, some of them medullary cells, and some of them bony lacunæ. He calls special attention to the fact that in the neighbourhood of medullary spaces cartilage cells with thickened walls are left, and that their outline contrasts plainly with the matrix. He does not, however, regard these as preparatory to the formation of genuine healthy bone. Virchow, on the other hand, looks upon this change as the mode in which bone-cells are formed, the cell wall becoming thickened, the cavity being gradually narrowed, and offshoots remaining to form the canaliculi.

To sum up, the intra-cartilaginous growth of rickety long bones varies from that of healthy bone in an enlargement of the layers of cartilage destined for the growth of bone, in deficient calcification of the bone and cartilage, in the incomplete absorption of the latter tissue, in the formation of more numerous and narrower medullary spaces even in cartilage not yet ossified, and in the continuance of thickened cartilage cells in the imperfectly ossified layers.

The *periosteal* growth of bone, such as takes place in health in the long bones whilst they are undergoing increase of thickness, occurs also in rickets with some modifications. The periosteum, with its cellular elements, becomes considerably thickened; beneath it is a congested trabecular network with medullary spaces. This network abuts on a layer of more compact whitish material, under which is a new layer of softer reddish pumice-like substance, with stronger trabeculæ, and upon this a denser cortical substance follows. These layers are repeated in rickety bones with a constant increase of the trabeculæ of the porous layer, which is characterised by deficient calcification, whilst the inner cortical layers, on the other hand, become more compact.

According to H. Müller, in the *interior* of rickety bones whole pieces of spongy substance are liable to be absorbed, and this circumstance explains the fact that a bone which was well formed before the disease set in, often loses its firmness in a comparatively short time. The flat bones, such as those of the skull, the pelvis, and the scapula, exhibit in like manner layers of deposit, which are vascular, easily cut, and for the most part coarsely porous. These layers are closely united on the outside with the thick succulent periosteum. Beneath the porous layers are found a thin, firm table of bone, nearly or quite normal in structure.

What is the explanation of the morbid nutrition which takes place in rickety bone? It is conceivable that it might depend either on increased absorption, or on diminished supply of earthy salts, or on a combination of both. Theories have been propounded based on each of these hypotheses.

Experiments have been instituted of keeping pigeons without any lime salts. It has been found that the bones of birds thus treated contained much less earthy salts than normal, the fat was increased, the specific gravity diminished, and the quantity of carbonic acid was 25 per cent. less than in the bones of healthy animals. The bones were thin and brittle, their surface rough, they were all very bloodless, and the cavities of the long bones were abnormally wide. *There was nowhere a trace of new formation in the bone tissue, everywhere loss of substance.* In this respect there was a total dissimilarity between these bones and those of rickets. Dr. Dick has produced great flexibility of the bones and consequent deformity in the skeletons of puppies, by depriving them of milk at a very early period and feeding them exclusively on meat. There is some doubt whether the condition thus produced has been identical with rickets.

It has been supposed that, owing to an increased amount of some acid (lactic or oxalic) circulating in the blood, the earthy salts of bone are absorbed in an undue quantity. To support this theory it is stated that an excess of phosphates is constantly secreted in the urine of rickety patients. Lehmann (Schmidt's Jahrbuch, 1843, Bd. 39, s. 8) has found in the urine of rickety children four times more phosphate of lime than in healthy ones. Marchand * and Beneke † also found in the urine of a rickety child considerably more earthy phosphates than are usually found in health. Some chemists have also found a quantity of lactic acid in rickety urine. According to Lehmann it also often contains oxalic acid, the urea is dimi-

---

* " Journal für Prakt. Chemie," 1842, p. 83.
† " Zur Phys. und Path. des Phosph. und Oxal. Kalkes." Göttingen. 1850.

nished, and uric acid, which is not absolutely increased, bears a larger proportion than usual to the urea.

On the other hand, Friedleben found no increase of phosphates at all. In none of these observations has any note been taken of the amount of phosphates supplied to the patient whilst the experiments were being carried on. No note has been made as to whether the disease was in an active condition when the urine was examined.

It is difficult to collect the whole urine of a child under two years of age, which is the usual period for the active stage of rickets. So far as my observations have gone, I have not been able to detect any decided excess of phosphates in rickety children.

To prove the existence of an acid which unduly promotes the absorption of the earthy salts, it should be found in the bones themselves. The bones, however, are found to have a faint alkaline reaction.

No theory, based on an excess of absorption, would explain the changes observed in the growth of rickety bone, which are characterised as much by an excess of soft material as by a deficiency of the harder substances. Great stress has been laid on the fact that disturbances of digestion, with catarrh of the stomach and bowels, nearly always occur amongst the very earliest symptoms of rickets, and that there is an excess of acid formed in the stomach under these circumstances. Guérin and Trousseau assert that they can produce rickets in young animals by almost any diet which is unsuited to them at the time. They have not, however, described the bones of animals thus treated with sufficient accuracy to satisfy one that they were really rickety, and not ill-nourished in some other way.

All these theories proceed on the assumption that the bone-changes constitute the whole of the disease in question; they make the mistake of regarding a general disease as one which is local, or at any rate which affects but one tissue. Rickets, no doubt, is quite as much a constitutional disease as tuberculosis or syphilis, and affects the nutrition of other tissues as well as the bones. Organic chemistry has not yet explained the exact changes on which these disorders of nutrition depend.

The following may be taken as an illustration of rickets in a well-marked degree, leading to muscular weakness, contracted stature, and various deformities.

Emily E. Copus, æt. 5½ years, has never been able to walk. Head rather large, measuring 20¼ inches. The height of child is only 28 inches, not much more than the normal height at 14 months of age. Has lost six incisor teeth, and several other teeth are broken and coated with tartar. Anterior fontanelle closed, the bone in this situation depressed. Clavicles, humeri, fore-arms and femora much deformed. Chest deformed, flattened laterally, with beading of anterior extremities of ribs. Abdomen large, resonant. Heart's apex beats a little outside of left nipple. Spleen not enlarged. Lymphatic glands of groin and neck hard and freely movable. Lower half of dorsal spine much curved backwards, not laterally. At the end of about 3 months' treatment, chiefly with cod-liver oil and a nutritious diet of well-chopped meat, milk, and other suitable food, she was able to walk, and began to grow.

More or less complete arrest of growth in height occurs in rickets when the disease is at all aggravated; in the majority of diseases this does not take place, and in many growth is accelerated. At 3 years old a healthy child is from 33 to 36 inches high, a rickety one seldom more than 30 and often less than 28 inches.

The long bones in rickets become either generally thickened or irregularly knobby, whilst the edges are rounded off and the cartilaginous extremities are enlarged. The circumference of the head bears a much larger proportion to the length of the body than is usual in health, in this respect resembling the fœtal condition.

The long bones are liable to simple flexion, or to incomplete or complete fracture; the last occurrence is rarest. When the disease is at its height the bones are so flexible that they scarcely ever break, but when the disease is at an early stage or is on the decline, fractures are more apt to occur. In addition to the flexibility of the bones the ligaments are unnaturally lax, so that patients can often easily bring their toes as high as their foreheads. In the case of incomplete fracture, the periosteum remains uninjured in consequence of its softness and tenacity, and the broken fragments remain in apposition, or else only the thin bony lamella which is nearest to the medullary cavity is broken, and this often only on the concave side. In many cases the medullary cavity becomes obliterated by the displacement of the fragments and the formation of callus.

From the peculiar nature of rickety fractures it is not usual to find any great swelling from callus outside the bones at the seat of fracture as in the case of healthy bones. It is sometimes not easy to distinguish between the movability of the fragments on each other and the natural flexibility of rickety bone. It is exceptional to be able to detect crepitation between the fragments. Recovery takes place more slowly than in health, the callus sometimes does not become firm under 18 or 20 months, especially if the rickety condition is much developed.

The order in which different parts of the skeleton become involved varies in different cases. It has been stated that the disease always spreads from below upwards, but this is entirely erroneous, no constant order is observed. The skull and the thorax often show signs of disease before the lower limbs. The degree and seat of deformities are determined by the age and habits of the child, the lower limbs are not much bent unless the child attempts to walk, deformity of the chest is much aggravated by intercurrent attacks of bronchitis; as a general rule the younger the patient when rickets sets in, the more does the skull become altered.

Alterations in the Skull.—In *size* it is relatively larger than in health, the growth of the face and the limbs being arrested. Unless there is hydrocephalus or hypertrophy of the brain it is not *absolutely* larger than in healthy children.

In *form* it is more angular, from the prominence of the frontal and parietal protuberances. There is occasionally a want of symmetry from an undue prominence of the parietal protuberance on one side. The fontanelles remain open much later than in health; the anterior fontanelle, instead of being closed at 2 years of age, is often open for 3 or 4 years. The sagittal suture, which is usually closed at 12 months, remains open till the third year of age, and so with the coronal and lambdoidal sutures. There is often a depression in the course of the coronal suture and at the anterior fontanelle when the disease has run its course and the bones are united. Softness of the occipital bone is occasionally detected between the ages of 3 months and 2 years. The soft spots are about the size of a lentil or bean near the lambdoidal suture. The affected portions of bone are elastic, and feel like a bladder inflated with air. With this sign there are usually other symptoms of intense rickets, such as profuse sweating of the head, great restlessness, intolerance of being handled or moved. The back of the head is frequently bored into the pillow and thus denuded of hair. In more advanced cases the children lie most comfortably on their face or on the shoulders of their nurse, so that the back of the head is relieved of pressure.

Soft spots occasionally occur on the heads of newly-born children in whom there is no other sign of rickets. There is some doubt whether this is, in such cases, a sign of rickets or not.

The presence of soft spots in the skull cannot be explained by the pressure of the brain and the constant position of the child on its back. They are dependent on deficient ossification and irregular distribution of bony growth.

Amongst the signs of rickets is sometimes mentioned a blowing murmur heard over the anterior fontanelle, synchronous with the arterial pulse. It is to be heard in a great many cases of rickets when the fontanelle is open, but it is not confined to rickety heads, and it is absent more frequently than present in rickets. I found it present in 13 and absent in 29 rickety children whose fontanelles were open. It cannot be regarded as a sign of any value. It is most frequently heard between the 7th month and the end of the 3rd year. Various explanations have been given to account for the sound. The most probable theory is that it is produced in the arteries at the base of the brain. Some suppose that it is produced in the venous sinuses by a widening of the longitudinal sinus towards the yielding fontanelle. It is, no doubt, intensified by a watery condition of the blood.

Dentition is retarded in rickets, and the teeth are often deficient in enamel, so that they crumble and become dark in colour. The late appearance of teeth is sometimes the symptom which first excites the suspicion of rickets. Many of the ailments which are popularly ascribed to retarded dentition are really often due to a rickety constitution : such are spasm of the glottis ; unhealthy, pale, and offensive stools ; restless nights, and great intolerance of being handled.

Spinal curvature is another result of rickets, it is not angular as in caries of the spine, though the spines of some of the vertebræ may be very prominent ; the curve may always be obliterated by laying the child on its face and extending its legs, while the palm of the hand is pressed gently on the convexity of the spine. The natural flexibility of the spine is much exaggerated in rickets from the following causes : from the softness and compressibility of the intervertebral fibro-cartilages, the softness of the bodies of the vertebræ from imperfect ossification, and the laxity of the ligaments and muscles. The most ordinary curvature in rickets is an excessive bending backwards in the lower half or two-thirds of the dorsal region, whilst the upper half or third of this region is almost straight, and the lower lumbar vertebræ with the sacrum are brought forwards. If the child cannot yet walk, the posterior curve implicates the lumbar as well as the dorsal vertebræ. The natural cervical curve is exaggerated and the face is directed upwards. Lateral curvature is less common than the antero-posterior, but it is by no means rare. Its direction depends on the position accidentally assumed by the child ; as, for instance, when the patient is habitually carried on the same arm of the nurse.

The clavicles are usually deformed, sometimes very much so ; the inner portion is bent forwards and upwards at an angle, and the outer portion backwards. The curves are produced partly by the weight of the arm, the inner portion being supported by the sterno-cleido-mastoid and pectoral muscles, but chiefly by the force exerted on it when the weight of the trunk is thrown on it by the arms, when the child sits and helps himself along with his hands on the floor.

The deformity of the thorax is one of the earliest noted, and is most important in its influence on the patient's health. The enlargement of the cartilaginous ends of the bone is a striking feature; the back is flattened, the ribs form a well-marked angle at the junction of the back and sides, instead of a curve, as in health. From this point, at which the lateral diameter of the thorax is greatest, the ribs pass forwards and inwards to their junction with the cartilages, where the lateral diameter is least; the cartilages curve out before turning in to join the sternum. The sternum is thrown forwards, and the antero-posterior diameter of the thorax is increased, having somewhat the shape, on a horizontal section, of a pear with its narrow end forwards. In consequence of the direction of the ribs being inwards, and the cartilages outwards, the thorax presents a groove from above downwards on its antero-lateral aspect, from the first to the ninth or tenth rib. The groove extends lower on the left than on the right side, but is deeper on the fifth and sixth ribs on the right than on the left side; these differences are due to the liver and heart. The points of deepest grooving correspond to the fifth, sixth, and seventh ribs. A little below the level of the nipple, the chest expands considerably, the thoracic walls being borne out by the liver, stomach, and spleen.

The chief agent in producing thoracic deformity in rickets is atmospheric pressure, aided by the elasticity of the lungs, as explained by Dr. Jenner. If the chest walls were made of unyielding material, the diaphragm could only descend so fast as air could enter by the larynx and overcome the elasticity of the lungs. But the thoracic parietes are flexible; and in health there is a due relation between their rigidity, the power of the diaphragm and the rapidity of its contractions, the size of the orifice of the larynx, and the elasticity of the lungs. The proper relation between these several conditions may be disturbed in a variety of ways. For instance, in a healthy child the diaphragm may act suddenly with abnormal force, as in sobbing; there is then a recession of the most flexible parts of the chest walls during inspiration. Again, if the orifice of the larynx be narrowed by spasmodic muscular contraction, or inflammatory exudation, whilst the chest walls have their usual strength, and the diaphragm acts as usual, the soft parts of the parietes will recede at each inspiration. If, on the other hand, the diaphragm act normally, and the opening of the larynx be of normal dimensions, but the chest walls abnormally softened, there will be recession of those walls at the parts which are too soft. This is the case in rickets, and recession consequently takes place at the part of the thorax corresponding to the soft parts of the ribs, which is just behind their junction with the cartilages, giving rise to the longitudinal furrow above described. If there is chronic laryngeal obstruction in a healthy or tubercular child, the cartilages, being softer than the ribs, yield; and, as a consequence, the antero-posterior diameter of the chest is diminished and the lateral diameter increased.

The femur is usually curved forwards and onwards. Before walking this curve is produced by the weight of the child's legs and feet, when sitting with its lower limbs unsupported. The curvature thus produced will be increased by the weight of the body when the patient begins to walk. The tibia, if bent before walking, is usually curved outwards (an exaggeration of the normal curve). After the child has begun to walk the direction of the curve may be modified in different directions, according to the way in which the weight of the body is received. The ulna and radius are sometimes much bent; the curve is often increased by the child throwing part of the weight of its body on its hands and fore-arms to support itself in the sitting posture. These bones are often twisted as well as bent outwards.

The pelvis in rickets is often deformed; it is frequently stated that, whereas in mollities ossium the pelvis is triangular, in rickets it is oval. The fact is that the rickety pelvis is more frequently triangular than oval. Its form is determined by a number of conditions : the direction in which it is compressed by the spine from above, and the thigh bones from below ; this direction will vary with the usual attitude of the child, whether it can stand and walk, or only sit and crawl ; and the degree of ossification attained by the pelvic bones.

As a result of the thoracic deformity, the heart's apex in rickets often beats outside the nipple line, and the chest walls are brought into abnormally close contact with the heart's apex, so that the impulse often seems greater than natural ; unless this is borne in mind, one is likely to fall into the error of supposing that the heart is dilated and hypertrophied.

White patches are frequently seen on the visceral pericardium of rickety children, just above the apex, on the anterior aspect of the left ventricle. They seem to be produced by pressure of the heart against the cartilaginous extremity of the fifth rib. These white patches consist of fibrous tissue ; they are sometimes smooth and sometimes villous on the surface.

Another result of the thoracic deformity is vesicular emphysema of the anterior margin of the lungs, with a margin of collapsed lung tissue separating it from the healthy pulmonary structure ; the collapse is due to recession of the ends of the ribs during inspiration, and the emphysema to the sternum being thrust forward during inspiration.

Collapse of the posterior and inferior portions of the lungs is also very common in rickets as a result of bronchitis. All young children are exposed to this danger from bronchitis, owing to the flexibility of their chest walls ; the elasticity of the lungs and the abdominal muscles enables them to expel the air in expiration, but their muscles of inspiration are too weak to overcome the impediment offered by the accumulation of mucus in the bronchial tubes. In rickety children the efficiency of the inspiratory apparatus is further impaired by the softening of the ribs, and the weakness of the diaphragm and other inspiratory muscles, a weakness which they share with all the muscles of the body.

Other Symptoms of Rickets unconnected with the Osseous System.

General Nutrition.—Sometimes rickety children have a good supply of fat; in others they are extremely emaciated. The skin is almost always anæmic and flabby; and in severe cases of the disease emaciation is the rule, though by no means of constant occurrence.

Dr. Jenner is of opinion that the emaciation is almost always due to albuminoid infiltration of one or several organs. This opinion has been controverted, and, I think, with justice, by Ritter von Rittershain. There is no doubt that in a certain number of cases characterised by extreme emaciation, the lymphatic glands, spleen, and liver, are in a morbid state, as described by Dr. Jenner; but other cases equally emaciated do not present these morbid appearances; and some patients with these changes are not much emaciated. The changes, when present, are as follows :—The lymphatic glands of the body generally are implicated; they are hard and very movable, varying in size from a large pin's head to that of a horse-bean. The cut surface is pale and translucent, compact, smooth, tolerably moist, and, to the naked eye, homogeneous in structure. The substance is tough, and the specific gravity high. In rare cases the glands may be purplish in tint instead of being pale.

The spleen usually partakes of the same change; it may only be slightly larger than natural, or it may reach nearly to the middle line below the umbilicus and downwards to the crest of the ilium, measuring as much as eight inches from above downwards. Its capsule is not usually thickened; it is not adherent to adjacent structures; its anterior border is sharp; it is firm to the touch and smooth on the surface. Its substance is tough but elastic, and thin sections can be cut with ease; its section is very smooth and translucent, looking more or less as if infiltrated with glue. Only a little pale blood can be squeezed out of it. Usually the colour is pale red; occasionally it is dark purple. The splenic corpuscles are seen more readily than in health. The liver, when involved in the same process, is larger than natural, very tough; its cut surface smooth, its substance semi-transparent, its specific gravity increased. It may be accompanied with fatty degeneration of the hepatic cells. The albuminoid exudation sometimes infiltrates the portal canals and the interlobular spaces, and sometimes the circumference of the lobules.

The kidneys are less frequently involved; when they are affected, their size is increased; they are tough, translucent, and pale. When the disease is considerable all appearance of structure is lost to the naked eye. The thymus and the brain are said by Dr. Jenner to be subject to increase of size from the same kind of albuminoid infiltration.

The transparent material which infiltrates the various organs, is not subject to the reactions with iodine and sulphuric acid which are ascribed to lardaceous infiltration by Virchow.

The frequency of these changes in rickety children is not very great. During life both the spleen and liver are more easily felt and depressed below their normal position in rickety children, owing to the thoracic deformity and the flatness of the diaphragm. When there is actual enlargement of these organs, it is not uncommonly tubercular in the case of the spleen, and fatty in the case of the liver. When coincident with peculiar hardness of the lymphatic glands, it is no doubt usually of the character above mentioned.

One of the most frequent concomitants of rickets is a catarrhal affection of the air passages. After death the mucous membrane of the larynx and trachea seldom presents any considerable change; in eight cases out of forty-five non-tubercular rickety subjects examined by Ritter von Rittershain, there was inflammation of this membrane; of these, three were cases of croup or diphtheria, and one was syphilitic.

The bronchial mucous membrane usually presents different appearances in different parts. In those parts in which the pulmonary tissue is collapsed and dense, the bronchi have thickened, reddened walls, sometimes present saccular dilatations, and contain an abundant tenacious or purulent secretion;

in the anterior parts of lung which are more distended with air, the bronchial mucous membrane is usually pale, and contains a more fluid frothy secretion.

As a general rule, the anterior edges of the lungs are pale, distended with air, and contain but little fluid; whilst the lateral parts, especially of the right upper lobe, and the most dependent portions, are more or less condensed and carnified; contain no air, are either smooth on section or very finely granular, and contain bloody fluid, whilst they are beset with numerous open bronchial tubes, from which yellow puriform fluid can be pressed. The amount of pulmonary change varies in different cases from slight condensation with hyperæmia and compression of the lung tissue, to complete disappearance of the air cells and arrest of the capillary circulation.

The pale insufflated tissue occupies the anterior border of the lungs, extending backwards for about three-quarters of an inch. This emphysema is due to the sternum being pushed forward during inspiration, leaving a space into which the lung tissue can expand. Behind this is a groove of collapsed lung corresponding to the anterior enlarged extremities of the ribs, which are forced in at each inspiration owing to the softness of the bones. The collapse of the posterior portions of the lungs is due to bronchitis, involving those parts which are most usually the seats of bronchitis. In order to expel mucus from the bronchial tubes, a deep inspiration is necessary, then the glottis is closed, and by the aid of the muscles of expiration, the air in the lung is compressed; when a certain amount of pressure is attained, the glottis is opened, and air is expelled with a force proportionate to the compression existing at the moment the glottis becomes open. The greater the force exerted on the air, the more probability is there that the bronchial tubes will be cleared of their secretions.

Mucus in the bronchi impedes the entrance of air to the vesicular structure of the lungs, and in rickets the difficulty of inspiration is great, even without this impediment.* As a

---

* As Dr. Jenner says of the rickety child,—" The chief occupation of his life is to get the air into enough of his air cells to enable him to live."

consequence, the rickety child, with bronchitis, does not inspire air enough to expel the mucus; by the elasticity of the lung and his abdominal muscles, air is expelled in expiration, and the lung tissue beyond the mucus becomes collapsed; hence the great fatality of bronchitis, measles, and hooping-cough in rickety subjects.

In rickets that spasmodic affection of the respiration, known as laryngismus stridulus, is very common; this disease is, in fact, more common in rickety children than in any others. It is not unfrequently the precursor of general convulsions, and is often preceded by spasmodic contraction of the thumbs and toes. It was supposed by Elsässer that the softened occipital bone was the cause of these spasmodic affections; they no doubt more frequently depend on the general constitutional state which induces irritability of the nervous system.

Chronic hydrocephalus, with effusion of clear fluid into the lateral ventricles, is occasionally found in rickets. There is often an excess of fluid in the arachnoid, and the membrane itself is milky in appearance.

Hypertrophy of the cerebral substance is met with but rarely in rickets. It is said that the grey matter is but little involved; and Rokitansky states that the enlargement is due to an excess of granular matter between the nervous fibrillæ. The symptoms which have been observed are apathy, alternating with restlessness; occasional severe headaches, with attacks of feverishness and vertigo. The intellect becomes less active; death often results from intercurrent disease, or may occur from convulsions followed by coma. Idiotcy may ensue; this condition is met with in cretins. I have seen the brain of an emaciated rickety child, nearly three years old, much enlarged; the weight of the encephalon was nearly $3\frac{3}{4}$ pounds, about one-fifth of that of the whole bone. The grey and white matter seemed to be equally hypertrophied.

Course, Complications, and Termination of Rickets.—This disease has been described under the heads of *acute* and *chronic*. Like other diseases of nutrition it runs a very rapid or a comparatively slow course. When it appears at a very early age, when the food and general management are excessively bad, and when there is a strong hereditary predisposition the disease is often more intense and is more rapid in its progress than under other circumstances.

It is often very difficult to determine when rickets begins. A number of symptoms present themselves before the bones undergo any change. Amongst the earliest symptoms are usually seen gastro-intestinal disorders, such as sour vomiting, hiccup, diarrhœa alternating with constipation. The stools contain at first undigested milk, subsequently become greenish and have a very sour smell. The urine is said in some cases to be albuminous. The abdomen becomes distended, the child is restless, and cries when his limbs are rubbed, pressed firmly, or tossed about. The child becomes pale and the skin hangs loosely. In some cases bronchial catarrh occurs, with quickened respiration, cough, and moist râles in the chest. There is also usually profuse sweating of the head, and a tendency to get rid of all the bed-clothes at night. These symptoms can none of them be regarded as pathognomonic, although the last-named occur under scarcely any other circumstances. Changes in the bones soon make their appearance; the parts affected first are generally the ends of the ribs at their junction with the cartilages. The bones of the skull and the lower end of the radius are very usually the next which are noticed to be undergoing changes.

Curvatures of the clavicle, of the humerus, of the bones of the fore-arm, and of the lower extremities, are now seen; the time of their occurrence and the amount of deformity depend mainly on the positions usually assumed by the patient. The deformities of the legs will often continue to increase after the patient is getting better, in consequence of the greater activity of the child in walking.

Rickets may terminate in recovery more or less complete; or it may be followed by tubercular disease, or by caries of the bones. Of more frequent occurrence in rickets are bronchitis or broncho-pneumonia and convulsions as causes of death.

Measles and hooping-cough are also very fatal to rickety patients. If the child does not die from secondary disease he may become a strong man, always bearing marks of the disease in bony deformities of various kinds.

Treatment.—In the way of prophylaxis attention to the health of the mother during her pregnancy and whilst suckling the child is of paramount importance. A too rapid succession of pregnancies is a most frequent cause of rickets. If possible, the child should for the first few months be nourished by its mother's milk alone. When from any cause this is impracticable, the milk of a wet-nurse is to be chosen. Failing this, or if the mother's milk be scanty or insufficient, cow's milk, with the addition of from a third to a half of water, and a little white sugar or sugar of milk should be given. The addition of lime-water instead of pure water to the milk of London cows will be desirable to prevent acidity. Occasionally, even in very young children, the use of cow's milk thus prepared will not agree with the child. It may be necessary to give farinaceous food; but the only safe way to give it until the child is 7 or 8 months old, is in the way proposed by Liebig, in which wheat is given with malt and a little bicarbonate of potash. By this means the acid of wheat flour is neutralised, and the starch converted into dextrin. It may be prepared as follows:—An ounce of wheat flour, an ounce of malt flour, and $14\frac{1}{2}$ grains of bicarbonate of potash, are to be well mixed together, two ounces of water and ten ounces of milk are then to be added. The mixture is then to be warmed gently (not boiled) and constantly stirred till it gets thick; the vessel is then to be taken from the fire, the mixture to be stirred again for five minutes, then put again on the fire, and taken away as soon as it gets thick. It must now be stirred again until it is a thin liquid, and afterwards well boiled. The husks are then to be separated by means of a fine hair sieve. The preparation requires a good deal of care, and is not likely to be generally carried out by the poor.

When the child is 8 or 9 months old, the gravy of meat and beef-tea should be given as well as plenty of milk. When

12 or 14 months old, half-cooked mutton pounded in a mortar should be given; or the child may have a piece of meat enclosed in a muslin bag, from which he may suck all the juice, leaving the cellular tissue in the bag.

Pure air and plenty of light are of great importance as well as suitable diet.

It is also very important to attend to the skin, using daily baths with or without .common or sea salt. Baths of oak-bark have also been recommended. The child should lie on hair mattrasses and pillows, and never use those made of feathers.

Amongst drugs, the most useful are cod-liver oil and iron. In the early stages, and when there is a tendency to diarrhœa, the patients cannot always take the oil; but when it does not disagree, which it does only in rare cases, it is of the utmost value. Steel wine, or the wine of the citrate of iron, the syrup of the iodide or of the phosphate of iron, are convenient modes of administering iron. Lime-water may be given with milk or with the cod-liver oil to correct acidity and prevent diarrhœa; but I do not think it does any good by virtue of its lime. If the diet be well arranged it will contain a sufficiency of lime and of phosphates.

Constipation may be treated by aloes or rhubarb; diarrhœa by chalk and catechu at first, and if it persist tannin or alum, with occasional doses of Dover's powder.

In treating the bronchial and pulmonary affections of rickety children it is not allowable to resort to any lowering measures. A few drops of ipecacuanha wine, or squills with ammonia and chloric ether may be given.

Mechanical appliances for the prevention of deformity are seldom advisable, especially during the earlier periods of the disease. It is impossible to correct deformity in one direction without incurring the risk of aggravating it in the opposite. When the disease is arrested, and whilst the bones are still more flexible than in health, an attempt may be made to prevent the increase of deformity of the lower limbs, by means of leather, pasteboard, or wooden splints.

THE HOSPITAL FOR SICK CHILDREN.

Statistics, showing the Proportion borne by Number of Cases of Rickets to Total Number of Out-Patients.

Years.	Total Number of Out-Patients Treated.	Number of Patients suffering from Rickets.	Proportion of Cases of Rickets to Total Number.
1854	6721	233	3·5 per cent.
1855	8087	256	3·2 ,,
1856	9995	315	3·2 ,,
1857	9025	392	4·3 ,,
1858	9773	442	4·5 ,,
1859	9867	561	5·7 ,,
1860	8833	589	6·6 ,,
1861	10772	852	7·9 ,,
1862	9615	805	8·3 ,,
1863	11224	1039	9·2 ,,
1864	11523	906	7·8 ,,
1865	11685	978	8·3 ,,
1866	11536	1051	9·1 ,,

TUBERCULOSIS.

This pathological product, so common amongst adults, is even more so amongst children. During the active growth of infancy and childhood it is developed more rapidly than in later life, and is found in many more of the organs and tissues of children than of adults.

It may be produced in utero, during lactation, or at any later period of childhood. The order of frequency in which tubercles occur in different organs is given by Rilliet and Barthez, as follows : lungs, bronchial glands, mesenteric glands, small intestines, pleuræ, spleen, peritoneum, liver, large intestines, brain or meninges, kidneys, stomach ; other parts, as ovaries, heart, skin, may be involved. In adults over 15 years of age, as Louis pointed out, if there be tubercle in the body the lungs are scarcely ever free. Louis only found one exception to this rule, but it does not hold good in children.

Of 312 tuberculous children examined by Barthez and Rilliet, 47 had no tubercle in their lungs. Of 160 children with tubercle, 25, or 15·6 per cent., had no tubercle in the lungs, at the Hospital for Sick Children, London. At this hospital the order of frequency in which different organs contained tubercle was not the same as that ascertained by Barthez and Rilliet : the lungs came first, next the bronchial glands, then the liver, the spleen, the brain or meninges, the mesenteric glands, the intestines, the peritoneum, the kidneys, the pleuræ, other lymphatic glands, the heart, the larynx, the thymus, the stomach, the spinal cord, the skin, and the suprarenal capsules.

Where tubercles occur in one organ only, the lungs are most frequently the organ involved; next the lymphatic glands; and much more rarely the pleuræ, the brain, and the meninges.

Tubercle of the skin and subcutaneous tissue is not very common: it occurs in infants under 2 years of age, and is an indication of an intensely tubercular constitution. The appearance presented at first is that of an indolent swelling, varying in size from a pea to a nut. At first it is freely moveable, and there is no discoloration of the skin over it. Adhesions after a time occur, the skin assumes a reddish purple tint, and gives way, and softened tubercle with pus is discharged. Sometimes tubercle is in the substance of the skin itself. The whole tuberculous mass may be discharged at once, or only a small opening is found in the skin, through which portions of softened tubercle escape from time to time. The ulcers thus formed may heal, and deep scars are left. This form of tubercle is often accompanied or followed by tuberculous infiltration of the bones.

Children who are the subjects of well-marked tuberculosis, that is to say, who are of that constitution in which the organs are peculiarly prone to become the seat of tubercle, may be known by the following features, so well depicted by Dr. Jenner. They are very pretty, have a tall thin figure, straight limbs, a delicate transparent skin, beautiful bright eyes with large pupils and long eyelashes; their face is of an oval contour, and their hair silky. They are intelligent and often precocious, they cut their teeth young, run and talk before other children of the same age. Their bones are small in circumference, long and firm in texture; their cartilages are soft and yielding; this last quality may be measured by the ease with which the stethescope depresses the sternum.

From this contrast between the bones and cartilages of the rickety and tubercular child, the shape of the chest in rickets and tubercle is very different. In rickets, when any impediment arises to the entrance of air, the soft ribs are driven in, and to the greatest degree at the part where they are softest. In tubercle, under the same condition, the ribs maintain their form, and their position is altered, producing an entirely different shaped thorax. There are three typical forms of thorax in tubercular children : 1. The long, circular chest. 2. The long chest, with narrow antero-posterior diameter. 3. The long, pigeon-breasted chest. In all, the lungs are of proportionally small size ; as a result, if the cartilages be firm as well as the bones, the ribs become more oblique, the four upper intercostal spaces near the sternum are widened, the lower spaces are almost obliterated, the antero-posterior and the lateral diameters of the chest are diminished, and thus the first form is produced. If, on the other hand, the cartilages remain soft, the sternum is depressed ; in some cases the sternum may even sink below the level of the cartilages, and the second form is produced. The third form of thorax (long, with pigeon-breast) is produced by repeated attacks of catarrh affecting the lower lobes of the lungs. During these attacks air does not so readily enter the lower lobes ; and, as a consequence, the lower ribs are driven in by each inspiration, causing the lower end of the sternum to advance. The same kind of deformity may be produced in a child not the subject of tuberculosis who suffers repeatedly from bronchitis, giving rise to more or less collapse of the lower part of the lungs. This kind of pigeon-breastedness is distinguished from that met with in rickets, by its being limited to the lower part of the sternum, the upper part of the thorax being narrow in the antero-posterior diameter. Combined with this is often found a knuckling forward of the cartilages close to the sternum.

The proportion of children in whom tubercle is found in cases of post-mortem examination is very large. In many cases the existence of it during life was not suspected, and death resulted from other causes. In 446 post-mortem examinations at the Hospital for Sick Children, tubercle was found in 159, or $35\frac{1}{2}$ per cent. It is also certain that a very large number of children who recover are the subjects of tubercle. This is shown by the number of cases in which obsolete tubercle is found on post-mortem examination. It is very difficult to state positively that a deposit of tubercle has taken place in many patients in whom there is strong reason to suspect its existence. In the lungs there is often no physical sign on which to base a diagnosis, as the deposit may be of the miliary form, and be quite insufficient to cause a change in the percussion note or in the respiratory sounds; or if there be evidence of consolidation, the diagnosis between chronic pneumonia and tubercular infiltration is at times next to impossible.

In a case like the following, in which recovery has taken place, we may venture to say that there was tubercle in the bronchial glands.

CASE I.

A. K., æt. 8 years, a delicate boy, whose mother died when he was very young, so that he was brought up by hand. His mother's family are many of them phthisical. Has frequently had attacks of dry hard cough, especially when cutting teeth. Twelve months ago he had an attack of rigors, followed by fever, which was thought to be ague.

On 12th June, 1863, he came home from the sea-side, and appeared pale and poorly. He was thought to be cutting his molar teeth, and was taken to a dentist, who said there was nothing in his mouth to account for his indisposition.

On the 15th June he was seen by his medical attendant, who found him very feverish, and thought he was going to have scarlatina. No rash however appeared, but a dry cough came on with some hurried breathing. He continued to be very hot at night, but seemed rather better, and was allowed to leave his bed on the 26th June.

On the 27th I first saw him. He had a pale face, large pupils. During examination flushing of face came on. Pulse 80, distinctly intermitting every 7 or 8 beats. Tongue moist, slightly furred. Skin perhaps a little warmer than natural. *Thorax;* left infra-clavicular and supraspinous regions presented a rather wooden and high-pitched percussion note. Respiration weak in left infra-clavicular region. Moist rhonchi posteriorly, more marked at upper part of left lung than elsewhere. Abdomen moderately full; small gurgling sensation in right iliac fossa; no tenderness. Bowels not relaxed. Appetite tolerably good. Pupils large, act freely to light, and oscillate a little.

30*th.* Rather better. Pulse at times intermitting. After this he got better and resumed his usual condition. Towards winter it became necessary to consider whether it was prudent for him to stay in London, as he is seldom free from wheezing, and catches cold on the least exposure to cold. On October 18th, I saw him in consultation with Dr. Jenner. His left infra-clavicular region looked a little more prominent than right. This region is decidedly duller than the right, and the dulness extends partly across the sternum. Respiration here is decidedly weak, vocal resonant fremitus is much diminished at this part. Left supraspinous region is now not dull, but there is some dulness below spine of scapula and towards the spine. There is general wheezing heard over the chest posteriorly. It appeared that there was tubercular deposit in the glands at top of sternum, and pressing on the left lung at its apex, so as to interfere with respiration and the vocal signs. He was advised to take cod-liver oil, to rub in unguentum iodinii under the left clavicle, and to spend the winter at Bournemouth.

In May, 1864, I went to see him. I found his general health improved; flesh firmer. In the preceding January had had measles, followed by a fine red rash with swollen tonsils, and enlarged lymphatic glands of neck (? scarlatina). These attacks did not aggravate his cough. On examining his chest I found that the left infra-clavicular region expanded on inspiration less than right; the vocal fremitus at this part almost abolished. Percussion note clearer than right, which is now decidedly dull. The expiratory murmur under right clavicle is now prolonged, and vocal fremitus and resonance rather increased. Posteriorly left interscapular region dull above level of spine of scapula. In right supraspinous region expiration prolonged. He has had occasional flying blisters under left clavicle; is now to have one or two under the right clavicle. He spent the summer at Clevedon, where he gained strength, and almost got rid of cough; occasionally wheezing when he had a slight cold. He again wintered at Bournemouth. In March, 1865, I again saw him. He was looking much better and had gained strength. Still wheezes at times and pants when he runs. The physical signs in thorax almost the same except that the dulness on *right* side was more obvious, and respiration was divided and expiration prolonged, under right clavicle and in right supraspinous region. He scarcely ever coughs now.

April, 1866. Has spent the year at Clevedon. Is much stronger. Very seldom has wheezing now. Left infra-clavicular region almost normal except that inspiration is a little harsh. Right infra-clavicular region is not so dull as it was; expiratory murmur too long; right supraspinous region also less dull; expiration prolonged.

Remarks.—In this case there can be little doubt that there has been tubercular deposition in the bronchial glands near the bifurcation of the trachea, at one time encroaching on the left bronchus, and also in the upper part of right lung. By the persevering use of cod-liver oil, the application of counter-irritation, and change of residence to a sheltered place on the south coast, with attention to hygienic measures generally, his health has been restored, and some absorption of tubercle has taken place, or else contraction has occurred by cretifaction. The intermitting character of the pulse after the acute attack, together with flushing of the face alternating with pallor, made me think of the possibility of cerebral mischief. I was not then aware how very frequent it is to meet with an intermitting pulse in children after acute disease of various kinds, such as the exanthemata, typhoid fever, pneumonia or pleurisy, even when the attack has not been a severe one.

In the following case we may fairly assume that there is tubercular disease which has become arrested, and the deposits have probably undergone retrograde changes.

CASE II.

J. Plumer, æt. 5 years, the youngest of four children, whose father is phthisical, and one of whose paternal aunts, as well as mother's sister, died of consumption. A sister died of tubercular meningitis. He is a delicate, fair-skinned, light-haired boy, with long eye-lashes, blue eyes and blue sclerotics. His bones and limbs generally are small. Two years and a half ago he had a blow on his head which made him insensible for half an hour, after which he continued drowsy for some hours. Has had measles and hooping-cough.

Oct. 23rd, 1863. For fourteen days past his sleep has been much disturbed; has been restless and moaning. Appetite has failed. His bowels habitually costive, have been more so than usual. For two days has had headache, and light hurts him. Yesterday was very drowsy. For the last two months has had a habit of sighing without apparent cause. During the last fortnight has been subject to frequent flushing of 'the face. For a week past has had cough and cold. His tongue is clean. Pulse 120 regular. Chest flattened from before backwards, with slight tendency to pigeon-breast. Moist rhonchi generally audible over lungs. No distinct dulness at any part on percussion. The last few nights skin has been hot. ℞. Mist. Ipecacuanhæ c. acid-nitromur.

30th. Is better. Does not complain of his head. Pulse 76 not quite regular. Sighs less and flushes but seldom. To take Ol. morrhuæ, ʒj. Ter die.

Nov. 11th. Looks still very pale. Sleeps better at night, and is not drowsy in the day. Bowels rather costive. Pulse 84 not quite regular.

Jan. 12th, 1864. Had been improving till last week, when he took cold. Has alternate chills and heats. Pulse 124. Sleeps badly, wakes easily. Complains that light or noise hurts his head. Right clavicle and infra-clavicular region certainly duller than left; expiration slightly prolonged, and a pause between inspiration and expiration. Left supra-spinous region duller than right; expiratory murmur prolonged. He sweats at night.

CASES OF ARRESTED TUBERCULAR DISEASE. 109

28*th*. Had an attack of scarlatina of the mildest possible description. Fever subsided on the 3rd day. Desquamation ensued, but there was no albumen in the urine. Since that time this boy has gone on at times improving, at others relapsing into a feverish state with cold and cough.

Remarks.—This case may serve as an 'illustration of a circumstance to which it may be well to call attention, that scarlatina occurring in the course of tubercular disease often runs a very mild course, and has no effect in hastening the course of the chronic malady.

The symptoms in this boy's case were such as to make one apprehend tubercular meningitis, and I think it very probable that there was a deposition of tubercle in the meninges.

The following case, in which there was no doubt tubercle in the mesenteric glands, with effusion into the peritoneum, is an interesting one.

CASE III.

H. W., the son of a farmer, aged 8 years, was brought to me in January, 1866. His father and mother were living, and not consumptive. Two of his brothers had died at the ages of 9 and 10 years of mesenteric disease, so that his mother was the more alarmed at his symptoms, which began like those of his brothers who died. For some time past he had been weak, at times losing flesh and then regaining it again a little. For several weeks his appetite had failed, and his abdomen had much increased in size. When he presented himself to me he looked pasty, with languid circulation, and large pupils. His abdomen was very large, and moveable masses of an irregular more or less rounded shape could be felt on pressure near the umbilicus. There was some uneasiness, but no distinct pain in the abdomen. I ordered him to take cod-liver oil; and twice a week an aperient of rhubarb and soda; to have the abdomen painted daily with tincture of iodine. Diet, milk, fresh meat, eggs, and bread.

Feb. 8th. His appetite had improved. But the abdomen was more swollen, now measuring 30 inches in circumference. The swelling is symmetrical, tilts up the ensiform cartilage, and there is distinct fluctuation from side to side; the percussion note is dull in flanks and resonant near the umbilicus. He has occasionally severe pains of a colicky character, rather relieved by pressure. His motions are pale and semi-solid. The last two nights his cod-liver oil has made him sick. Pulse 100 not very strong. Urine is scanty and high-coloured. To take three times a day 1 gr. of iodide of potassium and 4 gr. of bicarbonate of potash in 2 dr. of infusion of digitalis. And every other morning 1 gr. of calomel with 12 gr. of jalap. To use the liniment of iodine to his abdomen.

15*th*. Has improved the last two or three days. The liniment has blistered the skin a little. Abdomen now measures only 27 inches; there is tympanitic resonance in the left flank and up to the navel, but a dull note in the right flank and hypogastrium. Pt. c. medic.

March 1*st*. Appetite much better. Is more cheerful. Bowels regular. Urine abundant. Has had no pain lately in stomach. Pulse 80 stronger. The cold weather does not affect him unfavourably. His abdomen still measures 27 inches, has a more doughy feel; the percussion note is dull below umbilicus in front, clear in left flank. He is to reduce the dose of iodide of potassium to half a gr. and to add 20 minims of syrup of the iodide of iron to each dose. To take 1 gr. of calomel and 6 of rhubarb once a week. Still keep up slight counter-irritation with iodine. From this time he improved rapidly. His complexion became much clearer, and he gained strength.

May 24*th*. I noted that he looks robust and ruddy. Appetite good. Pulse (standing) 80, strong. His abdomen now measures only 24 inches. There is a rather hard doughy mass to be felt in front of abdomen more towards the right side and below navel. He has continued to take oil and mixture till the present time.

Remarks.—Mesenteric disease and tabes mesenterica are terms very frequently used, and are assigned as the causes of death in a very large number of children. The notion that atrophy in children is usually due to obstructed absorption through the lacteals from disease of the mesenteric glands, was formerly held as beyond dispute. There is no doubt that occasionally the lacteals are obstructed by rapid deposit in the glands, but this is not by any means a frequent cause of infantile emaciation; though when any large proportion of the mesenteric glands is involved, emaciation is a constant result.

Whenever a child is losing flesh and has a large abdomen, by many persons mesenteric disease is assumed to exist. This disease is rare under three years of age. An enlarged abdomen and emaciation are, however, very often met with in very young children, from rickets or from simple indigestion, arising from unsuitable food or unwholesome conditions surrounding the child. It often occurs in tubercular children without disease of the mesenteric glands. It is comparatively rare to find the mesenteric glands to be the main seat of tubercle, or to find them containing much tubercular deposit, whilst other organs are comparatively free ; these glands are liable to become the seat of tubercle in common with other parts when general tuberculisation exists ; many cases of " mesenteric disease " are really *general tuberculosis*, with but little tubercle in those glands. When the intestines are the seat of tubercular ulceration, the glands are liable to be enlarged and are frequently tubercular.

It is not safe to assume the existence of mesenteric tubercle unless the enlarged glands can be felt, although it is frequently present when the glands cannot be felt. The glands are sometimes to be felt as moveable rounded masses, varying in size from that of a nut to that of a walnut ; in other cases they are matted together into a large doughy mass near the umbilicus.

Tuberculous enlargement of the mesenteric glands may be simulated by fæcal accumulation, by tubercular disease in the omentum, or by cancerous disease, which is rare.

Fæcal accumulation occurs chiefly in the transverse and descending colon ; it assumes an elongated shape, the long diameter taking the course of the large intestine. When such an accumulation exists there is usually passed from time to time offensive flatus of a peculiar odour ; and fragments of fæcal matter, which are pale, lumpy, and of granular fracture, are usually at intervals discharged per anum. If there be any reason to suspect the existence of such a collection, enemata should be used, and the effect produced on the mass, as well as the character of the stools returned with the enema, carefully noted.

The injection, unless given properly, will very probably have no effect ; a good syringe should be used ; a few ounces should be introduced at first, then after waiting a few seconds whilst the child is kept quiet and encouraged to keep in the injection, a further quantity should be injected, making up if possible as much as a pint or more. This should be retained as long as the child can be induced to hold it ; but this will probably not be more than a minute or two. The enema may consist of gruel and olive oil, with a table-spoonful of oil of turpentine, or soap-and-water may be used instead of gruel.

Omental Tubercle is usually more superficial, the mass is less nodular, and its edge is sharper ; but the distinction between tubercle of the mesenteric glands and tubercle of the omentum is not always possible, and is not of practical importance.

Cancerous disease of the omentum or the mesentery is very rare in the child ; cancer of the kidney is more common, but ought not to lead to an error in diagnosis ; the position and form of the tumour should prevent mistakes. Very probably there will be hæmaturia to facilitate the diagnosis. (See Case VI., page 120.)

Mesenteric tubercular disease is sometimes accompanied by flat or even retracted abdominal walls, but more frequently there is distension from the accumulation of gas in the intestines. This is due to the unhealthy condition of the mucous membrane and glands of the small intestines, which have determined congestion and tubercular deposit in the mesenteric glands. From the same cause diarrhœa is often met with ; there may be pain and tenderness from enteritis or peritonitis. The latter condition often gives rise to serous effusion ; or ascites is sometimes caused by pressure on the portal vein near the liver.

Tubercular disease of the peritoneum is a frequent occurrence, either with or without tubercles in the mesentery ; when acute it is not attended with any symptoms ; when chronic, it is usually attended with more or less peritonitis, with tympanitic distension of the abdomen, with tenderness and griping pains.

Adhesion often takes place with the wall of the abdomen and between the coils of intestine ; so that a large doughy mass is felt ; sometimes fluid is encysted between the coils of the intestine, giving rise to a sense of fluctuation at different points.

The patient is usually very cautious in walking quickly or going down stairs, because shaking gives much pain, so that the abdomen is steadied with the hand ; he has a worn, anxious look, and emaciation proceeds with alternating diarrhœa and constipation. Twice I have seen fæcal discharge from the umbilicus in children, due to fistulæ established after perforation of ulcerated tubercular intestine. (See Case V., page 119.)

A very common seat of tubercle is in the subcutaneous lymphatic glands, especially those of the neck, the groin, or the axillæ. It is important to distinguish between enlargement of these glands from tubercle and from other causes. In healthy children the lymphatic glands are liable to undergo enlargement whenever the true skin is inflamed. In this case the enlargement is due to excess of blood, to effusion of serum and of lymph into the gland. If the inflammation be long continued the gland may continue enlarged for a long time, or permanently. The glands nearest to the source of irritation are usually most enlarged. At an early period the glands are soft and tender, and their outline is not well defined. When chronic, they are harder, moderately moveable, and not disposed to suppurate. A good illustration of these changes is afforded by common impetigo of the scalp, giving rise to enlargement of the glands behind the sterno-mastoid muscles.

If the child be *strumous*, the glands are prone to greater enlargement, and very liable to suppuration from very slight irritation. Even without irritation the glands of strumous children are larger than those of healthy ones, but are not hard nor very sharply defined, and nearly all the glands alike partake of the enlargement.

Another common change in the lymphatic glands, especially seen in rickets, syphilis, and protracted suppuration of bone, is the so-called albuminoid infiltration of them. The glands are then hard, sharply defined, and very moveable. They are usually larger than in health, but do not attain a great size, and are seldom the seat of suppuration.

In leucocythemia the glands also undergo great enlargement.

In a tuberculous child the lymphatic glands, unless congested, or unless the seat of local disease, are not usually enlarged. They are, however, sometimes albuminoid. When they become the seat of tubercle, they are enlarged, hard, irregular in form, oval or round, but more or less uneven on the surface. One gland alone is seldom involved, and often those symmetrically placed are attacked, though not often to the same extent on the two sides. Sometimes a large mass is produced by the union of several large glands by condensed cellular tissue. The skin over them is at first colourless and non-adherent. The diagnosis rests on their large size and their indolent chronic condition; the deposition of tubercle is, however, often determined to a special set of glands by congestion from irritation. When glands which have been long enlarged undergo softening, and the skin over them becomes red and inflamed, we may pronounce almost certainly that the enlargement was due to tubercular disease.

Sometimes glands containing tubercle are not very large, and cannot be distinguished from glands enlarged merely by an effusion of lymph, or from albuminoid infiltration. In a doubtful case the aspect of the patient and the hereditary predisposition will be of assistance in coming to a decision.

Tubercle in the bronchial glands is more frequent than in the mesenteric. It is the position in which obsolescent tubercle is most frequently met with, and tubercle is often seen there when no other organ of the body contains tubercle.

During life its presence may not cause any symptoms; its commencement can often be traced to hooping-cough, measles, or bronchitis with " infantile remittent fever." Sometimes a paroxysmal cough, almost resembling pertussis, may be caused by the enlargement of the glands. When the glands are enlarged to any extent, this condition will be indicated by dulness on percussion over the first bone of the sternum, which may encroach on one or other of the subclavicular regions, more commonly the right. Dulness in the interscapular regions is not frequent, but will be present when the glands of the posterior mediastinum are much enlarged.

The presence of tubercle in children is much less frequently indicated by physical signs than it is in adults, on account of the scattered nature of the deposit, many organs being attacked at once, and the amount of deposit at any one spot being small. Examination of the apex of the lungs is comparatively of little importance in diagnosing early stages of tubercular disease in children. Percussion of the top of the sternum is of much more value. Numerous small granules dispersed over all parts of the lungs cannot be recognised by percussion or auscultation. We must not expect assistance from the expectoration or from flattening of the chest. Crackpot sound in children does not always imply the existence of a cavity.

The *diagnosis* of tubercular disease is consequently rendered more difficult, and is more dependent on general symptoms. The probability of its presence will be increased if the child be of the build and appearance above described. The symptoms may be very acute, or subacute, or very chronic.

One of the most valuable aids to diagnosis is afforded by the thermometer, as insisted on by Dr. Ringer, who has shown that "there is probably a continued elevation of the body in all cases in which a deposition of tubercle is taking place in any of its organs." Dr. Ringer is of opinion that if in any case there be a daily elevation of temperature for more than a fortnight, unless it be a case of rheumatism, ague, or typhoid fever, or some local inflammation, the diagnosis of tuberculosis may be made. I have made a good many observations since Dr. Ringer published his results, and so far my results tend partially to confirm his observations. I have, however, certainly met with some cases of tubercular disease in which the morbid process appeared to be making decided progress, in which there was no distinct elevation of temperature; this appeared to be especially true in cases of tubercle of the brain; so that I am not quite prepared to admit as an invariable rule that "the temperature may be taken as a measure of the amount of tuberculosis and tuberculization."

The diagnosis between acute tuberculosis and typhoid fever is sometimes very difficult. The patient in both diseases may have headache, be languid, dull, and heavy, the skin hot with morning remission, the pulse quick, the tongue coated, the lips dry, the appetite deficient. In the chest there may be signs of slight catarrh or some coarseness of respiration; and the child loses flesh. Usually in tuberculosis the abdomen is of normal form and free from tenderness, so it may be in typhoid fever; the stools are pale or parti-coloured and offensive. There may be diarrhœa or constipation in tuberculosis; the stools may not be characteristic in typhoid, and diarrhœa may be absent. The eruption may be wanting, or so scanty and ill-defined, as not to be sufficient for diagnosis. The diagnosis will rest mainly on the stools, the abdomen, and the temperature. In tuberculosis the highest temperature will not be usually above 102°, and at some part of the twenty-four hours it will fall to 100° or lower. The persistence of the symptoms after the third week, with progressive emaciation, the existence of a marked tubercular aspect and hereditary predisposition, will, when present, confirm the diagnosis.

In typhoid fever, unless there be local inflammation, or a relapse, the fever subsides soon after the 28th day, if not before; there will be much debility, but the appetite will return, and the stools regain their normal character, unless the ulceration of intestines have been severe and the ulcers have not healed.

In other cases of acute tubercular disease, the symptoms may lead to the suspicion of cerebral mischief; there may be headache and drowsiness, with dislike of noise and light. Another set of cases is mainly characterised by a furred tongue and deranged secretions from the alimentary canal, so that one is inclined to ascribe all the fever and other symptoms to disordered stomach and liver. At another time the chief symptoms appear referable to the bronchial tubes, and the child is believed to be suffering merely from a protracted catarrh. After a period of a month or six weeks the child may recover, or one organ is so far implicated as to cause death within a short period. The organ involved is usually one or both of the lungs, the brain, meninges, or the intestines.

In chronic tubercular disease, the symptoms are extended over a much longer period, are subject to remission or intermission, and are not so well defined until an advanced period, when there may be, in addition to emaciation and fever, signs of local disease in the lymphatic glands, lungs, or brain.

The *treatment* of the tubercular constitution in children resolves itself mainly into prophylaxis and attention to hygiene. Hereditary transmission is the most prolific cause, and over this, unfortunately, the physician seldom has any control. In contracting marriage, the probable health of the offspring is not usually a subject of consideration, and medical advice is not taken. If we are to have a healthy race, persons of strongly marked tubercular tendencies must not marry. It is specially important that there should not be a tubercular predisposition in both parents. These considerations, however, are not regarded; tubercular subjects do marry, and children are born. If the mother is tubercular, a healthy wet nurse is desirable; or, failing this, the child should be brought up on cow's milk, with one-third water and a little sugar of milk. Attention must be particularly paid to ventilation of the rooms occupied by the child; he should be warmly clad, and sent much into the fresh air. A dry, well drained house, on a porous soil, should be chosen. It has been recently shown by Dr. Buchanan's researches, that soil has more to do with the causation of tubercle than almost any other single condition. A cold bath in the morning, after the child is six or twelve months old, according to its strength, and a tepid bath in the evening, are of great value. Under twelve months of age, or if the child is very delicate and its circulation very languid, a tepid bath should be given twice a day. The addition of sea-salt to the baths renders them more bracing. The diet must be light and nutritious; good milk should enter largely into its composition till the child is eight years old; fresh meat should be given from the age of eighteen months, and animal broths and gravy from the age of twelve months. To correct the alvine secretions, an occasional dose of rhubarb and soda should be given; rarely a dose of calomel and rhubarb will be needed. A simple antacid saline, such as citrate of potash, with slight excess of potash, followed by calumba and soda, is occasionally of service for dyspepsia. When there is a tendency to anæmia, the vinum ferri, or vinum ferri citratis, is a good preparation. In the majority of such children, cod-liver oil, as an article of diet, especially during the colder months of the year, will be found of great benefit. When acute tubercular disease sets in, the symptoms and general condition must be treated, but there is no drug with which I am acquainted that is of much use.

I have tried the hypophosphites and pancreatin, both with suet and with cod-liver oil, for children, but have not met with any encouraging results from the use of either remedy in acute or chronic tubercular disease.

CASE IV.—*Tubercular Peritonitis and Pleurisy. Recovery.*

Alma Duensing, æt. 10 years. Mother phthisical. Had tolerably good health until July, 1861, when she had chronic suppuration of glands of neck, remaining open until February, 1862.

March 6th. Had pain in lower part of thorax on both sides and in the abdomen, was restless, hot, and feverish during the night, with some cough and dyspnœa. The next day lay about, looking pale and ill, at times flushing. After a compound jalap powder the bowels became very relaxed.

11th. Brought to hospital. Suffering much from paroxysms of pain in left side of chest and below the epigastrium. Bowels acting nearly every hour. Skin cool, hands and feet cold. Pulse 104 small, weak. Respirations 28.

Thorax.—Some dulness on percussion at lower part of left chest, back and front, with friction sound and feeble respiratory murmurs. No difference in vocal fremitus on the two sides.

Abdomen.—Considerably swollen, measuring at umbilicus 26¼ inches. Percussion resonant, except in lumbar regions, below umbilicus, and left iliac fossa; fluctuation marked from side to side of abdomen. No enlargement of any viscus, or any tumour to be felt; superficial veins not enlarged. Urine sp. gr. 1029, free from albumen; deposits phosphates on boiling. To take Potassii iodidi, gr. ij; potassæ acetatis, gr. x; misturæ potassæ citratis, ℨiij. Ter die.

15th. Bowels open only once a day. Is now free from pain. Abdomen a little softer.

19th. Bowels act two or three times daily, copious semi-relaxed frothy motions.

April 4th. Child's general aspect does not improve. Bowels still act two or three times daily. Motions semi-solid, light coloured. Pulse 92. Thorax much the same, but no friction sound. Abdomen now measures

CASE OF TUBERCLE WITH FÆCAL FISTULA. 119

only 24¾ inches. Is resonant throughout and free from fluctuation. After this the child went on to improve; the abdomen remained tympanitic and large, but there was no sign of fluid or any enlarged glands to be felt. Measurement 23½ inches. There were scattered rhonchi audible in chest, with some coarseness of respiration under right clavicle, and dulness on percussion.

CASE V.—*Rickets and Tubercle. Suppuration in peritoneal cavity; communication with intestine. Abscess discharging at umbilicus. Recovery.*

Ann Galligan, æt. 20 months. Father's family consumptive. She was tolerably well until four months ago, when she had hooping-cough, which lasted two months. Seven weeks ago she suddenly complained of pain in epigastric region; towards evening she seemed quite ill. The bowels were relaxed, and motions of a dark colour. She had powders given her for ten days under the impression that she was only suffering from dentition. The pain continued, and at times was very violent; she had also a slight dry cough, and lost flesh. Gradually the abdomen became enlarged and very hard; fourteen days ago the navel was observed to project, and at length it protruded more than an inch beyond the abdominal wall, and of a shining white colour.

March 27th. It gave way, and more than a quart of pus is said to have escaped. During this fortnight the child remained very ill, had much pain in abdomen, and her bowels were relaxed. On admission a linseed meal poultice was applied to abdomen, and 2 minims of laudanum were given to the child every four hours. Beef-tea and milk. She is a pale child, with straight limbs, large joints, and rickety chest.

April 7th. The following notes were taken:—A varying amount of thick pus has, since admission, been daily discharged from umbilicus, some days a very large quantity. Bowels have been open three or four times a day. Motions relaxed and offensive; no pus has been seen in them. Appetite has been good. Abdomen full and soft, every where fairly resonant, except below and to left of umbilicus for 2½ inches. An indistinctly defined swelling is felt below umbilicus extending to pubes, about the size of a small orange; pressure on the sides of the tumour causes an escape of pus at the umbilicus.

Thorax.—The right side looks a little contracted; there is a dull percussion note from clavicle to base, and also behind to a less extent, except at the lower part, which is clearer. From clavicle to nipple, inspiration is high pitched and expiration loud, almost blowing; below nipple inspiration is weak and expiration rather loud. Similar respiratory sounds behind. The child cries frequently as though in pain, but does not suffer from examination of the abdomen.

10th. The child was ordered syrup of the iodide of iron 15 minims, with a grain of iodide of potassium, in water, three times a day.

15th. The patient seemed much worse, weaker, losing appetite, frequently flushed. Discharge from navel appears to be fæcal mixed with pus. The bowels act three times daily, and the motions now contain pus. There are signs of effusion in the right pleura. Four days later she was rather better; appetite improving. Discharge from umbilicus purulent, with bubbles of gas which escape on pressure. Has taken 1¼ gr. of Dover's powder every night; lately a thickening is felt in abdomen below the umbilicus, and a mass about the size of a large chestnut is felt deeply in the right iliac fossa. Meat diet and wine are now given.

May 1st. The child's condition had improved. Discharge from navel abundant, of thin purulent character, not particularly offensive; some gas was occasionally observed.

14th. All discharge from the umbilicus had ceased, the fistula had closed, the abdomen was flat and soft, except below the umbilicus. Pulse 108 improved. Right lung still imperfectly expanded; fluid in pleura absorbed.

25th. She was discharged convalescent.

Remarks.—This case was probably one of tubercular disease of the intestines, with some mesenteric glandular disease; perforation of bowel took place, together with suppuration around the seat of perforation, limited by adhesions. There was also tubercle in the right lung, with pleurisy on the same side. This and the preceding case present many points in 'common; in both there was tubercular disease, peritonitis, and pleurisy. The symptoms were in many respects the same. Both cases looked unpromising; both recovered.

CASE VI. (referred to at page 112).—*Cancer of Kidney.*

M. H., the daughter of healthy parents, except that on the mother's side there is a tendency to phthisis. She came under my notice in March; I was informed that she had enjoyed good health until six months before, when she was 1 year and 9 months old. She then had some difficulty in passing water, amounting to retention for some hours; when passed it was of the colour of blood, and she was in much pain for a little time. She was seen by the family doctor, and at the end of a few weeks she was tolerably well again, until a few days before I saw her. She then seemed to have caught cold, and lost her appetite. Three days before my visit there was blood in her urine. On the morning of my visit she passed some blood in her water, but that passed at midday was of the normal colour. I examined the early urine and found it very acid, containing a trace of albumen, and a sediment of blood. Under the microscope were seen numerous blood discs and a few cells of

CANCER OF THE KIDNEY.

spheroidal epithelium, one group of 5 or 6 side by side. These symptoms passed off, and she was again in her usual health, though somewhat thinner, until the end of April. She then, after a fit of screaming, passed several ounces of nearly pure blood by the urethra. The blood continued to colour her urine for a few days, and then it resumed its normal aspect.

May 12*th.* I made the following notes :—Has been losing flesh and strength lately. Her complexion is much the same. Has lately complained of pain in the lower part of her back after passing water ; the urine is not passed more frequently than usual. For a week or more has had a short cough. This morning for the first time was noticed a swelling in the left hypochondrium ; when lying on her back it is visible ; it is of rounded outline, extends up under the ribs and back to the loin ; it feels smooth and hard ; it is dull on percussion, is not tender, nor is the skin over it discoloured. It measured 4 inches by 3, and dipped down into the renal region. When placed on her face there was bulging of the left loin, and the lumbar muscles could not be made out distinctly as on the right side. It appeared to be an enlargement of the left kidney. After this she had at intervals paroxysms of pain in the back and left side, also paroxysmal attacks of cough sometimes almost like hooping-cough. The tumour grew rapidly. On the 18th May, in consultation with Dr. Jenner, we found the lower part of right lung and right interscapular region dull on percussion, and respiration weak over lower part of this lung. Dr. Jenner agreed with me that there was cancer of kidney, with probable secondary growths in right lung and bronchial glands.

June 16*th.* Has lost flesh. Complexion not sallow or waxy. The tumour has increased, especially in the anterior direction. Her bowels are rather costive. She has not lately passed much blood, nor suffered much pain.

27*th.* I noted that she had very rapidly emaciated for four days past. Her finger-nails are blue, and there is general pale lividity. She has suffered a good deal from dyspnœa and restlessness. She died the next day. No post-mortem examination was allowed.

Remarks.—There can be little doubt that there was cancerous disease of the left kidney, with secondary deposits in lungs. Had there been a more careful examination of the abdomen before the 12th May, no doubt the enlargement would have been detected sooner. The disease appears to have begun nine months before death, and was marked by hæmaturia. For the first six months there was no other symptom, except slight failure of strength ; there was certainly no notable enlargement of the kidney until the disease had existed seven or eight months. The chief cause of hæmaturia in children is renal calculus; this leads to attacks of hæmaturia at intervals, generally following active exercise, and attended with renal pain passing down in the direction of the ureter. An examination of the urine will often indicate a tendency to lithic acid deposits. Another cause of hæmaturia is the existence of that condition which has been lately described, both in adults and children, by Dr. Dickinson, Dr. Pavy, Dr. Harley, and Dr. Greenhow, giving rise to intermittent hæmaturia after exposure to cold. The colouring matter of the blood is found, but not unaltered blood discs; oxalate of lime is also present; but in the intervals of the attacks there is nothing abnormal in the urine. Until there is enlargement to be detected, such a case as I have narrated might be mistaken for one of this kind; but if the urine be examined, it is found to contain numerous blood discs, and no oxalates.

The absence of the cancerous tint, and the very gradual failure of strength and flesh, were remarkable features in this case.

When the diagnosis is made in such a case, it is obvious that, in the present state of our knowledge, treatment can be only palliative.

DIPHTHERIA.

M. E. S., a fair, delicate-looking girl, aged 3 years and 3 months, was brought to the Hospital for Sick Children in the month of July with the following history.

She had measles about 7 months before, and had not been strong ever since. Was subject to swollen glands under the lower jaw. Three weeks ago she suffered for 14 days from diarrhœa, which left her weak. About the time at which the diarrhœa ceased, namely on 6th July, she became feverish and her neck was swollen; she had no pain in swallowing. Two days later had a croupy cough; her throat was less swollen. Her voice also was suppressed and husky, and there was some difficulty in her breathing. Mustard poultices and poppy-head fomentations were applied to her throat, and a "white powder" was given night and morning. She did not improve, the breathing became more difficult, causing lividity of lips and hands. Her urine was observed to be scanty.

July 12*th*. She was taken into the hospital. Her skin was cool and moist; temperature of axilla 99·4°. Some lividity of cheeks, lips and fingers. Pulse small, so quick, and the child so restless, that it could not be counted. Respiration 66 in the minute; inspiration dry, hissing; expiration rather crowing. Nares dilated with respiration. On inspiration the base of thorax and the root of the neck recede. She makes frequent ineffectual attempts to cough (is unable to draw the preliminary deep inspiration). Voice reduced to a faint whisper. Restlessness excessive. She drops off to sleep, but quickly wakes up with greater dyspnœa. On examining the fauces, there was found to be a large accumulation of mucus; the mucous membrane was swollen and dusky; and on the tonsils and uvula there were patches of adherent greyish white exudation. Her tongue was coated with a yellowish white fur. The glands at the angles of the jaw were not swollen.

This case is a good illustration of the course often run by diphtheria. A child somewhat out of health becomes feverish, complains of headache and a little pain in swallowing. In this case both of the last-named symptoms were absent, or so slight as to be unobserved, as they not unfrequently are. There is at first swelling of the neck; the throat is not examined. In two days there are loss of voice and laryngeal cough, whilst difficulty of breathing sets in and increases so gradually that she is not brought to the hospital for six days, when there are well-marked signs of deficient aëration of the blood; in fact, asphyxia is imminently threatened. On examination of her throat, a false membrane is seen lining the tonsils and the uvula.

Tracheotomy was performed the same evening with entire relief to the dyspnœa; the child however was very slow in rallying; the pulse for many hours was scarcely perceptible. The next day she seemed better, there was no lividity, and she was actually lively. The pulse was, however, weak and thready, beating at the rate of 180 per minute. The urine was found to be clear with a light flocculent sediment, which consisted of scaly epithelium, pyoid cells, spheroidal epithelium, and small fragmentary casts highly granular, in one or two cases enclosing renal epithelium. It contained a considerable amount of albumen, occupying $\frac{1}{3}$ of the tube when allowed to settle after boiling, and the addition of nitric acid. The following mixture was to be given every 2 hours: Iodide of potassium, 2 gr.; sesquicarbonate of ammonia, 1 gr.; syrup and water, of each 1 drachm. And a few drops of a solution of 2 gr. of chlorate of potash in an oz. of water were to be dropped into the trachea every $\frac{1}{4}$ of an hour.

Towards evening she refused food, her pulse was still weaker, and lividity had returned. The canula was removed, and by means of a probe a false membrane was detached, forming a cast of the trachea, the bronchi, and their primary divisions. This gave but little relief, and the next morning about 6 o'clock she died.

On post-mortem examination the tonsils were found coated with a yellowish white deposit dipping down into the follicles, detached with difficulty, leaving a smooth surface beneath. Uvula and epiglottis covered with a similar membrane. Fauces generally, and root of tongue, not injected. From epiglottis downwards the membrane extended continuously, covering the true and false vocal chords, and the whole larynx. Below the vocal chords it is rough on the surface, of considerable thickness, and tough. The wound was found to implicate the 2nd and 3rd rings of trachea. Mucous membrane of larynx, denuded of deposit, is seen to be fibrous, and to have lost its epithelium. There is a thin layer of deposit below the wound, which can be stripped off as a pellicle as low as the bronchi to their primary and secondary divisions, leaving the mucous membrane beneath unchanged. Very little collapse of pulmonary tissue; some vesicular emphysema; no capillary bronchitis. Lobules of liver rather distended. Kidneys rather opaque in cortical portion. Many tubuli were found to be distended with granular epithelium. Peyer's patches of the ileum were some of them swollen and bloodstained.

INSIDIOUS LARYNGEAL DIPHTHERIA.

Another little girl, aged 5 years, came under my care on the 17th November with the following history :—

E. A. B. had been a tolerably healthy child, occasionally subject to a cough. She had a cough of an ordinary character for 3 weeks before any distinctive symptoms manifested themselves.

Nov. 12*th*. Her voice was observed to be a little hoarse. The next day she had a noisy hollow cough "like a consumptive person." The following day she was seen by a hospital physician, who wrote on her prescription paper: "Consolidation right apex. Wasting." Cod-liver oil was given. No recent acute disease appears to have been suspected.

16*th*. The cough had become husky and dry; respiration difficult. All her symptoms rapidly became worse, and the next day she was taken into the hospital. When I saw her I found her lying, with anxious expression, livid face and lips. Her mouth was opened widely during inspiration, and her nostrils dilated. Her respirations were 28 in the minute, expiration twice as long as inspiration; it was faintly audible, dry, and of hissing character. On inspiration there was some recession of the lower part of the chest in front. Cough dry, husky, and somewhat ringing. Pulse 148 weaker during inspiration. Tongue moist, coated posteriorly with a dirty-white fur. Tonsils much swollen; no false membrane visible on the fauces. No swelling of glands of neck. The temperature of axilla was 99·8°.

Here, then, we have another case of laryngeal obstruction, setting in gradually with very little febrile disturbance, without pain in swallowing or headache. In this case there is no glandular swelling of the neck—no deposit on the mucous membrane of the throat. Diphtheria is not epidemic at the time; no other children with whom she has associated have been similarly affected. Is this a case of croup or of diphtheria? Most persons would probably call it croup, and according to the descriptions laid down in books, and the diagnostic differences between the two diseases which are described, they would be justified in this nomenclature. Whether called croup or diphtheria, there can be no doubt that this child was suffering from a disease essentially the same as the one whose case was first related.

There was no deposit of diphtheritic membrane visible in the throat, but the insidious character of the laryngeal symptoms and the slight amount of febrile disturbance were distinctive. The urine could not be saved, being mixed in every case with stools. Emetics were administered, but she became worse, so that tracheotomy was performed the same evening, but the child only lived 24 hours after. There was found after death adherent false membrane on the posterior surface of the epiglottis and lining the whole larynx; below the wound (which implicated the cricoid cartilage and 1st and 2nd tracheal rings), the trachea was injected and lined with tenacious mucus, which extended into the bronchial tubes to their secondary ramifications. The kidneys looked quite healthy. Peyer's patches were swollen and opaque, and the mucous membrane around them was much injected.

In the following case the attack was sudden, the fauces were not involved; there was albuminuria after the ninth day. Tracheotomy was performed on the third day, but the patient died after sixteen days of lobular pneumonia in a state of great exhaustion.

R. W., æt. 6 years, a strong healthy child, was quite well on Sunday, the 21st March, went to church, and took her meals as usual; towards evening her voice was slightly hoarse. The next day she seemed quite well, the hoarseness had disappeared; she went to bed at 7 p.m. and about 8 o'clock woke up and vomited. She had then some difficulty of breathing, and a croupy cough set in. She passed a very restless night, vomited 8 or 10 times; after lying down a short time she would start up, vomit and gasp for breath; then lie down and go off to sleep for a short time, breathing with a loud noise. She did not complain of sore throat or pain in swallowing. The mother ascribed the attack to her sitting in a hot church and then walking home in the cold without fastening her cloak. No cases of croup or diphtheria had been heard of in her neighbourhood.

March 23*rd*. She was admitted to the hospital; she walked from home. Her breathing was distinctly laryngeal and her cough croupy. There was slight recession of the soft parts of the chest during inspiration. She was not obviously distressed. Her skin was not hot. Pulse 140. Respiration 30. Tongue moist, pale red. No redness, swelling, or exudation about the fauces, and no external swelling in neck. At three o'clock in the afternoon the dyspnœa had much increased, the pulse was 160 small and rather hard. Six leeches were applied to the throat and 1 gr. of calomel was to be administered every 2 hours. By half-past five the dyspnœa had become so urgent that it was decided to perform tracheotomy. A moderate quantity of blood was lost during the operation. Great relief was afforded, the patient being afterwards quite comfortable. For three or four days she progressed favourably; after this prostration set in and occasional vomiting. Albumen was detected in the urine for the first time on 29th March; it was constantly present after this and in increasing proportions. The canula was withdrawn on the 30th, and the patient breathed through the glottis. Lobular pneumonia set in, and the patient lingered till April 8th, when she died, having been for some time much exhausted.

After death there was found lobular pneumonia in the 3rd stage in both lungs, and there was effusion of turbid opaque serous fluid in the left pleura and some adherent lymph on the visceral layer of this membrane. In the trachea and bronchi the mucous membrane was of a dark red colour, and studded with points of dirty-grey granular lymph. The lymphatic glands were somewhat enlarged, congested, and friable. The soft palate and tonsils were healthy. .

Remarks.—This case affords an excellent example of pseudo-membranous croup. It may be said, " It certainly resembled croup at the commencement ; but the subsequent occurrence of albuminuria, and the great debility, show that it was not croup, but diphtheria." This appears to me a very unfair way of regarding the case. I would rather say, " Here is a case of undoubted croup, accompanied by albuminuria after the ninth day."

Whichever way the case is regarded, can any one maintain that the three cases which I have just sketched are not instances of one and the same disease ?

It is possible that pure sthenic laryngitis, not at all infectious, may in some places be attended with an exudation of lymph taking the form of false membrane ; but I must say that in London I have not met with any cases having this anatomical character, which did not appear to be closely allied to diphtheria.

I can detect no distinction between membranous croup and laryngeal diphtheria. There are two diseases sometimes called croup which are totally different from diphtheria; one is laryngismus stridulus—a purely spasmodic affection, seen most commonly in rickety children ; the symptoms here are intermitting, frequently relapse, and are often accompanied by tonic contraction of the thumbs and toes, with a tendency to convulsive movements ; the other is a simple inflammation of the larynx and trachea, which induces thickening of the mucous membrane and muco-purulent secretion. The inflammatory and spasmodic elements may be combined in various proportions, giving rise to mixed symptoms, and to cases of disease which form the connecting links between pure spasm of the glottis and pure inflammatory laryngitis. In children the reflex nervous system is very active, and laryngitis, especially when severe, is attended with more or less spasm.

Besides these two classes of cases, a large number of cases are designated croup which I should prefer to call laryngeal diphtheria. "Epidemic croup" is always diphtheria. It is a very frequent occurrence for a patient to be brought to the hospital suffering from diphtheria, with the statement that one or two members of the family, or children in the same house, have recently died of croup. It is by no means an uncommon thing for patients to be here, whose case one physician considers to be croup, whilst another calls it diphtheria; or a case is admitted, and at first called croup, but when albumen is found in the urine, or a slight patch of exudation seen in the throat, it is called diphtheria.

The points of distinction insisted on by Dr. Jenner are, that croup is a local disease, not contagious; that it does not occur as an epidemic; that it does not affect any large proportion of adults; that there is no albumen in the urine, and that there are no symptoms of disordered innervation consequent upon it.

Let us look at these points seriatim. It appears to me as impossible to maintain that croup is merely a local disease, as that pneumonia is merely local, or catarrh, both of which are generally indications of a morbid constitutional state. As to contagion; diphtheria is only observed to be contagious when an epidemic prevails, or when a number of cases are brought together in ill-ventilated hospitals. Croup, as well as diphtheria, has been described as epidemic; sporadic diphtheria is not uncommon. Albumen has been found in the urine of patients with croup; it is only quite recently that it was found in diphtheria. The other two points (its frequent occurrence in adults and the symptoms of disordered innervation) have most weight in establishing a distinction; but in reference to these, it must be remembered that when diseases become epidemic they are more liable to attack adults, who escape when the disease is only sporadic; and that a certain set of symptoms often prevails in one epidemic which has been absent in other epidemics of the same disease. This is illustrated in the history of scarlatina, in different epidemics of which renal complications may be very general, or may be almost unknown. Even when epidemics of diphtheria prevailed in former times, the nervous sequelæ were not noted; we have no record of these phenomena till a comparatively recent period. It is quite probable that even if symptoms of disordered innervation had followed sporadic croup in as large a proportion of cases as they follow epidemic diphtheria, they would not have been connected with the previous illness.

Though epidemics of diphtheria have been described from the time of Aretæus downwards, Orillard, in 1834, was the first to recognize that a distinct connexion exists between paralysis and diphtheria. Previously to this isolated cases of paralysis following diphtheria had been reported by Chomel and Ghisi in 1749, by Bard in 1771, by Sedillot in 1810, and Guimier in 1826; they were lost sight of by later writers, or regarded as mere accidental coincidences. The term "diphtheritic paralysis" was introduced by Trousseau. The first complete account of diphtheritic paralysis was given by M. Faure, in "L'Union Médicale de Paris," for February, 1857. A valuable account of the same subject is given by Maingault, in an essay, "De la Paralysie Diphthérique," published in 1860, at Paris. Since then many English authors have written on this symptom of the disease. M. Roger, in the "Archives de Médecine," 1862, vol. i. ("Recherches sur la Paralysie Diphthéritique"), estimates the frequency of paralysis at a third of the cases in which life was prolonged sufficiently to allow of its occurrence. This is certainly a much larger proportion than is seen in London. At the Children's Hospital, instead of one in three, the proportion has not been higher than one in ten, or ten per cent. In private practice, and amongst adults, the proportion has been a little higher, but nothing approaching the ratio of one to three.

In Reynolds's "System of Medicine," Mr. Squire, who writes the articles on croup and diphtheria, maintains that there is an essential distinction between the diseases, adopting Virchow's views. The distinction is based mainly on the condition of the mucous membrane upon which the false membrane is deposited. In diphtheria, he says, there is interstitial necrosis, or ulceration of the mucous membrane immediately beneath the exudation; in croup the exudation is free, and there is no such lesion of the membrane. He also states that diphtheritic exudation is pure fibrin without albumen, whilst croupal exudation consists of effused lymph, in which albumen can always be detected. This distinction cannot be maintained. In cases of undoubted diphtheria there is sometimes lesion of the mucous membranæ of the pharynx, and no such lesion in the larynx and trachea. The tracheal exudation frequently contains albumen; in undoubted croup there may be ulceration.

He attaches much importance to the presence or absence of false membrane on the tonsils or soft palate. "One spot of diphtheritic deposit brought into view clears up all doubt." "The presence of albumen in the urine is," he says, "conclusive in the diagnosis." On the latter point I have already spoken. In a case (R. W., page 112) which, according to Mr. Squire's description, was undoubtedly croup, albuminuria was a prominent symptom.

On the condition of the fauces I have something more to say. If we refer to accounts of croup which prevailed before diphtheria was generally recognized, we find that it was not unusual to see it accompanied with exudation on the velum and tonsils. Dr. West gives his experience of croup in ten years from 1839 to 1849. Of 23 cases, 11 were primary and 12 secondary; of the 11 primary cases, two had false membrane on the velum and tonsils, and of the 12 secondary seven had such a formation. On referring to an excellent account of croup by Dr. Charles Wilson, in the "Edinburgh Medical Journal" for 1855-56, you find that he describes, as one of the most constant symptoms, a swelling of the tonsils of a pallid red colour. Accompanying this, in about 11 per cent. of the cases, he found "traces of exudation on the amygdalæ or pharynx. Now and then the uvula also is slightly tumid, and a flush of red tinges the whole palatine arches, or the upper portion of the gullet."

Wherever you read accounts of croup, there is a certain varying proportion of cases in which exudation involves the tonsils and soft palate. The degree to which this occurs appears to vary much in different epidemics. So much is this the case that some French writers fall into the opposite error, and assert that membranous croup is always accompanied with pharyngeal false membrane.

In Cheyne's account of Cynanche trachealis, he speaks of redness and swelling of the tonsils and soft palate, but not of the occurrence of false membrane.

Mr. Squire says that whilst more boys die of croup than girls, from diphtheria more girls die than boys. In this hospital this rule has not been found to hold.

Up to the end of 1865, from croup 11 boys died and 15 girls; from diphtheria 30 boys died, and 27 girls. The ratios were here exactly reversed. From all causes there died 370 boys and 354 girls. The numbers are not, however, sufficiently great to be of much value in a statistical point of view. It must be remembered, too, that the cases in this hospital do not include many patients under two years, and it is at this age that the mortality amongst males from all causes is considerably higher than amongst females; in the ratio of 84 to 66 under twelve months.

From croup it appears that 38 per cent. of the deaths occur under two years of age; from diphtheria not quite 20 per cent.

Of the cases of diphtheria which have occurred in the Children's Hospital, two-thirds of the cases have suffered from laryngeal complications. The laryngeal symptoms set in on the 1st, 2nd, or 3rd day, in a very large proportion of cases within the 1st week. I have seen them occur once on the 12th and once on the 19th day of the disease.

The rate of mortality amongst those with laryngeal symptoms is very high, about 80 per cent.; amongst those without laryngeal complications it has been between 50 and 60 per cent.* It would not be so high as this probably if all cases coming to the hospital were reckoned, because many of the milder cases are treated as out-patients, and are therefore not available for these calculations.

Dr. Jenner has described the following varieties of diphtheria: mild, inflammatory, insidious, nasal, primary, laryngeal, and asthenic. These terms indicate different phases of the disease, which commonly present themselves. They may often run into each other; the nasal variety is usually asthenic or insidious. The primary laryngeal variety was formerly called croup, and still goes by that name unless diphtheria of some other form is prevailing at the time. The case of R. W., just described, is a good illustration of this variety.

In a medical man, whose case was reported in the "Medical Times and Gazette," February, 1861, the disease began with rigors, pain in swallowing, redness and swelling of the mucous membrane, of the pharyngeal arches, of the palate, and uvula. Laryngeal symptoms very quickly supervened, afterwards a little lymph was seen on the arches of the palate, more abundant on the base of the arch than above, as if proceeding from below upwards. Towards the latter half of the second day his voice was gone; he had the utmost difficulty in respiration; frequently expressed in writing and by signs his urgent desire to have the windpipe opened.

Laryngotomy was performed on the third day of his illness, just in time to save life, for immediately before the operation he was on the point of asphyxia, and without relief, could not have survived five minutes.

When the disease commences in the nares its true nature is liable to be overlooked until false membrane is detected on the fauces, or laryngeal symptoms occur to indicate that the exudation is spreading to the glottis. Beyond slight impairment of health, such as loss of appetite and want of spirits, there may be nothing to excite apprehension. A slight muco-

---

* M. Roger's statistics for the Children's Hospital in Paris, in 1859 and 1860, show a mortality of 77 per cent. for laryngeal diphtheria, and 46 per cent. for cases limited to the fauces and pharynx.

purulent or sanious discharge from one or both nostrils, with little or no enlargement of the lymphatic glands of the neck, may be the only sign of local mischief. A closer examination of the nose will sometimes reveal a deposit of false membrane within the nares, but more commonly this will not be detected, and the disease may have lasted two or three days before the exudation is seen in the throat, or the occurrence of laryngeal symptoms throws light on the true nature of the case. Very often, however, the fauces, uvula, and tonsils, will be found to be red and swollen, and covered with tenacious mucus.

A case of this form was under my care at the Children's Hospital in May, 1865.

W. P., a rickety boy, was admitted on account of being generally poorly and having a slight cough. When admitted, on 22nd May, he presented the ordinary symptoms of slight bronchitis in a rickety child ; at night there was a considerable amount of dyspnœa, not at all laryngeal in character. On the 25th May, a muco-purulent discharge was observed from his nose, and the lining membrane of the nares was seen to be covered with a thin pellicle of false membrane. His tonsils were slightly reddened and not much swollen. There was a little enlargement of the lymphatic glands of neck. The next day his urine was found to be decidedly albuminous ; this condition lasted 10 days. He was ordered potassæ chloratis gr. ij. ; potassii iodidi gr. ss. ; liquoris cinchonæ $\mathfrak{m}$vj. ; syrupi, aquæ āā. 3j. in a mixture every three hours. Wine 3 oz. in 24 hours. As much beef-tea and milk as he could be induced to take.

On the 2nd June there was considerable swelling at the right angle of the lower jaw ; the cellular tissue became brawny and suppurated. It was opened on the 9th June. He then took quinine and chlorate of potash, with meat diet and 4 oz. of wine. The abscess healed, he regained his strength ; and on 19th June the rash of measles appeared. He passed through this easily, and left the hospital well on the 7th July.

In this case the insidious character of the disease was well illustrated. The exudation not spreading beyond the nares is an exceptional circumstance ; still more rare is the suppuration round the lymphatic glands which ensued. Swelling of the glands is nearly always present in diphtheria, but it very seldom leads to suppuration. The occurrence of albuminuria in such a mild case is noteworthy, and may be regarded as an indication of the whole system being implicated.

Diphtheria is sometimes met with on the skin, either near mucous orifices or on abraded surfaces of cutis elsewhere. This is of rare occurrence in London, but appears to be much more frequently met with in Paris.

A good illustration of it was afforded by an eminent surgeon, whose case I have reported in the " British Medical Journal " for September, 1864, in his own words.

The day after performing tracheotomy on a child supposed to be suffering from croup, a puncture which he had on his right thumb became painful, the next day a pustule was noticed, which was punctured and poulticed. A day or two later, on removing the epidermis, the subjacent cutis was found in the condition of a peculiar dark slough; there was an entire absence of suppuration and excessive pain. This was followed in six days by diphtheritic deposit on the tonsils. The wound on the thumb was a month in healing. A month later there was paralysis of the soft palate, partial paralysis of the fingers and legs, and some impairment of sensibility.

Another case of the inoculation of diphtheria is reported in the "Medical Times and Gazette" for December, 1866, by Dr. Patterson.

Experiments attempting to propagate the disease by inoculation have, I believe, always failed. Negative results are, however, of slight importance compared with positive ones such as those mentioned.

The rarity of cutaneous diphtheria in London may be estimated by the statement that Dr. West has only met with three cases of it in his practice, and I have only seen it once.

Description of the Exudation in Diphtheria.—Most frequently the tonsils are first affected, they are enlarged, usually one more than the other; their colour is of a deeper red than in health; in a little time swelling and glistening are observed on one part of their surface; spots are seen on this surface, at first semi-transparent, then opaque; these detached spots soon coalesce. In other cases there is but one centre of exudation, the mucous membrane around being elevated and of a violet-red colour. Sometimes the back of the pharynx is first attacked, deposits being observed in the orifices of the follicles, which soon coalesce, or lines of opaque tenacious secretion are found. The false membrane, however commencing, spreads at the margins, whilst the central portion is thickened by additions on the under aspect; on its surface it assumes a yellowish-white appearance. At the same time there is more abundant secretion of mucus mixed with fibrinous material from the adjacent mucous membrane.

The cervical glands enlarge, and the cellular tissue of the neck becomes more or less infiltrated with serum. The upper layers of false membrane, when thick, become partially separated, and from exposure to damp warm air are liable to undergo decomposition and to become fœtid.

The false membrane, examined microscopically, is found usually to consist of altered epithelial cells, of granular corpuscles, and nuclei. In the deeper layers pus globules and blood discs are often seen. Fibrillation, such as occurs in fibrinous exudation, is sometimes seen on the under layers of the deposit.

Chemically, the exudation has an alkaline reaction, swells, and is rendered transparent by acetic acid, and is dissolved by strong alkalies. It is not affected by maceration in water, and yields no gelatine when tested by tannin. When boiled it sometimes gives no indication of albumen. It is sometimes soluble in an aqueous solution of nitre (half a drachm to the ounce) like pure fibrin. The exudation in the bronchi usually becomes disintegrated by maceration in water, and gives indication of albumen on boiling.

In some epidemics of diphtheria, amongst the sequelæ have been observed erythematous, papular, or vesiculo-pustular eruptions on the skin. Ecthyma has been observed by Dr. Greve in Norway, and is considered by him as of the gravest signification.

Seat of Paralysis.—The cases in which paralysis is limited to the palatine arches and the pharynx are by far the most numerous, constituting at least two-thirds of the whole number (27 out of 36 of Roger's cases). Paralysis which involves other parts is rarer in children than in adults. Maingault collected 90 cases of diphtheritic paralysis, excluding all those in which the paralysis was limited to the arch of the palate and the pharynx, of these 29 (less than one-third) were children. Considering the much greater frequency of diphtheria in children than in adults, this is a remarkable fact, which it is not easy to account for. It may perhaps be partly explained by the larger proportion of recoveries from the disease in adults than in children; partly by supposing that in very young children slight degrees of paralysis, especially of the lower limbs, or impairment of vision, may be overlooked. These explanations are, however, insufficient to account for the great relative frequency of paralysis in the diphtheria of adults.

Maingault has given the order of frequency in which different parts were involved in his 90 cases, as follows:—paralysis of the soft palate, 70; general paralysis, 64; amaurosis, 39; paraplegia, 13; strabismus, 10; paralysis of muscles of neck and trunk, 9; paralysis of sensation without muscular weakness, 8; loss of virile power, 8; paralysis of rectum, 6; paralysis of bladder, 4.

It is quite exceptional for the palate to escape when other parts are involved. The symptoms of this are a peculiar nasal character of the voice, return of fluids by the nose, and, on examining the throat, the palatine arches are found to hang down towards the tongue; the uvula is elongated and flaccid, the palate does not move when touched with a feather or pen, or when the vowel *a* is uttered. Power of suction is also impaired. The voice is peculiarly thick and indistinct: it is really due to air passing more through the nose than the mouth, and in this respect is exactly the opposite of what is commonly, but incorrectly, called " speaking through the nose," which is, really, speaking when the nose is obstructed. The French have two words for the two varieties of voice, "voix nasillarde," when the nares are partially obstructed, and "nasonnement" for the change due to paralysis of the velum palati.

When the pharynx is paralysed there is difficulty of swallowing, there is inability to gargle, or to distend the cheeks by blowing. The inability to swallow may be absolute or it may extend only to fluids, whilst solids can be swallowed. In one case recently in the hospital nothing could be swallowed, and an unsuccessful attempt was made to nourish the child by the stomach-pump tube. The patient had stomatitis after alleged scarlatina. After five months she had again a scarlatiniform rash with diphtheria, not very well marked in its characters. There was very great prostration; the throat symptoms disappeared till after about a month; paralysis of the pharynx set in, causing an accumulation of mucus in the throat; and total inability to swallow.

In other cases there is loss of sensibility above the glottis, so that fluids swallowed, partially enter the larynx, this happens more frequently after tracheotomy than in other cases.

A great many theories have been proposed to explain the occurrence of paralysis as a sequel to diphtheria, none of them entirely satisfactory. When diphtheritic paralysis was believed to be confined to the palate and pharynx, it was very natural to suppose that it was a result of the inflammation of the mucous membrane extending its effects to the subjacent muscular tissue. It has been found, however, that this paralysis may occur in cases of cutaneous diphtheria, when there has been no exudation or inflammation of the throat, and the subsequent occurrence of paralysis in the limbs and elsewhere point clearly to a more general cause being in operation. Neither the amount of exudation and its localisation, nor the severity and the duration of the primary disease bear any constant proportion to the occurrence or extent of subsequent paralysis. Severe general paralysis may follow slight pharyngeal, nasal, or cutaneous diphtheria. Diphtheria confined to the larynx is less frequently followed by paralysis than when limited to the pharynx or fauces. Then, on the other hand, the most severe attacks of diphtheria are not at all more frequently followed by paralysis than the more mild attacks. The occurrence of albuminuria is not by any means constant in cases followed by paralysis. The gradual and progressive accession of these nervous sequelæ appear to me to suggest their dependence on some nutritive change in the nerves or muscles, which is dependent on the same general condition (whatever that may be) which gave rise to the primary symptoms. Dr. Cumming has suggested that diphtheria, in its course, leads to the destruction of some element in the blood which is essential to the nutrition of the nerves. This is pure hypothesis.

Its dependence on extension of the disease from the pharynx to the meninges of the brain or cord is neither borne out by post-mortem examination nor by the character of the symptoms. The paralysis is also quite different from the atrophic paralysis of infancy, which attains its maximum at once (see chapter on Paralysis).

A very grave symptom of disordered innervation sometimes occurs which has not yet been mentioned. Dr. Jenner says that in frequency it is next to paralysis of the pharynx. The symptom alluded to is an extreme infrequency and weakness of the pulse, the pulse falling to 40, 32, or 20 in the minute. This is generally a fatal symptom when very well marked. It is often attended by vomiting, without furring of the tongue, which is usually of itself a bad symptom as a sequel of diphtheria. It is sometimes due to renal mischief.

In the following case there was gradual failure of the heart's action without infrequency; there was also obstinate vomiting, albuminuria and scanty secretion of urine.

G. Edwards, a delicate boy, aged 6 years, became feverish and disinclined to sit up on the 10th December; he also complained of sore throat and pain in swallowing; two or three days later he was hoarse, but less feverish; on the 16th December he had a bad headache, complained more of his throat, which was swollen outside, and his mother noticed "some thick white skin" on both his tonsils. He was not brought to the hospital till December 19th. His complexion was rather pale and pasty. Pulse 108 soft and weak. His voice was not hoarse. There was a thick loose white deposit on both tonsils, arches of palate, and uvula. A trace of albumen in his urine. Tincture of sesquichloride of iron was applied freely to the throat, and he was to take 10 drops of the same in water every two hours, with wine, beef-tea, and milk as diet. Two days later, on the 21st, he was about the same, pulse 124 weak. Appetite bad ; tongue clean. Deposit on fauces seemed to be clearing away from tonsils at one or two points. Very slight hoarseness. Urine in 24 hours 16 oz., specific gravity 1015, with a considerable trace of albumen.

Dec. 23rd. Takes solids tolerably well, and fluids at first, but does not seem able to take a long draught. Pulse very weak. In other respects much the same, but more cheerful. Urine 15 oz., specific gravity 1019, with a considerable amount of albumen.

24th. Sick this morning before breakfast. Less deposit on fauces. Appetite improving. To have 6 oz. of wine daily. Urine only 7 oz., specific gravity 1026.

25th. Better. Drinks without difficulty. Appetite tolerably good. Pulse 108 not so weak. Urine 8 oz., specific gravity 1025.

26th. Throat free from deposit, tonsils slightly swollen. Is quite lively. Pulse 104 feeble. Urine 9 oz., specific gravity 1024, highly albuminous.

27th. Appetite good. Pulse 88 of moderate volume weak. Urine 6½ oz., specific gravity 1020, with rather less albumen.

28th. Seems better. Pulse 88 soft. Sick after breakfast. Urine 7 oz., specific gravity 1029. Throat quite well.

29th. Urine 6 oz., specific gravity 1024, less albuminous.

30th. Cheerful, but more languid. Pulse 104 very weak. Throat well. Urine about 8 oz., specific gravity 1024. Under the microscope it exhibited crystals of uric acid, granular epithelial cells, medium-sized waxy casts, also some small and large granular casts.

31st. Weaker. Sick after everything he takes. Urine only 3½ oz. in 24 hours, specific gravity 1035.

Jan. 1st. Almost constantly sick after taking anything. Very listless and weak. Pulse very weak, not quite regular, about 120. Urine only 1½ oz. ; the deposit of albumen occupies about ⅜ of the tube. To have brandy and beef-tea enemata every 4 hours, with brandy-and-egg mixture by the mouth.

2nd. Sickness has continued. Was at times very restless through the night. Is now moribund ; restless with pale livid countenance, cold extremities, at times feebly vomiting. Tongue moist, cool. Is quite sensible. Pulse cannot be felt. About noon he breathed quickly and feebly for a few times and then ceased to breath, with the exception of a few gasps, accompanied with spasms of muscles of jaw. Urine passed in 24 hours 1 oz., almost solidified by heat. After death only 2 drachms of urine were found in the bladder.

On post-mortem examination, 25 hours after death; the blood was found loosely coagulated. About 7 oz. of clear yellowish fluid in peritoneum ; about 3 oz. in each pleura ; no signs of serous inflammation. Very little fluid in pericardium. Heart large and flabby, weighing 3 oz. 15 drachms. No valvular disease. A patch of atheroma under the ventricular endocardium of the great tongue of the mitral valve. Muscular fibre of left ventricle was found under the microscope to be the seat of considerable fatty degeneration ; the normal striation and fibrillation of muscular fibres were at points entirely obscured by clustered oil particles ; some muscular fibres looked quite healthy, whilst others had but a few oil globules attached to them. The lungs presented but little amiss ; there was very slight collapse at the edge of the middle and lower lobes of the right lung. In the larger bronchi of the left lung was a little glairy mucus and some injection of the mucous membrane. Tonsils, uvula, arches of palate, and epiglottis, were perfectly healthy. Kidneys weighed—the left, 2 oz. ; the right, 1 oz. 14 drachms. Capsules separated readily, leaving a smooth surface with ramiform injection ; cortical substance pale, of uniform appearance in texture. It is wide in comparison with pyramids, which are congested towards the base, and contrast strongly with cortex. The tubuli from cortex, examined microscopically, were found much altered ; in the majority they appeared full of granular matter, no cell or nucleus being visible. In some there was a faint indication of the cell walls of epithelium, the nuclei being irregular and granular. The granules were not dissolved by ether.

Liver weighed 26¾ oz. On section it presented a mottled appearance from the pale colour of centre of lobules, as compared with the darker colour of their periphery. The liver cells contained an excess of fat globules.

Fatty degeneration of the muscular fibre of the heart has been met with by Dr. Bristowe (Path. Soc. Trans., vol. x.) in a case of diphtheria which died after five days' illness, and I have met with it in two other cases. May not this condition be the cause of the extreme weakness of the heart's action in some cases of diphtheria?

The period at which symptoms of disordered innervation first appear varies from the 7th day of disease to the end of the 4th week. An altered tone of voice and return of fluids through the nose occur earliest ; these are accompanied with loss of sensibility of the velum pendulum palati. The weak, infrequent pulse is noticed towards the end of the 2nd or the beginning of the 3rd week. Complete inability to swallow does not often occur till the 3rd or 4th week of illness.

The first symptoms of paralysis usually appear before the end of the 4th week ; there may be a progressive increase of symptoms till the 7th or 8th week, occasionally for 2 or 3 weeks longer. They have usually disappeared before the end of six months. There is often observed with paralytic symptoms considerable pallor and opacity of skin.

The special senses occasionally suffer, vision most frequently. The pupils are at first sluggish and somewhat dilated, there is a difficulty in reading near objects ; next, more distant objects become obscure, double vision is not infrequent, with or without strabismus. Sometimes one eye only has been affected. Taste is now and then impaired ; and, less frequently, hearing.

As these affections of special sense disappear, numbness and tingling of the fingers and toes have been noted, extending gradually to the hands and arms, as well as to the feet, legs, and hips. These sensations, with some formication and various degrees of anæsthesia, sometimes occur before any decided loss of muscular power. They have been known to disappear for a while, and to return with increasing muscular weakness. Hyperæsthesia of the paralysed parts is said to have been observed. As a general rule, the lower limbs suffer more than the upper. The arms are never paralysed without the legs. Sometimes there is complete loss of voluntary power; at others a loss of muscular sense, and of the power of co-ordinating muscular movements. Reflex action is sometimes diminished. Paralysis of the bladder is but rarely met with. In a case reported by Dr. Gull ("Lancet," 1858, vol. ii. p. 5), the " breathing became entirely thoracic. The diaphragm was unmoved in inspiration and depressed in expiration, indicating a loss of power in the phrenic nerves."

The way in which the secretion of urine became more and more scanty was very striking in the case of G. Edwards. Several cases of the same kind have come under my notice. Dr. Sanderson pointed out that when albuminuria sets in there is often no decrease in the amount of urea secreted; but that, on the contrary, there is often an increase. Mr. Squire has insisted on the same thing. In my experience, however, it is usual, when albuminuria exists to any notable extent, for the amount, both of water and of solids, passed from the bladder, very markedly to diminish, as in the case just reported.

Albuminuria is generally present in severe cases after the 5th day. It may be met with on the 1st or 2nd day of the disease; this, however, is, in my experience, very rare. It may set in as late as the 19th day. It is present in the majority of severe cases that live beyond the 4th or 5th day. It is a grave symptom, but not by any means a fatal one. It is not attended, as in the nephritis of scarlatina, with dropsy, and but rarely with serous inflammations. It is rarely accompanied by stupor or convulsions. I have not met with any cases in which it has been the foundation of chronic Bright's disease. Why there should be this difference in the course of diphtheritic and scarlatinal albuminuria is not very obvious, because the earlier renal changes are almost identical. There is, however, in scarlatinal nephritis a more abundant desquamation of renal epithelium. Albuminuria is proportionally more frequent in diphtheria than in scarlatina. Of 51 cases examined, 45 had albuminuria at some period; in 29 cases I have no note on this point.

The most common cause of death in diphtheria is the extension of disease to the air passages: this has been the cause in about four-fifths of my cases. Of 58 deaths of which I have notes, 47 were cases in which diphtheritic deposit had extended to the larynx. Of 20 recoveries there were symptoms of laryngeal obstruction in 11. Laryngeal symptoms may occur on the 1st day of illness. In the majority of cases it is present during the 1st week of illness. I have seen laryngeal symptoms commence on the 12th day, and once on the 19th day. Muco-purulent fluid very commonly accumulates in the smaller bronchial tubes, and leads to pulmonary collapse, and often to congestion and inflammatory exudation. This fluid is partly drawn into them from above, and partly secreted in them; the morbid secretion passing down from the trachea irritates their mucous membrane, and leads to secretion from that membrane.

Death is favoured after tracheotomy by the ease with which

all morbid secretion from the larynx, trachea, and larger bronchi, is inspired into the smaller tubes and air cells, and the difficulty which exists in expectorating any accumulation from the smaller bronchi when the windpipe is opened. This difficulty is vastly increased if there is, at the same time, unusual flexibility of the chest walls, as in all young children more or less, but especially so in rickety subjects; this condition acts prejudicially, by rendering it impossible for the patient to take that full inspiration which is essential to the production of a cough, such as will at all effectually expectorate accumulations from the air cells and smaller tubes.

Death occasionally, but rarely, occurs in diphtheria after only a few hours' illness from the virulence of the general disease, before there is time for the characteristic anatomical lesions. The same thing occurs in scarlet fever, small-pox, and other acute specific diseases.

Amongst the rarer sequelæ of diphtheria may be mentioned pleurisy, endocarditis, peritonitis, and erysipelas. Membrane has been passed per anum, and also from the bladder.

Mr. Bridger, of Cottenham, in a paper read before the British Medical Association in 1864, speaks of a change in the auriculo-ventricular valves as one met with in all the post-mortem examinations which he had made—24 in number. It consisted in roughness, redness, and thickness, as if from interstitial deposit. The symptoms denoting this affection of the heart were an anxious countenance, hurried respiration, rapid pulse, and tenderness in the præcordial region.

Diagnosis.—The earlier stages of diphtheria may be confounded with *catarrh*. The redness of fauces in diphtheria is more intense, but less diffused; the cervical glands and those beneath the sterno-mastoid are usually swollen; there is more debility and often premonitory indisposition. The peculiar deposit, when it is seen, is pathognomonic.

Tonsillitis.—The febrile disturbance is great in proportion to local inflammation. Pain in swallowing is more intense, and the glands are not enlarged at first. If exudation occurs, it is at the orifices of the glands, is yellowish, soft, and semi-transparent, not of any extent laterally. Occasionally tonsillitis assumes a slightly diphtheritic character.

Scarlet Fever.—Heat of skin and quick pulse usually precede the sore throat. The fever is higher and more durable. The redness of throat is more uniformly diffused, extending to both tonsils and forwards to the tongue. Both tonsils are sometimes covered with a milk-white layer, which extends to the soft palate, and does not become thicker on its under surface. The subjacent mucous membrane is reddened and sensitive. In scarlet fever the tendency is to deep ulceration and sloughing of tonsils, and there is no tendency to extension to the air passages. Serous inflammations much more frequently follow scarlet fever than diphtheria. Dropsy, also, is very rare after diphtheria; whilst paralytic symptoms are rare after scarlet fever. Albuminuria is common to both, hæmaturia to any extent is rare in diphtheria.

One disease may complicate the other, or either one may follow the other.

Diphtheria may follow typhoid fever or other exhausting disease. It is necessary to distinguish sordes, and the product of muguet from diphtheritic exudation. Sordes occur in patches of unequal thickness, not of great extent or tenacious; and they may be removed without injuring the subjacent parts, which are red and tender. They chiefly collect in front of the upper part of the mouth.

The deposit in thrush or muguet is at first adherent, afterwards more easily detached; it consists of little roundish collections, slightly projecting, and of acid reaction. If scraped off, the mucous membrane beneath is slightly depressed, red or grey. It occurs chiefly on the membrane of the cheeks, and is never of any great extent or tenacity.

There is a form of sore throat, of a herpetic character, in which a thin, whitish pellicle is formed, surrounded by a red margin on the tonsils, the palate, or pharynx. This may be mistaken for diphtheria at first, but its course is quite different.

Treatment.—No specific is known by which the disease can

be cut short. It is necessary to study in what way death is threatened, and endeavour to avert that. Diphtheria kills either by asphyxia, from obstruction to the entry of air into the lungs ; by asthenia, from the intensity of the general disease ; from gastro-intestinal catarrh with diarrhœa, by disordered innervation affecting the heart and stomach, or by defective elimination from the kidneys. To avert asphyxia, which is the most frequent cause of death, three different classes of remedies are used. One is to attempt to limit the extension of the disease below the fauces by the application of caustic on the edges of the false membrane and the surrounding mucous membrane. This is only suitable in cases where the whole of the false membrane is visible, and its margins can be reached by the caustic. In such cases it is useful to paint the surrounding mucous membrane on to which the disease is spreading with strong hydrochloric acid and honey in equal proportions. I prefer this caustic to nitrate of silver. This application sometimes arrests the local spread of the disease. It does not, however, act on the general constitutional state, which continues for a time and may give rise to fresh exudation, either on the same part or elsewhere, as well as to the other symptoms above described. It does good, however, because the morbid process generally spreads by continuity of tissue rather than from separate centres.

Another way in which asphyxia is sometimes averted is by the use of emetics, which relieve the congestion of the larynx, and the spasm, and may loosen and bring up the false membrane which is occluding the glottis. This last result is but seldom obtained. The emetic which I prefer is ipecacuanha. In sthenic cases, limited to the larynx, tartar emetic in an emetic dose followed by nauseating doses of the same is most efficacious. If this drug is given to delicate children or in asthenic cases, it causes too much depression, and leads to troublesome and exhausting diarrhœa.

Another drug which has a decided effect in checking plastic exudation is calomel. This is a remedy which is now almost discarded in the treatment of diphtheria in this country. I

L

must say, however, that I am not prepared entirely to give it up. In thirteen of my worst cases I gave calomel, and in seven recovery ensued. In some of the worst cases in which recovery has occurred, calomel has been the remedy. It is not a drug to be used indiscriminately in all cases. I limit its use to children of moderately good constitution, and to cases in which the exudation is firm and thick, or causing laryngeal obstruction with sthenic symptoms. I give it until the bowels become relaxed with greenish stools. The dose is half a grain or a grain every two or three hours, with or without a grain of Dover's powder, or half a grain of ipecacuanha. I at the same time give abundant fluid nourishment, and sometimes wine. I have been surprised to find that patients thus treated have made as good, if not a better, recovery than those who have been treated only with tonics, or with salines and chlorate of potash. In a few cases, where small doses of calomel caused diarrhœa, and the false membrane was very tenacious, and the pulse good, I have used a flannel belt smeared with mercurial ointment around the abdomen. I have not much confidence in the benefits to be obtained from this proceeding. In cases not treated with calomel, chlorate of potash in full doses is given, and frequently combined with liquor cinchonæ. This seems to be of service, though not so obviously as in ulcerative stomatitis. Tincture of muriate of iron, in full doses, is strongly recommended by many. I cannot speak so highly of it from my own experience. I have occasionally used with success iodide of potassium in one or two-grain doses, combined with the chlorate of potash. I have also used it in large doses much diluted, as recommended by Dr. Wade of Birmingham, who says it invariably prevents renal mischief. It has seemed sometimes to depress the patient without always preventing the accession of, or arresting albuminuria. In cases where there is much heat of skin, much pain in swallowing, intense redness, and swelling of tonsils and soft palate, hot fomentations to the neck, the inhalation of steam, and a simple saline, such as citrate of potash, with a mild mercurial aperient, are the best remedies. If ice is preferred by the patient, I recommend that it be given to be sucked *ad libitum*. A strong solution of chlorate of potash may be used as a gargle ; or, if there is fœtor, which is rather exceptional in children, Condy's fluid diluted in the proportion of two drachms to six ounces of water, makes a useful gargle ; a solution of carbolic acid has also been strongly recommended as a local application.

To avert death by asphyxia, when other measures fail, or if death is imminent, recourse is to be had to *tracheotomy*. I would recommend this operation whenever there is decided and persistent distress from want of air, with laryngeal respiration, and increasing recession of the chest walls and root of neck in inspiration, if it is not relieved by an emetic. It is not well to wait till lividity sets in ; at the same time, however near death the patient may appear to be, if laryngeal obstruction is the probable cause, the operation is to be recommended, with the understanding that the case is a desperate one which cannot be made worse by an operation, and that there is the remotest chance of success from an operation. The circumstances which render the operation inexpedient, are extreme rickets (on account of the great flexibility of the chest walls and the certainty of pulmonary collapse), the existence of consolidation in the lung from pneumonia, and the child being less than twelve months old. In a very young child, under twelve months old, it is impossible to get a tube, of sufficient size to maintain life, into the trachea. In such cases, it might be well to try to keep the wound open with wire, bent like an eye speculum. In the first three successful operations recorded in this country, no tracheal tube was used ; two are recorded in the third and sixth volumes of " Medical and Chirurgical Society's Transactions ;" and the third was by Mr. Carmichael, of Dublin, in 1820.*

The proportion of recoveries to deaths, when this operation has been adopted, has been higher in France than in England. M. Trousseau has saved one in four in hospital practice, and one half amongst private patients. Dr. Buchanan, of Glasgow,

---

* " Transactions of the King's and Queen's College of Physicians in Ireland."

has saved seven out of 21 cases. In the Hospital for Sick Children, in my own practice I have had five recoveries out of 22 cases operated on. I believe that in well-selected cases at least 25 per cent. may be saved. Even when the patient ultimately dies after tracheotomy, the death is generally a much easier one than when he is left to die without operation. No death can be more distressing than that which is due to a slowly advancing laryngeal and tracheal obstruction.

The causes of death after tracheotomy for diphtheria are various. One of the most frequent is extension of the disease down the trachea and bronchi. Pneumonia, especially of the lobular variety, is a frequent cause; from the opening in the trachea, the patient is unable to cough, and the secretions accumulate in the smaller bronchi and air cells, and give rise to inflammation in the small lobules. Bronchitis is another cause. The patient sometimes dies from asthenia with renal complication, leading to almost complete suppression of urine. Suppuration in the mediastinum, ulceration of the trachea, hæmorrhage either at the time of the operation or some days later, the tube not fitting and consequently slipping out of the trachea, diarrhœa from drugs previously administered, gangrene or erysipelas round the wound, pyæmia, and scarlatina supervening, have each of them been causes of death which have come to my knowledge. Hæmorrhage has been known to occur from ulceration of a large vessel below the wound, or from the surface of the wound or the mucous membrane of the trachea; an hæmorrhagic condition seems to be induced, as is sometimes manifested by purpuric spots on the skin and ecchymoses of the serous membranes found post mortem.

The operation is not in itself a painful one, and, if it is likely to be so, chloroform may be generally administered without any disadvantage. I prefer not to use it, though some surgeons advocate its employment. I have seen it in one or two cases aggravate the dyspnœa; so that it has been thought prudent to suspend its administration, and to operate without it. In other cases, it seems to relieve spasm, and thus to be of benefit for the time. When it is remembered that nearly every case in which tracheotomy is resorted to would certainly die if left to itself, if a much smaller proportion than one in four, even one in fifty, could be saved, the operation would be justifiable. The chances of success are much diminished by delaying the performance of the operation until the patient has been suffering for some time from insufficient aëration of his blood. Those practitioners meet with most success who resort to the operation early.

In the adult, laryngotomy through the cricothyroid membrane answers very well; but in the child this space is not wide enough for the canula. The cricoid is sometimes cut through, but this is undesirable. It is better that the first and second, or second and third, rings of the trachea should be divided. The operation should be performed slowly, the patient's neck being raised on a pillow, and the head depressed. Great care must be taken to keep in the middle line; if possible, the isthmus of the thyroid should be avoided, by being drawn down, and care must be taken of the large veins in its neighbourhood. When the trachea is exposed, it may be kept steady by a hook, and an incision made from below upwards of the requisite length. It is sometimes necessary to cut through the isthmus of the thyroid; the hæmorrhage in such cases is not usually excessive; the trachea may sometimes be cut through below the isthmus instead of above it.

For insertion into the wound at first, it is convenient to have a double canula, the outer one of which is bivalved and opens laterally, the fissure between them being vertical. The outer tube is inserted by itself with the valves closed, its lower end then being quite narrow; it is opened out by the insertion of the inner tube. On the first occasion that the canula is withdrawn for any purpose, it is advisable to substitute a double canula which is not thus valved, because the swollen mucous membrane is liable to be pinched between the flaps of the valves, if this form of canula is retained.

The lower aperture of the inner tube is circular, and should have a diameter equal, at least, to the narrowest part of the air-passages, which is the cricoid cartilage.

Mr. Marsh has measured the larynx of a number of children, and published the results in a valuable paper on tracheotomy in children, in the St. Bartholomew's Hospital Reports for 1867. For children from 1 to 4 years old, the diameter of the opening in the canula should be about $\frac{9}{40}$ths of an inch; from 5 to 8 years old, $\frac{11}{40}$ths of an inch; from 9 to 12 years, $\frac{12}{40}$ths of an inch, and from 14 to 16 years, $\frac{14}{40}$ths of an inch.

The advantage of the double canula is, that the inner one can be taken out from time to time without distressing the patient.

It is very important that the interior of the canula should be kept clean and smooth; otherwise a dry cake of mucus soon lines it, and narrows the aperture gradually, thus interfering seriously with the oxidation of the blood, almost without attracting the notice of the attendants.

An attempt should be made to dispense with the canula at the end of four or five days, although it is frequently impossible to do so under ten or twelve days, or even a fortnight. It may at first be left out for a few hours at a time, or during the day only, and replaced at night. In some cases the tube has been worn for months. There are dangers in leaving the tube in the larynx longer than is necessary; one is that ulceration frequently results at the point of the trachea to which the lower end of the canula reaches; another is, that there is a tendency in the larynx to contract if no air passes through it. In a case under the care of Dr. Steiner, of Prague, there was complete obliteration of the cavity of the larynx by firm cicatricial tissue. In a case under the care of Dr. West it was necessary to dilate the laryngeal aperture by introducing daily a piece of sea-tangle, and leaving it for some hours until it had swollen considerably. In this way the patient, who had worn the tube for some months, was enabled after a time to dispense with it. When a tube has been worn for a long period, the muscles which regulate the admission of air through the glottis are sometimes disabled, or act irregularly so as to cause dyspnœa.

Tracheotomy has no power to arrest the progress of the

disease either locally or in its general course. It is therefore important not to suspend constitutional treatment after the operation. If the patient is tolerably strong, and the exudation is very tenacious and membranous, calomel may be given with the same precautions as before. In a larger number of cases chlorate of potash, with cinchona, is indicated, or tincture of sesquichloride of iron. Fluid nourishment should be given, and stimulants according to circumstances. I believe that it is possible to give too much alcoholic stimulant; it appears, when given in excess, to favour albuminuria. It should be administered largely diluted.

There is often great difficulty in inducing the little patient to take nourishment; the success of the operation will often be secured by the tact of the nurse in coaxing or compelling a patient to take food. Where it is impossible to induce the patient to take enough by the mouth, it is well to give enemata of beef-tea and wine, not exceeding four or five ounces at a time, and repeating them every six or eight hours, being careful not to set up diarrhœa.

The disorders of innervation which follow diphtheria are treated with tonics, such as quinine and iron. Strychnia is sometimes given. In cases of complete paralysis of the pharynx, it may be necessary to feed the patient with the stomach pump; this is difficult in young children. Injections per anum of beef-tea and wine may also be given. These patients often die of starvation.

Immediately after tracheotomy, we usually surround the patient's bed with blankets and introduce steam into the tent thus enclosed. If the weather is very cold and dry this is useful; otherwise I believe it is better omitted. Care should be taken not to raise the temperature too high, not above 65° or 68° Fahr.; for we thereby exhaust the patient's strength.

In cases in which there is great prostration from the first, in cachectic subjects, or subjects debilitated by some recent disease, and in all cases during some epidemics, the main indication is to keep up the patient's strength. This should be attempted by the use of beef-tea, eggs, and milk, and wine or brandy. The tincture of muriate of iron, with chlorate of potash, may also be given with advantage. In these cases all lowering remedies must be avoided, and aperients, if required, must be of the mildest character.

The *prognosis* in diphtheria should always be cautiously made. No case, however mild, is entirely without risk. If fresh exudation is taking place, there is danger of extension to the larynx; during the first week laryngeal complication is always to be apprehended. When the larynx is implicated, the chances of recovery are not more than one to five. Albuminuria adds to the gravity of the case; it is by no means a fatal omen; if the quantity of albumen increases rapidly from day to day, it becomes almost certainly fatal.

Vomiting is an unfavourable symptom at any period, but especially so after the first week of disease. It often denotes renal mischief, or is a sign of disordered innervation of the pneumogastric which may lead to syncope.

The prognosis will be more serious in a family, several of whose members have already suffered severely from diphtheria, than in a family without such a history. The character of the prevailing epidemic will also influence the prognosis.

One attack of the disease does not afford much, if any, protection from a recurrence of it at a subsequent period; relapses are very frequent, and second attacks not very rare.

ACUTE HYDROCEPHALUS AND MENINGEAL TUBERCLE.

THIS disease depends, in the majority of cases, upon tuberculisation involving the cerebral meninges. For these cases, whether there be any considerable amount of *water on the brain* (hydrocephalus) or not, the name tubercular meningitis is now generally employed.

This term is based on the notion that inflammation is an essential part of the process, which is not really the case, any more than in acute phthisis. Acute deposition of tubercle in the meninges may occur with or without inflammatory exudation. Moreover, just as in the lung it is sometimes difficult to draw the line between tubercle and exudation in lobular pneumonia, so it is not always easy to say in the encephalon what is simply tubercle and what is inflammatory lymph.

Dr. Bastian has recently shown that the meningeal granulations take their origin in the epithelial lining of the peri-vascular sheaths which surround the vessels of the pia mater and the brain. In this respect they differ from tubercles of other serous membranes which take their origin in the connective tissue.

MM. Barthez and Rilliet have pointed out very clearly the distinctions between idiopathic meningitis and tubercular disease of the meninges. The latter exhibits different anatomical characters from those presented by meningitis occurring in non-tubercular patients. In the vast majority of cases the tubercular disease is preceded by a set of premonitory symptoms indicating the commencement of tuberculisation.

The anatomical characters are—

I. Meningeal tubercles seated in the meshes of the pia mater, commonly yellow granulations or miliary tubercles. They vary in size from that of a grain of sand to a line or more in diameter. They may be few or many. They are found on the convolutions or on the sulci; on the convex surface of the brain, or between the hemispheres, or at the base, and especially in the Sylvian fissures; and follow the course of small arteries.

II. Meningeal inflammation. This bears no proportion in amount to the quantity of the granulations; nor is there any relation as to position. It may be almost absent with many granulations; or there may be much inflammation without granulation. The Pacchionian bodies are usually large, and the arachnoid near them opaque. The general surface of the arachnoid is usually dry and sticky. The signs of inflammation may be mere congestion of the pia mater, or there may be a limpid fluid or a greenish gelatinous fluid infiltrating its meshes, or solid lymph. More rarely the meshes of the pia mater are thickened, red, and brittle, and give exit on pressure to a bloody fluid. All the signs of inflammation are much more frequent at the base than at the convexity. In idiopathic meningitis the signs may be met with at the base as well as at the convexity, but never *at the base alone.* In tubercular meningitis the exudation is *concrete*, in simple meningitis it is *fluid* pus. In idiopathic meningitis there is frequently pus, serum, or false membrane in the cavity of the arachnoid, which is not the case in tubercular meningitis.

III. Lesions of the Brain. (1) The grey or the white substance may be congested. (2) The cortical layers of the brain may be softened, so that they are torn on removing the pia mater; they are sometimes reddened, sometimes pale. (3) Effusion into the ventricles. This is the lesion which has excited most attention, and in many cases is the most striking one. The amount of it varies from a drachm or two to many ounces. The extent of effusion does not hold any constant relation to the severity of the symptoms, or to the degree in which other lesions exist. The fluid is usually clear and colourless, containing very little albumen, and a larger proportion of potash salts and phosphates than exists in the blood. (Vogel's Kinderkrankheit, 1863, p. 280.) There is commonly combined with it softening of the central parts of the brain, especially of the commissura mollis, the velum interpositum, and the fornix. (4) A sandy roughness of the membrane covering the corpora striata, and forming part of the wall of the lateral ventricles, is common. (5) Tubercular masses, dipping down from the pia mater, are often met with in the brain itself, if carefully looked for. In the great majority of cases tubercle will be found in other organs of the body. The disease characterised by anatomical signs, such as those just described, is no doubt a well defined and common one. It may be regarded as acute tuberculosis, developing itself mainly in the meninges of the brain. The inflammatory element may enter largely or but slightly into the case. Serous effusion in the ventricles coming on acutely in children is generally, but not always, due to tubercular disease. The following case is an illustration of non-tubercular acute hydrocephalus.

Case of Hydrocephalus, non-tubercular, probably due to venous obstruction.

F. H., aged 9 months, was admitted into University College Hospital on the 16th July. No evidence of hereditary tendency to tubercle. He is his mother's first-born. She is only 20 years of age. At birth the child seemed healthy, and continued so for 5 months, except that till 3 months of age there was apparent weakness in the muscles of the neck. The child cut 2 teeth when 6 months old, and was very strong for his age. At the age of 7 months his mother was putting him to bed one evening, when he shivered all over and was very sick. The sickness lasted about a week ; at the end of this time he had convulsions which lasted 48 hours. The mouth was drawn down to the left ; the left eye was turned upwards and outwards ; there was paralysis of the right arm and leg. From this time the child remained unconscious. For 3 weeks before admission there was much screaming and groaning. The head gradually enlarged and became very markedly hydrocephalic. Before admission he had taken iodide of potassium without benefit. On admission this treatment was continued, and a blister 12 × 8 was applied to the head.

Aug. 2nd. The following notes were taken :—Lying on right side with head retracted, the body curved with convexity forwards; feet and legs extended; great toes drawn in beyond the line of other toes, hands rigidly flexed, arms extended. The child constantly uttering a kind of grunt with expiration, and working left arm about. Eyes half-open; tenacious mucus about lids; internal strabismus of right eye; pupils medium-size, sluggish. Has 4 teeth. Gums are clenched firmly on the finger being introduced into his mouth. He appears to be quite blind. Anterior fontanelle distended; no pulsation. Head warm, not hot. Sometimes takes fluid nourishment eagerly, at others refuses it. Bowels usually act once a day; motions slate-coloured, very offensive. Spirit lotion to be applied to head. The child became gradually weaker and died on 12th August.

Autopsy 6 hours after death. Weather dry. Convex surface of brain pale, except posteriorly, where the larger vessels are injected (from position). Convolutions flattened, sulci opened out. In removing the calvaria, brain substance was ruptured, and a quantity of clear fluid allowed to escape.

Lateral ventricles much distended. Roof of the ventricles rather tough. Foramen of Monro large enough to admit thumb readily. Septum lucidum thin and quite perfect. Ventricle of the septum large. Surface of corpora striata presents a mammillated aspect; the elevations look soft, and can be felt by finger. It would appear, however, as if the corpora striata were depressed by sulci, and hence elevations produced, rather than as if there had been an abnormal deposit on them. There are some large vessels ramifying on the outer surface of these bodies. There is no softening of the floor of the ventricles, of the fornix, or of the commissures. There is more difficulty than usual in drawing back the velum interpositum, owing to the firmness of the large veins entering into its structure. The corpora striata and optic thalami have a less rounded outline than usual. The left thalamus has a transverse depression near its centre from before backwards; the posterior portion is pinker than normal. On removing the brain from cranium, a portion of the membranes, with a thin layer of brain attached, is left adherent to the neighbourhood of the left sphenoidal fissure. On the posterior surface of the middle lobe on left side is a vessel distended with what appears to be clotted blood, undergoing the changes commonly met with in blood effused during life. The coats of the veins are not thickened; the clot is partly yellow and partly pink. The vein leads to one of the sinuses near the torcular Herophili, where is also a decomposing clot. In the other sinuses there is much clotted and some fluid blood. The membranes at the base of the brain do not tear readily, but their toughness appears to be entirely due to vessels entering into their composition, which are tougher than usual; the intermediate parts are thin and transparent. There is *no granular appearance at the base or in the Sylvian fissures,* or any *inflammatory exudation around the optic tracts, the origins* of the cranial nerves, or between the middle and posterior lobes. On the wall of the posterior cornua of the lateral ventricles there is a peculiar yellow discoloration, having an irregular outline, reticulated and irregularly stellated ; it appears a little raised above the surrounding portion of the walls, and is surrounded by a slight vascularity.

Fourteen ounces of serum were collected from the ventricles ; some was lost. Cerebellum, pons and medulla oblongata healthy.

Lungs. Left.—Free from tubercle, emphysematous.

Right.—The seat of lobular pneumonia in the middle and lower lobes. No signs of secondary deposits.

Spleen small, firm, and healthy.

Liver healthy.

Kidneys very pale but healthy.

Heart.—Right side full of black clot ; clot in left auricle and ventricle. Valves and muscular structure healthy.

No trace of tubercle in any part of chest or abdomen.

This case was mistaken during life for one of tubercular disease affecting the brain and its membranes. The sudden accession in perfect health is remarkable ; the occurrence of rigors, which are rare in young children, is also noteworthy. The obstinate sickness was much like the early history of tubercular meningitis, and the occurrence of convulsions at the end of a week was quite consistent with this hypothesis ; hemiplegia, though rarely met with, was not incompatible with this theory. The slow after progress of the case rendered the diagnosis of tubercular meningitis more unlikely ; a duration of 3 months is quite exceptional in fatal cases ; it is usual for the progress of tuberculisation to put an end to life within that period. The pathology of this case is not quite clear ; it seems however probable that coagulation took place in the veins, and that serous effusion occurred from this obstruction to the circulation. What led to the coagulation it is not easy to say, whether phlebitis, or simple coagulation such as often occurs in the course of phthisis or after fever.

The hardness of the central parts of the brain in this case is evidence, if any be needed, that mere soaking of those parts in serous fluid will not produce softening such as is commonly met with in tubercular meningitis.

A case such as the one just described would be a favourable

one for puncture. This plan of treatment was proposed, but the mother would not give her consent.

In the great majority of cases of tubercular meningitis there are premonitory symptoms indicative of failing health, especially loss of flesh, loss of colour, loss of spirits, or some change in natural disposition. Loss of appetite is a very frequent premonitory symptom. In the following case there were scarcely any prodromata.

H. C. S., aged 7 years. His father (a surgeon) and mother were healthy, and there was no hereditary tendency to tubercle so far as could be ascertained. He is the 2nd of a family of 5 children. The eldest boy has been rather delicate, and a younger brother died 12 months before of hooping-cough. He had himself been in very good health till about the 1st May. When an infant, and being suckled by his mother, his father had a severe illness, which gave his mother much anxiety; at the age of 18 months he did not seem to be thriving, did not grow much, and had a large abdomen. Since that time he had become a strong and apparently healthy boy. Twelve months ago it was observed that he had a habit of frequently sighing; and priapism was noticed to an unusual degree. For some months past he had slept with his head drawn back, and he frequently put up his hand to the side of his head to support it, but never complained of headache. He was not a precocious or peculiarly excitable child. He usually studied about 2 hours a day. For about a week before the end of April he complained of soreness of his head when it was brushed or combed. About the 1st May he was sick in the morning on getting up; his tongue was furred; he had 2 aperient powders given to him, which acted freely; the sickness however continued for several days; he had *no headache*. The appetite failed, and on the 3rd there was a little feverishness. From the 4th to 7th May the bowels did not act, but were freely moved by medicine on the 3rd. On the 7th his pulse was observed to be under 70 beats in the minute. On the 8th he became drowsy, his pulse was slower and a little irregular; towards night he became restless and somewhat excited. His bowels were freely opened by a mercurial powder and senna. On the 11th, when I first saw him, he was lying in a very listless condition, but not liking to be disturbed. Head warm, not hot. Had been sick once during the day. Pupils of medium size act moderately to light. Pulse regular, about 60 in the minute. No strabismus and no rigidity of any part. In the left infraclavicular region of thorax the percussion note is slightly dull, and expiration is prolonged. He knows his father, and at times speaks spontaneously. He is very slow in replying to questions, and sometimes does not answer simple questions. Small doses of calomel were given frequently, and tartar emetic ointment was applied to the head.

12th. Worse; quite unconscious. Passing motions in bed very freely. Slight strabismus. During the night he was very weak and low; some stimulant was given.
13th. Worse. Comatose. Some rigidity of limbs and subsultus.
14th. He died.
No post-mortem examination was made.

Remarks.—There is no doubt that this was a case of tubercular meningitis, although occurring in a child not known to be tubercular. All the symptoms were characteristic. The absence of headache is worthy of note, though not very rare; constipation was not a marked symptom. The duration (eleven days) was short considering the absence of prodromata and the comparative slightness of the febrile symptoms.

The following case affords an illustration of a class in which, although no tubercle is met with in the body, the symptoms are almost identical with, and the *post mortem* appearances, though without meningeal granulations, are similar to what is met with in tubercular meningitis. The duration of the case was somewhat protracted.

C. V., aged 2 years, brought to the Children's Hospital on 16th January, 1860. Seen by the house-surgeon, who noted on his card, "Debility," and ordered Ol. morrhuæ, ʒj. Ter die. He is the youngest of 8 children; 4 others living said to be healthy; 2 died of inflammation of chest at the ages of 21 and 6 months respectively; 1 of small-pox. He was a healthy baby; never had any illness till the present, except a little diarrhœa in the summer. About a month before Christmas last, he fell out of bed on his forehead. He was not insensible, but cried a good deal. Nothing more was thought of it, as he seemed in his usual health, until a fortnight later. At this time he was seized with vomiting and purging; the sickness occurred both before and after food. There was also thirst and loss of appetite; he rolled his head about on his pillow a good deal. About a week later than this, his head was drawn back on his shoulders. It was called "rigidity of the cervical muscles," and rubbed with liniment. This lasted for 3 weeks. About the same time he had screaming fits regularly every other night; these were accompanied with great pallor, alternating with flushes.

Jan. 23rd. I saw him for the first time as an out-patient, and noted on his paper, " Screaming fits, followed by vomiting and diarrhœa. Rolling of head." He did not present himself again until 6th February, when his mother stated that there was constant sickness, and that he had convulsive fits every other night. His head was not hot, and the pupils were equal. I gave him a purge of calomel and jalap; and a saline mixture with a grain of iodide of potassium three times a day. On 27th February his mother brought word that he was dying. He had had 2 strong fits of convulsions; he was violently screaming every 5 minutes and constantly sick.

March 19*th.* There has been for the past month rigid contraction of the hands; he has had a number of convulsive attacks, in which his arms and legs are drawn up and his face works about. Has been quite blind for more than a week. Still takes nourishment, though he appears unconscious, and there is constant twitching of the muscles, and frequent moaning.

22*nd.* Unconsciousness continues. Is in a constant succession of convulsive seizures.

26*th.* He died.

Autopsy (26*th*).—Lungs healthy, free from tubercle.

Liver, spleen, and kidneys healthy.

Head.—Fontanelles closed. Dura mater partially adherent on right side of longitudinal fissure over a patch the size of a crown piece. The convolutions are a good deal flattened. On slicing the brain the lateral ventricles are found distended with clear fluid; about 15 oz. of fluid was measured. The walls of the ventricles are rather tough, and the vessels of these membranes appear considerably injected. The corpora striata are darker in colour than usual, but not distinctly injected. The commissura mollis and the fornix are softened. The membranes at the base of the brain are tough and opaque in patches. No granulations are seen on them. Over the central part of base of brain there is a patch of membrane nearly 1½ inch square, which is infiltrated with lymph of a yellow colour and almost cheesy consistence. There is no undue injection about the membranes. The intestines were not opened.

Remarks.—The insidious way in which this case began, so that cerebral mischief was scarcely at first suspected, was very unlike idiopathic meningitis, but characteristic of tubercular disease. Whether the fall was the cause of the disease is doubtful.

In young children it is very important to bear in mind that cerebral symptoms are frequently due, not to disease in the brain, but to disease of other organs, causing sympathetic disturbance of the cerebrum. This is often a cause of incorrect diagnosis; pneumonia is constantly treated as cerebral by persons who do not examine the chest; so also is pleurisy. Both these complaints are very often comparatively latent in children.

It would not be compatible with my present subject to enter

fully into the diagnostic marks of these affections, which are indeed sufficiently obvious when attention is called to the matter. (See chapter on Pneumonia and Pleurisy.)

Dr. Wilks ("Guy's Hospital Reports," 1860), states that he has never met with tuberculous meningitis in any case which did not present tubercle in the lungs. I have more than once met with such a condition.

The following case is an example :—

E. J., æt. 9 months, was brought to the Hospital for Sick Children on *Dec.* 14*th*, 1857. Suffering from frequent and troublesome vomiting. There was heat of the head, retraction of the neck. She had suffered from convulsions several times.

21*st*. Child had cough and signs of bronchitis. Sickness of an obstinate character continued.

Jan. 4*th*. Had a convulsive attack which lasted 36 hours. From this time the child grew weaker and lost flesh, convulsions returning at intervals ; the thumbs and the great toes were constantly rigidly flexed.

18*th*. Her mother wrote :—" From 8 *a.m.* to 4 *p.m.* my child was in strong convulsions. She rolls her head and her eyes also ; they are very dull and deathlike ; her breath is very short and quick. I have given the medicine when I could get the mouth open. Nothing has passed her bowels for several days." She died a few days later.

On post-mortem examination the upper surface of the hemispheres were not much congested ; the membranes over the sulci were opaque and studded with yellowish white spots about the size of pins' heads. Both lateral ventricles were distended with clear fluid. The brain substance was not much softened except at the posterior part of left hemisphere. The fissures of Sylvius were obliterated by firm adhesions between the cerebral lobes ; the membranes were tough and opaque. Round the optic tracts and the roots of the nerves generally, the membranes were tough and studded with opaque spots. There was not much extra-vascularity at any part.

Lungs quite free from tubercle. The middle and lower lobes of the right lung were solidified by pneumonia. The bronchial and mesenteric glands were free from tubercle.

It has been usual, since the days of Whytt, to describe the course of tubercular meningitis as running through three stages. The first is one of febrile excitement, with quickened pulse, headache, vomiting, and constipation, pain in the head and intolerance of light. The second stage is characterised by diminished frequency with irregularity of the pulse, slow

M

respiration, a tendency to stupor, dilatation of pupils, strabismus, and impaired vision. During this stage remissions sometimes occur in the symptoms, which encourage delusive hopes of recovery. The third stage is characterised by a very rapid pulse, tonic spasms of the limbs, subsultus, complete loss of sight, coma, convulsions; the face is often of a dull violet tint, and the eyes become suffused and covered with a film of tenacious mucus from the complete loss of sensibility and reflex irritability.

The course of the disease is very varied in different cases, and does not by any means constantly observe the stages above described.

The disease may make its appearance under three distinct classes of circumstances. 1. In the course of chronic phthisis. In this case the symptoms are more latent. 2. It may break out suddenly without any premonitory symptoms. This is rare (see case quoted page 158). 3. The most frequent case is that the alarming symptoms are preceded by other premonitory signs, which indicate a failure of health. For a period varying from a fortnight to three months there is a loss of flesh and of colour, the muscles become flabby, the features may be drawn, there is a dark spot under the eyes and a loss of lustre in their expression. There is often at the same time a loss of spirits, a dislike to company and to play, and sometimes a disposition for sleep during the day. There may be some other change in the natural disposition. Mental occupation fatigues more than usual. The patient may have quiet nights, or he may be restless, and there may be grinding of the teeth and deep sighing. Some children become excessively passionate. Headache is not frequently noted. The appetite is usually capricious. Sometimes there are pains in the stomach; there may be alternations of diarrhœa and constipation. These symptoms are naturally referred, by parents and doctors also, to a variety of causes, such as teething, rapid growth, errors in diet, or want of fresh air. There is but little fever, and this only at night, when the child is not observed. One of the most constant premonitory symptoms is loss of flesh; next in frequency is an irregularity of appetite, or loss of spirits and a change in the natural temper and disposition of the child.

These premonitory signs are dependent on incipient tubercularisation. When tubercular meningitis is preceded by premonitory symptoms it seldom lasts less than ten days or more than three weeks. When occurring in the course of advanced tubercular disease the cerebral symptoms do not usually precede death by more than from 8 to 12 days, sometimes by not more than 3 days.

The symptoms of tubercular meningitis are very varied, and may be conveniently described in order under the heads of—
I. Pyrexia and organs of circulation. II. Organs of respiration. III. Digestive organs. IV. Organs of nervous system.

I. At the commencement there is often but little fever, but there is usually a certain amount of nocturnal pyrexia. The pulse is either quickened, or unaffected, or from the first it is retarded. When quickened at the onset it usually becomes slower and often irregular after 4 or 5 days, continuing so for several days. During the course of the same day there are often great changes in the frequency of the pulse. After a while the pulse is permanently accelerated. A day or two before death it is usually above 140 in the minute. The *irregularity* of pulse is in *force* or in *rhythm* only, or in both. Sometimes there is a peculiar vibration in the pulse, and the size of it is out of proportion to the age of the child, from complete loss of arterial tension before the heart's force has failed. These characters are observed during the periods of slower pulsation. The temperature usually rises with the pulse; but there is no constant ratio between the two. The elevation of temperature is not so marked as in many other forms of tubercular disease, and at advanced periods is quite absent.

The face is alternately flushed and pale; in the earlier stages the face often becomes flushed when the patient is spoken to or any attempt is made to fix his attention. Towards death there is often a dusky violet tint seen on the cheeks.

There is often seen what is called by Trousseau the "tâche cérébrale," which he thus defines:—" In making gentle friction on the skin with a hard substance, such as a pencil or the finger nail, there is rapidly developed on the points touched vivid redness, which lasts a longer or shorter time, from 8 to 15 minutes." This condition is best observed on the front of the thighs and the abdomen. Trousseau regards it as a diagnostic sign of great value, as indicating cerebral or meningeal disease, in distinction from pneumonia, typhoid fever, or other febrile diseases with cerebral symptoms. It is a sign frequently met with in tubercular meningitis, but it is also often absent; and is certainly found in other diseases of a febrile character.

Profuse sweating often occurs during the last day or two of life.

II. *Respiratory functions.*—At first there is nothing abnormal, but as the pulse becomes slower the respiratory movements also become irregular. Deep sighs of a peculiarly sad kind are uttered. At an early period of the disease, slow gaping often repeated is not an unusual symptom.

At a later period, a suspension of respiration for a few seconds is not uncommon.

III. *Digestive functions.*—The appetite is usually diminished at the outset, and completely absent at the end. Thirst is commonly absent. The teeth are generally moist, till towards the end. The gums are generally moist, often a white fringe appears at their free margins. The tongue is usually moist; it is rarely brown and rough in the centre, on the other hand it is rarely clean throughout. Most frequently it is whitish or yellowish in the centre, red at the point and the circumference. At other times it is coated with a thick lemon-coloured coating, as in gastric disorder.

Vomiting is very seldom absent; it appears at the outset and lasts 2 or 3 days. In some cases it has lasted 9 or even 25 days or 2 months; when once it has ceased, it does not often return. It is very readily produced by making the patient sit up. Once in 15 times, Barthez and Rilliet have seen this symptom wanting. Vomiting, combined with constipation, is a most important sign of tubercular meningitis. Constipation is present at the outset of three-fourths of the cases. There is often diarrhœa from 7 to 12 days after the commencement. Diarrhœa of some standing, dependent on tubercular ulceration, is sometimes checked for a few days at the outset of meningitis.

Retraction of the abdominal walls usually comes on 5 or 6 days before death. Abdominal pains may be observed at different periods; they are most frequent at the outset; they are often increased by pressure.

IV. *Cephalalgia* is nearly a constant symptom. When the child is very young he often indicates it by taking his head between his hands and pressing it firmly. It is a symptom usually of the 1st or 2nd day; it often lasts until the appearance of coma.

The intellectual functions are perfect at first; or if they are affected there is only a little slowness of answers. At the same time there is often somnolence and irascibility. When meningitis is secondary to confirmed phthisis grave disturbance of intellect may occur at the outset. Delirium may set in from the 10th to the 15th day; it varies much in intensity. In about one-third of the cases it is intense, and accompanied with cries and attempts to get out of bed, and general restlessness; this kind of delirium seldom lasts more than a day or two. When the delirium is tranquil, the patient mutters some inarticulate sounds. The delirium does not seem to have any constant relation to the cerebral lesions. The peculiar hydrocephalic cry is not very common. Somnolence and coma are seldom absent. There may be a little drowsiness at the outset, which passes off, to return at the end of a few days, and to be succeeded by coma. Coma may occur without any antecedent drowsiness. Profound coma is not necessarily dependent on abundant effusion into the ventricles.

Convulsions.—Tubercular meningitis, without tubercle in the brain, seldom or never *begins* with convulsions. When convulsions appear at the outset, or by their frequency and intensity constitute an important symptom, it is said by Barthez and Rilliet that there is commonly tubercle of the brain.* Convulsions may be very partial, or quite general;

* My observations do not confirm this.

when general they have a variable duration. They commonly occur towards the close of the disease. They are more frequent the younger the child. Rigidity and contraction of muscles are symptoms which exist at a somewhat advanced period of the disease ; they are usually partial. They last a day or two and disappear, again to return. Subsultus tendinum, picking of bed-clothes, catalepsy of limbs, grinding of teeth, twinkling lids, fixed gaze, are all common symptoms.

Paralysis, partial and transitory, is not rare, but scarcely ever general and permanent. Hemiplegia sometimes occurs ; the tongue may be drawn aside, or there may be ptosis of one lid. Retention of urine is rare. In young children hyperæsthesia at an early stage is common. More commonly there is a blunting of sensation, varying from a slight diminution to entire absence. Frequently a great intolerance of being disturbed.

Diplopia is an occasional symptom between the 1st and 2nd periods. Oscillation of pupils on exposure to light is an important symptom. At the outset there is an intolerance of light and sound. Strabismus usually exists after about the 8th day. Dilatation of pupils exists usually before death, it may be unequally marked in the two eyes. Sight usually fails some days before death, smell and hearing at a later period.

Prognosis is always unfavourable. When the disease is fully developed, and has advanced beyond the middle of the 2nd stage, the hope of recovery is lost. Cases have been met with in which the diagnosis of tubercular meningitis seemed undoubted, from which recovery has taken place. In these cases the symptoms had not gone beyond the earlier stages, such as vomiting, constipation, headache, with some slowness of the pulse and tendency to drowsiness.

Recovery may be complete or only partial, leaving some permanent symptom (such as amaurosis, local paralysis, or impaired intellectual powers) which is indicative of cerebral tubercle or effusion into the ventricles. The recovery is often only temporary ; the tendency remains, and a relapse occurs.

Causes.—Age. It is rare under 12 months, less rare between

1 and 2 years, then more frequent between 2 and 7 years; from 8 to 10 it is rather less common, and rarer still between 10 and 15 years of age.

Dr. West says, of 61 cases, five were under 12 months, fifteen were between 1 and 3 years, thirty between 3 and 6, five between 6 and 9, two between 9 and 10, one between 10 and 11, and one between 12 and 13 years of age.

Of 23 cases under my own care, six were between 2 and 3 years, one was under 12 months, three between 1˙ and 2 years, three were 3 years old, four were 4 years old, one 6 years, four 7 years, and one 9 years.

It must be borne in mind that children are not often taken into this Hospital under 2 years of age, and as my cases have been chiefly amongst the in-patients at the hospital, the number of cases that occur under that age is not fairly represented.

The tuberculous diathesis is the great predisposing cause. Hereditary predisposition is of frequent occurrence; several members of the same family in succession may be cut off by this disease. The marks of the tuberculous diathesis are activity of mind and body, the nervous system being highly developed; skin is thin, veins distinct, complexion clear, eyes bright, pupils large, eyelashes long; bones small, muscular, and fatty; development scanty; face oval, and hair silken. Subjects of this diathesis often cut their teeth, run, and talk early.

Undue intellectual exertion, exposure to the sun, and blows on the head, are possible exciting causes in predisposed subjects.

Diagnosis is often very easy, but in some cases very difficult. It is important not to be too hasty in coming to a decision. The patient may require to be seen several times in the day. Particular inquiries should be made as to the patient's health for several weeks previously. In hospital practice the information thus obtained is often of little value, because many of the parents of poor children do not observe very closely symptoms of slightly impaired health.

Loss of flesh, with want of appetite, are symptoms to be asked for. Headache at irregular intervals, pains in the limbs, giddiness, a change from the natural disposition, a loss of spirits, and unusual irritability of temper, are symptoms which alone or in combination are of some significance. Great constipation, restlessness at night, starting in sleep, and grinding of the teeth are often premonitory symptoms. If after such a class of symptoms there occurs a sudden accession of vomiting, with headache and slight drowsiness, a whitish tongue and pulse not quickened, there is great probability of the case being one of tuberculous meningitis. Gastro-hepatic disorder may cause similar symptoms; in such case the headache is probably not so severe, vomiting is after food, and not so often on an empty stomach; the tongue often has a yellowish tinge, and constipation yields easily to aperients. The patient has not an astonished, fixed, or uncertain look. Sometimes vomiting is not so severe, and occurs only after food; even in the cerebral disease constipation may not exist, but there are headache and some drowsiness, with possibly some peculiarity of aspect. These symptoms should rouse your suspicions. At the next visit you will probably find some other symptoms, such as an abnormally slow pulse (a very important sign, but still more so if also irregular). There may be some other signs, such as a vacant stare, or a dazzled expression, sighing, alternate flushing and pallor, great slowness in answering questions, or frequent application of the hand to the head.

In some cases the symptoms may resemble those which are occasionally seen in *typhoid fever*. Drowsiness, sighing, headache, and vomiting, may be common to both diseases; slowness of the pulse and constipation have also been seen in typhoid fever. The abdomen in typhoid is usually convex and not retracted, but even this symptom is not invariable; the "tâche cérébrale" has also been seen in typhoid, as Trousseau himself admits. In such cases as these (fortunately very rare) the diagnosis would be very difficult. The temperature will probably be higher in typhoid than in hydrocephalus, the vomiting is not likely to be troublesome, and the pulse is very rarely irregular or persistently slow throughout the day.

Trousseau mentions a case of typhoid fever in which there

was retraction of the abdominal walls, a slow pulse (56 in the minute), and constipation. There had been vomiting and at an early period diarrhœa; the pupils were dilated one more than the other, and there was a "tâche cérébrale." Such a case would very readily be mistaken for tubercular meningitis.

The diagnosis of this disease is most difficult in children during the period of dentition, when slight cerebral symptoms are frequent; such as vomiting without obvious cause, drowsiness, flushing, excessive irritability, starting in sleep. The impossibility of properly feeling the pulse at this age from the child's irritability will often increase the difficulty of diagnosis. In such cases the gums should be examined, and if necessary lanced, a mild aperient given, and the child must be seen again soon. If constipation exist, it will probably yield readily, and in a few hours there may be an absolute remission of symptoms.

To distinguish *idiopathic* from the tubercular meningitis, it must be remembered that the former is much rarer, that it attacks patients without tubercular tendencies. It is generally said that convulsions occur at the outset, and that fever is more intense, that there is more severe vomiting and more active delirium, that headache is more intense, convulsions more constant, and the course of the disease altogether more rapid. Such cases as these are extremely rare, and have never come under my observation. Cases are recorded by Gölis* under the name of water stroke, and by Abercrombie in his work on "Diseases of the Brain."

Cases of cerebro-spinal meningitis, quite independent of tubercle, running a course of two or three months, have come under my notice.

The prominent symptoms have been more or less rigidity of limbs with retraction of the head, tetanoid spasms and great hyperæsthesia; vomiting has been an early symptom, and coma has preceded death. Purulent fluid has been found in the.

---

* "Praktische Abhandlungen über die Krankheiten des Kindlichen Alters." Vol. ii.

lateral ventricles and in the arachnoid at the base of the brain, also in the spinal cord, but not a trace of tubercle in the body.

Death has in other cases occurred within a fortnight of the first symptoms.

A case in which recovery occurred was under the care of my colleague, Dr. Gee. The diagnosis made was, cerebrospinal meningitis.

A boy, aged 4 years, in good health, after a short walk, complained of languor, headache, and sickness. He became cold, and soon afterwards feverish. (On the previous day he had been out a good many hours in very hot sunshine.) The next day was delirious and noisy, with occasional vomiting. The two following days he was very low.

On the 5th day was very restless, crying out unintelligibly. Pulse weak, very irregular. Cheeks mottled, with scarlet flushes. Pupils small and unequal, slight squint. Great heat of head. Ice applied, and grey powder given.

On 6th day was better; mouth was sore, squint continued. To take iodide of potassium.

On 8th day, pulse 100-120. Sleeps more naturally. Head was less retracted.

On 9th day squint persists, was more irritable; feet, which had been cold, became warmer. Does not yet know one person from another.

On 11th day began to talk a little rationally. Cheeks still flushed. Pupils large, equal. Head not retracted. From this time he slowly recovered. His intellectual power returned in a few weeks. During convalescence it was found out that he was deaf, a symptom which remained permanently.

The diagnosis made in this case was probably correct. The occurrence of an acute febrile attack, with delirium, vomiting, great prostration, strabismus, inequality of the pupils, flushing, heat of head, coldness of extremities, and retraction of head, leading to permanent deafness without discharge from the ear, renders this diagnosis in the highest degree likely. It would seem as if the mercurial course was of service.

These cases are probably sporadic cases of the disease which occurred on the shores of the Baltic and elsewhere as an epidemic three years ago.

Hydrocephaloid disease, caused by exhaustion from diarrhœa or some other debilitating disease or lowering treatment, should always be kept in mind whilst making a diagnosis. The child is irritable and restless, the skin may be warm or cold, and pulse frequent; the patient starts at the slightest touch or at a sudden noise, sighing and moaning in sleep, often screaming. The bowels usually have been loose, and there is frequent vomiting. The arms are thrown about, or used to rub the head. If not properly treated the child becomes pale, the pulse almost imperceptible, the eyelids half closed, the pupils sluggish, the eyes not noticing anything; breathing becomes irregular and sighing, the voice husky or almost suppressed. Complete collapse follows, and death. I have seen this condition following typhoid fever mistaken for tuberculous meningitis. This class of cases may generally be distinguished by the history of the attack, the previous state of the bowels, and by the pulse, which is not slow in the earlier period. (See the diagnosis of Typhoid Fever.)

Vomiting, if present, is usually only after taking food or drink. The fontanelle if not closed is depressed. In such cases a warm bath, light nourishment, with stimulants and opiates, will usually bring about an improvement.

Pneumonia often sets in with vomiting and convulsions. Dyspnœa, pain in the chest, and cough, may not be marked symptoms. There may be headache, and delirium, with constipation.

In all cases of dubious character examine the chest carefully. The heat of the skin, with some movement of nares in respiration, will often rouse suspicion.

Treatment.—This must be chiefly prophylactic. When the disease has once fairly set in the chance of doing good is small. There is, however, no doubt in my mind that tubercular deposition may have commenced in a child's organs, even in the meninges, and be arrested.

In children with hereditary predisposition to tubercle, or of the aspect which is distinctive of tuberculosis, the rules of hygiene should be strictly enforced. Such children should breathe pure air, have plenty of light, with plenty of nutritious, easily-digestible food, especially milk and animal food. They should use saline or other tonic baths, and avoid great excitement and fatigue. They should have plenty of sleep in well ventilated rooms. Their physical development should be studied much more than their mental, until they are 8 or 9 years of age, and even then their hours of study should not be long.

If the child is thin, cod-liver oil should be given occasionally for a month or two, especially in cold weather, as an article of diet. If the digestion is impaired the child's appetite must not be indulged by unsuitable and tempting articles of food; but by the use of calumba and soda, or citrate of potash, with tincture of orange, the stomach may be strengthened, and the food must be of the most digestible and nutritious character.

When the disease is declared, ice or cold lotions to the head, mercurial aperients, followed by iodide of potassium, with absolute rest in a quiet darkened room, are the measures to be recommended. The treatment is adopted rather with the possibility of the case not being tubercular but idiopathic, than with the expectation of curing the disease if it is really what it is believed to be.

I have not seen any benefit from blisters or tartar emetic ointment to the scalp or nape; and as it adds to the discomfort of the patients I have ceased to recommend counter-irritation.

Short Notes of Cases of Tubercular Meningitis.

CASE I.

Bridget B., æt. 2 years. Doubtful hereditary predisposition to tubercle. Has a tuberculous aspect. One of 9 children, 5 living. Was bitten 4 months ago by a dog; from that date her health has failed. Abscesses and pustular eruption on skin. Her nights have been very restless, and she has been more nervous.

Sept. 27. Vomiting after food; next day drowsiness; delirium, biting at her clothes. Sickness lasted for a week.

Oct. 6th. Convulsions frequently repeated until death; biting at everything which came in her way.

9th. Death.

Autopsy.—Granular thickening at base of brain; 1½ oz. clear fluid in lateral ventricles. Tubercles in Sylvian fissures. Central portion of brain a little softened. No tubercle in any other organ.

Remarks.—The connection of the bite of a dog with the

child's illness (whether it can be regarded in any way as an exciting cause, acting on a subject predisposed to the disease), is a point on which there is room for difference of opinion. The case by itself will prove nothing, it may be borne in mind and compared with others.

CASE II.

Mary Wood, æt. 7. No history of tubercle in the family. Health began to fail 4 months ago. Went to the country for 2 months and came home much improved. A week after her return attacks of vomiting after food set in. Since this time has often vomited on an empty stomach and lost flesh. For the past fortnight sickness worse; great constipation, restless nights with moaning. Admitted November 8th. Has an expression as if exposed to a strong light; eyes nearly closed; corners of mouth drawn up. Pupils act well. Pulse 72 regular. Saline mixture and purgatives.

Nov. 19*th.* Bowels confined. Sick once in 24 hours. Has the same dazzled expression.

20*th.* Delirium in the night. Pulse 96 regular. Candle light hurts her eyes.

21*st.* Delirium at night. Flushing, gaping, sudden paroxysm of headache. Pulse 120; skin hot. Hair cut, and ice to head.

24*th.* Is rational but drowsy. Likes the ice. Frequently gaping and sighing. Pulse 116 regular. Abdomen full, resonant, and rather tender. Tongue clean at tip and edges, coated white on dorsum. Complains at times of headache.

26*th.* Increasing stupor. Tenderness of limbs and abdomen. Pulse 108 regular. Two leeches to each temple. Hydr. chlor. gr. j every 4 hours.

27*th.* Wandered during the night; occasional twitching of left arm. Early this morning, was very pale, her heart beating tumultuously. Pupils large, contract sluggishly on near approach of candle-light, relapsing into state of dilatation even with candle close at hand.

28*th.* Becoming more drowsy. Pupils act less. Pulse 132.

29*th.* Seemed to know her mother better than for several days. Right pupil larger than left. Pulse 140.

30*th.* Free perspiration. Slight convulsive movements of limbs. Face flushed, rather dusky.

Dec. 1*st.* Last night strongly convulsed at intervals for 10 hours. Since then in a state of coma, with eye-balls in constant motion. Towards noon she had strong convulsions and died.

Autopsy.—Lateral ventricles contained 2 oz. of clear fluid. Fornix softened. Lining membrane of ventricles unusually firm and finely studded with sandy granules. Membranes at base thickened, opaque, and beneath them yellowish gelatinous fluid. Small granulations in Sylvian fissures. No other part examined.

CASE III.

W. J. C., æt. 1 year 9 months. Father not strong. Child always delicate ; eczematous.

June 1st. Nights restless, bowels confined, skin perspiring ; occasional vomiting. After this could not walk for a week. He remained more or less poorly for a whole month. His temper was irritable ; he regained flesh towards the end of the month. In July his health failed again ; vomiting every day for a week.

July 24th. Much constipation. Change in temper and habits. Appetite bad. Speaks in an abrupt manner, and with peculiar intonation.

26th. Pallor and flushing alternately. Sickness of undigested gooseberry skins.

July 30th. Intolerance of light. Very irritable when disturbed. Pulse 65, very slow.

Aug. 3rd. Violent screaming at night. The sight of one eye is lost.

5th. Had a severe attack of convulsions in the night. Pulse very small, rapid. Respiration short, sighing irregular. Pupils dilated, insensible to light. Muscles of neck rigidly contracted.

6th. In a state of coma.

8th. Death. Opacity of membranes at base of brain and fluid in ventricles. Tubercles in lungs, liver, spleen, bronchial and mesenteric glands.

CASE IV.

Eliza Chapman, æt. 4. Had scarlatina 4 months ago, and measles with pneumonia 7 weeks ago. Had been observed to grind her teeth for some months and start in her sleep. Fourteen days before had headache, pains in legs and arms ; sweating at night. Bowels regular ; appetite bad. Drowsy in the day time. For a week had been delirious at intervals. Admitted Oct. 6th. She was then languid and fretful, Pulse 108 weak.

8th. Restless and feverish.

9th. Delirious and drowsy. Answers questions very slowly. Cries out at times with pain in head.

13th. Irregular contractions of right foot. Does not answer when spoken to.

15th. Bowels confined. Forearms contracted, especially right. Sight gone. Pulse 160. Slight ptosis ; dilatation of right pupil.

17th. Lies unconscious, but without natural sleep.

18th. Occasional spasmodic irregular action of diaphragm. Pulse 190. Pupils both insensible to light. Lay in this state until the night of 20th, when she died.

Autopsy. — Tubercles in lungs, bronchial glands, liver, spleen and kidneys. Membranes opaque near longitudinal fissure. Ventricles contained 10 oz. of fluid; centre of brain softened. Surface of corpora striata granular. Membranes at base tough, granular, yellowish, matting the roots of nerves together.

CASE V.

Thomas Graham, æt. 22 months. Has been teething rapidly; losing strength. For several days has been sick, grinding his teeth, and moaning.

Nov. 3rd. Admitted. Looks dull and sleepy. Eyes half shut; right more than left. Right pupil larger than left. Rigidity of one arm and leg. Pulse 144 irregular. Had a fit lasting 5 minutes.

4th. Several convulsive attacks. Takes no notice of people around him.

5th. Pulse 160. Head hot.

6th. Teeth clenched. Both pupils insensible to light. Pulse 200 irregular. He died the next day.

Autopsy.—Tubercles in lungs, liver, spleen, kidneys, and glands.

Head.—Typical characters of tubercular meningitis.

CASE VI.

H. Plumer, æt. 6. Twelve months ago was feverish, had headache, cough, and dyspnœa. Was ill 3 months and lost flesh. She got better till 2 months ago, then again began to fail. Took cod-liver oil, and her appetite improved.

Nov. 27th. Signs of consolidation of apex of one lung. Last night was sick on an empty stomach. Bowels not much confined.

Dec. 1st. Sickness persists. Pain in head; dislikes light and noise. Pulse 68 irregular. Sighing and flushing. Has a vacant look. Two leeches to temples. One gr. of calomel 3 times a day.

3rd. Convulsions. Pulse 80. Apathetic. Ice to head.

4th. Was much more conscious after ice. Arms were stiff and legs cold for 5 hours.

5th. Constant movement of hands, picking bed-clothes and face. At times quite unconscious. Strabismus; cannot see. Pulse 114. In the nights of 5th and 6th, convulsions, succeeded by coma, from which she did not rally.

8th. Death. No autopsy made.

CASE VII.—*Tubercular Meningitis, accompanied with some Tubercle in Brain and Capillary Hæmorrhage. General Tuberculisation.*

C. C., æt. 3¾ years. A rickety boy. He had measles 2 months ago, which left him weak. The first symptoms were on 1st February, when he ate very little and vomited. After this did not seem well; on the 19th February his bowels were much confined, his appetite was bad, and he was sick. His face frequently became flushed. The next day (20th) he became unconscious and remained so.

He was admitted to the hospital on 22nd in a state of stupor; head hot; pupils medium-sized, inactive. Pulse 180 irregular. Right forearm rigidly bent; abdomen retracted. He lived 3 days, never recovering consciousness; convulsive twitching of left side of face, left eye, left arm and leg. Moaning in the night. No general convulsions.

Autopsy.—Arachnoid moist; clear fluid in subarachnoid space. About an ounce of fluid in lateral ventricles; fornix softened. In the grey matter of left posterior cerebral lobe is a *nodule of tubercle* ¼ of an inch in diameter. On the anterior lobes between the olfactory nerves backwards to the optic commissure the surface of the brain is injected with capillary and punctiform injection. There is also minute capillary hæmorrhage which extends for a short depth into the grey cerebral substance. The Sylvian fissures contain small tubercles, with yellowish lymph and injected pia mater. Lungs, bronchial glands, spleen, liver, kidneys, small intestines and mesenteric glands also contain tubercle.

CASE VIII.

John Cornish, æt. 4 years. One of a family of 9 children, 7 living; 2 are said to have died of "water on the brain." He "cut all his teeth with fits;" has had no convulsive attacks since. Two months ago he had a mild attack of scarlatina; but has not been so well since. Soon after recovering from that illness he became frightened when lifted up, clenched his hands, and said he should fall; he was also afraid of falling when carried downstairs. Two or three weeks ago he became sick, especially after his meals.

Sept. 14th. He was violently sick without anything in his diet to account for it. The next day, however, he was sent to school, but after being there 2 hours was extremely sick and was sent home. He now for the first time complained of his head aching; he had previously complained of pain in his stomach on several occasions, even when not sick. No new symptom was observed until the 19th, when he became a little drowsy.

22nd. The sickness continues and constipation; his tongue is furred and lips are dry. Still has headache; at times he wakes up screaming. Pupils rather small, not quite equal in size. In the afternoon of this day he became a little delirious.

24th. Is very drowsy; does not speak unless spoken to; when asked says that his head aches. Bowels acted 8 or 10 times last night after some calomel; the motions nearly black. Still vomits every few hours; the retching is rather violent. Violent convulsions set in about 7 *p.m.*, after which he went off into a comatose state. He lived until 29th September, having been frequently convulsed during the last four days of his life, and almost constantly comatose.

Autopsy.—The usual appearances in the membranes of the brain at the base, with a moderate quantity of fluid in lateral ventricles, and softening of the fornix. Tubercles in the lungs.

Remarks.—In this case the sickness was a very prominent symptom for several days without headache; and it usually followed eating. Previously to this, however, he had suffered from a kind of vertigo which is not a usual premonitory symptom in this disease. The exciting cause appeared to have been scarlatina in this as in several other of my cases.

CASE IX.

H. Millard, æt. 9. Ailing 6 weeks. First symptoms loss of flesh, and feeling cold when washed. Ten or twelve days before death, headache, which was increased by noise or light. Unconscious about 36 hours.

Autopsy.—The usual appearances at base of brain. Tubercles in lungs and other viscera.

CASE X.

Elizabeth Bickley, æt. 4. Scarlatina 3 months before death, followed by hooping-cough. The first symptom, violent vomiting with constipation 16 days before death. Nine days later drowsiness and delirium; grinding teeth, knocking her head with hands. Coma and convulsions during the last 36 hours.

Autopsy.—The same as preceding.

CASE XI.

William Taylor, æt. 2. Family tuberculous. When 12 months old had attacks of stupor, preceded by pain in head. For 10 days had pain in head, aggravated by noise. For 4 nights he *lay awake* quite quietly; he was *not sick*, and his *bowels were relaxed*. Five days before death strong attack of convulsions lasting 5 hours. After this he lay in a semi-comatose condition, very irritable when disturbed; his face alternately flushed and pallid. Head not at all hot. Noises cause flushing of the face. The last 2 days in a state of stupor, with arms, legs, and neck rigid; pupils dilated. He was only sick once, and that 3 days before death.

No autopsy.

Remarks.—In this case the symptoms ran a different course from what is usual. The absence of vomiting and constipation at the outset, the early occurrence of convulsions, the sleeplessness without any screaming, and the great susceptibility to be disturbed by noises, are exceptional.

CASE XII.

M. A. C. Stockley, æt. 7. Tuberculous glands in neck. Violent sickness, paroxysmal headache, loss of appetite, constipation 16 days. Drowsiness, soon followed by unconsciousness 4 days. Rigidity of limbs, dilated pupils 2 days. Convulsions during the last 24 hours.

No autopsy.

178   ACUTE HYDROCEPHALUS.

CASE XIII.

Emma Watson, æt. 1 year 9 months. Father consumptive. Not a strong child. Sickness, thirst, and heat of skin 9 days before death. The 2 following days bowels loose, slept badly, screaming. Four days before death restless, very passionate, biting hands, squinting occasionally. From this period at times restless, at others drowsy, sight failing, and sickness frequently returning. Convulsions 8 hours before death.

Autopsy.—General tuberculisation. Fluid in lateral ventricles; central parts of brain softened. Membrane in sylvian fissures opaque, studded with white opaque granules of the size of small pins'-heads. A little opacity in velum interpositum. Lungs contained crude tubercle which was softening. The liver and spleen contained grey tubercle. The mesenteric glands were free.

Remarks.—In this case the symptoms were those of general tubercular disease, with an accession of cerebral symptoms only four days before death. There was looseness of bowels, instead of constipation, and this without ulceration of the intestines.

CHRONIC HYDROCEPHALUS, TUBERCLE OF THE BRAIN, AND OTHER CEREBRAL AFFECTIONS.

TUBERCLE in the brain is rare in the adult, but is sufficiently common in the child. Of 117 adults who died of tubercular disease, examined by Louis, only one had tubercle in the brain; of 312 tubercular children examined by Barthez and Rilliet, 37 had tubercle in the brain. In our hospital, of 160 children with tubercle, 40 had tubercle in the brain.

Tubercle attacks many organs in children which it usually spares in adults—the liver, kidneys, spleen, and peritoneum, for instance.

It is rarely that tubercle occurs in the brain without tubercle in some other organs. Barthez and Rilliet met with this condition twice only. Tubercle in the brain, when in large masses, is often found without any tubercle of the meninges. In small patches it is often seen as an accompaniment of meningeal tubercle by extension of the deposit from the membrane into the superficial layers of brain.

The deposit is usually circumscribed with a roundish outline, in size varying from a millet seed to a pigeon's egg, or larger. There is sometimes but one mass, more frequently there are several small masses, as well as a large one, or a number of medium-sized masses. All parts of the encephalon are liable to be involved. The deposit most frequently begins near the pia mater, from which it extends deeper into the brain substance. The mass can often be turned out, leaving the cerebral substance but little roughened and torn. The crude yellow form of tubercle is the variety met with. It is sometimes softened.

The brain around is usually quite normal in appearance; sometimes it is of a rose colour and a little softened when the tubercle has begun to be softened. Cretaceous change has been met with in cerebral tubercle. Barthez and Rilliet met with it once in 37 cases, Dr. West once in 19 cases.

Instead of being round in outline, tubercle is sometimes seen in the form of a patch not more than two lines deep under the pia mater, and half an inch long or more, by two or three lines broad, the brain around being softened. In other cases tubercle is infiltrated into the brain, which is of a rose colour and softened.

Tubercle in the brain is more common under than above the age of 5 years. Of 67 cases, 37 were under 5, 22 between 5 and 10, and 8 between 10 and 15 years of age.

The *symptoms* which are caused by tubercle in the brain vary according to the part in which the morbid product exists, and the rapidity of its growth.

Paralysis may be due to pressure on the origin or course of separate nerves, or to pressure on the central ganglia or the crura cerebri. Rigidity or convulsions depend on the peripheral portions of the brain being involved.

Owing to the slow growth of the tubercular product in many cases, and the absence of congestion, the disease may exist for some time before its presence is suspected.

A considerable mass of tubercle is sometimes present in the brain without any distinct cerebral symptoms, when something occurs which suddenly induces an attack of convulsions; or the progress of the tumour induces local paralysis. There may be ptosis of one eye from paralysis of the third cranial nerve, or internal strabismus with double vision from paralysis of the sixth nerve, or one arm may be weak. There is generally headache, subject to paroxysmal aggravation. It is usually frontal unless the cerebellum is the seat of tubercle, then the pain is occipital. After a while, amaurosis with dilated pupils and more or less complete paralysis, with enfeebled intellectual power, often occur.

The duration of cerebral tubercle is said to vary from six

weeks to two years; but it may, when the deposition of tubercle is arrested, last for many years. We know that tubercle in the brain may undergo cretifaction as in other organs, and if this be so, this condition may last many years, till a new outbreak of tubercular disease, or some other distinct affection, carries off the patient.

There is one symptom often observed in cerebral tubercle, namely, slowness of the pulse, with or without irregularity. I do not attach much value to this symptom. It is often noticed in children during convalescence from slight illness without any cerebral symptoms.

The usual terminations of tubercle in the brain are somewhat varied; it may be severe convulsions ending in coma, or gradual emaciation and death from asthenia, or some internal disease, such as pneumonia or one of the acute specific diseases. It is very rarely that recovery ensues. The tubercular constitution of course remains, and very generally manifests itself ere long.

It is very important to be cautious in giving a diagnosis and prognosis. Keep the case under observation for some time and see it frequently at different hours of the day before giving a distinct opinion; otherwise you may be led into giving too serious a prognosis in a case of functional disorder, or too favourable a prognosis in a case of tubercular mischief. Whilst your prognosis should be guarded, it is better not to make up your mind that the case is one of tubercle and consequently almost hopeless, until the symptoms are very decided. In any case, fortunately, the treatment may be nearly the same. Quinine and cod-liver oil with counter-irritation and attention to the bowels are the chief points to be attended to. I would not recommend mercurials or depletion in any of the cases.

In the following case, a patient who had previously suffered from empyema, been tapped, and made a good recovery, was seized with cerebral symptoms, vomiting, pain in the head, and listlessness. She had flying pains about her body, slight pericarditis, strabismus and double vision, followed by amaurosis. Her previous illness is recorded on page 55.

Jessie T., a girl aged 7 years and 9 months, was admitted under my care on the 5th May, 1862. Her father and mother were living; she had several brothers and sisters living; one had died of diseased knee. She was of a somewhat tuberculous aspect. In the previous autumn she had been a patient for empyema on the right side, for which I had performed paracentesis. A fistulous opening was established, which had closed at the end of about three months. She remained in good health all through the winter.

April 24th. She lost her appetite for the first time, complained of pain in her head, and seemed listless. During the following week she suffered more pain in the head, took no food except a biscuit occasionally; she was very listless and apathetic, and was once sick. The bowels were confined.

30th. She took powders of rhubarb and soda every night, since which the bowels have been open each day; but pain in the head has continued, she has also complained of pain in the eyes and in the bones, and has been occasionally sick after taking tea or small quantities of food. At night she has screamed frequently after being asleep for about half an hour; cries out for water to be applied to her head, and talks in an excited way about dying.

May 3rd. Said she saw things double.

5th. The following notes were taken on her admission :—She is fairly nourished, with fine brown hair, long eyelashes. Hands kept over eyes and forehead; she complains of pain in these parts. Objects to having her eyes open. Head not hot. Says she sees double. Pupils equal, act thoroughly and equally. Face has rather the appearance of a person looking at a very strong light; at times, however, the child assumes her natural smile and an arch expression. Answers questions pettishly. Observes accurately and quickly. Tongue moist, clean at tip, coated on dorsum. Pulse varies from 64 to 72, feeble and regular.

Thorax.—The right side is very moderately contracted from the old empyema. The auscultatory signs are almost normal, except at the right infraclavicular region, where percussion is dull and the expiratory murmur prolonged and separated from the inspiratory. She cries out occasionally with flying pains about loins, hips, and elsewhere. Two leeches applied to each temple. Ice applied to the head. ℞. Calomel, gr. ij ; pulveris jalapæ, gr. viij. Statim sumend.

6th. Leeches bled well. Passed a very restless night. Bowels open twice since taking the powder. Tongue moist, cleaner. Pulse 64 to 68, not quite regular. Pupils as yesterday. Very slight strabismus. No diplopia. Expression at times pettish, with occasional look of distress. From time to time she cries out, complaining of pain in head, eyes, or some part of the body. ℞. Hydrargyri chloridi, gr. j. Ter die sumend.

7th. Slight pericardial friction sound heard over the heart, for which a leech was ordered. Pulse 72 regular. General condition as before. No abnormal heat of skin.

8*th.* Better in all respects. Bowels open once. No pain to-day. To omit calomel. To take a citrate of potash mixture with an excess of bicarbonate.

9*th.* Pain in head has returned. Pulse 84 not quite regular. Strabismus more marked. Slight friction sound still audible over 4th left cartilage.

10*th.* Complains much of pain in left hip and left orbit. Pupils equal, rather small. Pulse 98 intermits about every 10th beat.

12*th.* Much the same, but pain is less.

13*th.* Pain in left ear. Strabismus less marked. Pulse 120 more regular. Bowels regular. Sleeps well.

14*th.* Pain less. Pulse 80 intermits about every 10th beat.

17*th.* Seems more herself to-day, though at times complains of pain in both eyes.

19*th.* Seems better. Pulse 120 regular. ℞. Ol. morrhuæ, ʒj. Ter die. ℞. Syr. ferri iodid. ♏xv ; potassii iodidi, gr. j ; aquæ ad ʒj. Ter die.

22*nd.* Much the same. There is still slight internal strabismus of right eye, and now also partial ptosis of that lid. At times suffers from severe frontal headache. Appetite moderately good.

June 1*st.* Has been up and walking about the ward for 2 days. She walks with very uncertain step. Has occasional cramps in left leg. Is cheerful and quite intelligent. Strabismus as before.

4*th.* Slept very little last night, complained much of headache. There is more strabismus. Tongue, when protruded, points slightly to left side. The features are very slightly drawn to the right. Appetite failing. Bowels open. Pulse 96 regular. Child's manner is heavier than it was.

5*th.* Much the same. Pulse 96 not regular in force or rhythm.

7*th.* Still has frontal headache. Strabismus continues. Deviation of features also. Bowels confined. ℞. Pulv. cal. c. rhei. gr. viij. Statim sumend.

9*th.* No headache to-day. Deviation of features to right still observable. Sight is decidedly impaired. Is more lively and cheerful.

14*th.* Allowed to get up. Walks with feeble uncertain gait ; no dragging of either leg. Sight evidently much impaired. Intellect perfect. Right pupil a little more dilated than left ; it contracts readily on the application of a lighted candle, but soon dilates again whilst the light is still applied. The left pupil contracts, and remains contracted till the light is withdrawn. No pains.

19*th.* Had a very restless night. Complains of pain in head. She at times has a kind of shudder almost resembling a rigor.

20*th.* Does not seem worse this morning, except that the sight of right eye is less distinct than it was. She cannot distinguish a person's features at a few feet distance even with both eyes open. She is cheerful and amuses herself with toys. ℞. Hydr. bichlor. gr. $\frac{1}{24}$; tinct. cinchonæ, ♏x ; aquæ ad, ʒj. Ter die.

23*rd.* The child seems quite well, but absolutely blind. Signs of consolidation at upper part of chest on the right side, but no active disease. A little later than this her eyes were examined with the ophthalmoscope, and it was found that the optic disc was quite pale and atrophied. The strabismus and deviation of muscles of face quite disappeared. The child seemed in very tolerable health, and was not troubled with headaches.

Remarks.—This case I regard as one of tubercle in the brain; it began in many respects like a case of tubercular meningitis. The character of the pain in the head, the affection of sight (that is to say, double vision at the outset, due probably to paralysis of some of the muscles of the eyeball, and subsequent gradual loss of sight), the character of the pulse, the temporary paralysis of one side of the face, with partial ptosis on the other side, all these symptoms point to the growth of some substance exerting pressure within the cranium. From the appearance of the child, her previous history, and the condition of her right lung, there can be little doubt that she was of a tubercular diathesis, and that the growth within the cranium was of a tubercular character. I will not venture to speculate on its probable situation. Convulsions, a very common symptom of cerebral tubercle, but not a necessary one, were absent in this case.

Amaurosis with atrophy of the optic nerves is not an uncommon result of tumours in different parts of the brain and of the cerebellum, even when these tumours exert no pressure on the nervous structures concerned in vision. The most probable way of accounting for this condition is that suggested by Dr. Brown-Séquard, that by a reflex action the vessels nourishing the optic nerves are excited to undue contraction, the supply of blood is reduced, and atrophy ensues.

It is not a common occurrence for tubercle in the brain to undergo retrograde changes, and the disease to be arrested. That tuberculisation of the bronchial glands and lungs is *very often* arrested in children there can be no question. Nothing is more common than a deposition of tubercle in various organs as a sequel of measles or hooping cough, and if these cases be properly treated a very large proportion of them recovers.

CASE OF CHRONIC HYDROCEPHALUS. 185

That the same may occur in the brain there is both clinical and post-mortem evidence.

In regard to the treatment of this case, I do not ascribe the recovery to the use of calomel, but much rather to cod-liver oil and iodide of iron. If a similar case came under my treatment again I should give cod-liver oil at an early period, resorting to counter-irritation. I would give a nutritious diet, and act rather freely on the bowels by calomel and jalap.

The following case is probably another illustration of cerebral tubercle leading to chronic hydrocephalus, lasting for several years.

G. S., æt. 5 years, brought to the Hospital for Sick Children on 31st December, 1860. There was no history of hereditary predisposition to tubercle. He had, however, the aspect of a tubercular child. He had passed through scarlatina and measles in early childhood, without any obvious ill effects. He was believed to be in good health till three weeks before admission, when he was seized with morning sickness, which has continued daily till now. At the same time he complained of pain in the head, which was never absent, but paroxysmally worse. It was generally very severe after his fits of sickness. For a fortnight he has been unsteady in walking, like a drunken man; of late he has been so tottering as to fall down. One of his hands was unsteady, so that he could scarcely use it at all. The leg on the same side was also decidedly weaker than the other. No squinting. Sleeps well. His appetite not quite so good as it was. Bowels open daily. He is never sick after food, but only on an empty stomach in the morning.

Jan. 8th, 1861. Expression dull and heavy. When asked if he has any pain, points to the lower part of left side of chest. No frowning or knitting of brows. Left angle of mouth drawn up more than right, especially when laughing. Shuts eyes equally well. Tongue deviates a little to right when protruded. It is clean. He grasps more firmly with left than right hand; moves right with difficulty. Requires support to walk; moves both legs equally well; when left alone he totters, and separates his feet. Sleeps well at night. No vomiting since admission.

16th. Vomits occasionally.

18th. Last night vomited a dozen times, complained much of pain in his head. Nurse says he always does so when he vomits. Bowels open twice.

25th. Had a severe attack of vomiting yesterday, during which headache was very severe. Tongue moist and furred. Bowels open once. Pulse 120. No paralysis of face, eyes, or tongue. No twitching of muscles noticed.

Fontanelles and sutures closed. No convulsions since admission.

March 20th. There has been but little change since last note. Able to sit up in bed.

April 19th. Another attack of vomiting this morning at 7. Complained of headache all night. He now lies semi-unconscious, but when pinched groans. Lying on left side, legs drawn up, eyes closed, very pale. Some heat of head. No throbbing of carotids. Slight strabismus. Constant twitching of right eye inwards and left eye outwards. Tongue furred, does not deviate. Extending his legs hurts him; there is also some rigidity of arms. Arms and legs remain in the position in which they are placed by observer for a minute or more.

May 21st. Intellectually duller. Speaks less distinctly. Sight evidently much impaired; cannot distinguish colours. Strabismus well marked. Pupils large, equal. The right side of his face is partly paralysed. Tongue is clean, deviates to right. For three weeks has had no vomiting, but frequently complains of headache. Bowels regular. Passes everything in bed without giving any warning. Can move arms but not legs; cannot raise head from the bed; when head is raised complains of pain in neck. Legs flexed on body; slight but distinct rigidity of legs, more marked left than right; he cries out with pain when they are extended.

Head decidedly enlarged, especially posteriorly from side to side, and the forehead is pushed forward. Expression vacant, but answers questions at once, and to the point. Is but little emaciated. On irritating the inner parts of thighs, there is no retraction of testicles. Pulse 116 good. Not much emaciated.

June 17th. Yesterday, whilst laughing heartily, had some kind of fit. He turned blue, and had some difficulty in getting his breath. There was twitching of the mouth, and constant shutting and opening of his eyes. This lasted about five minutes, during which he was quite unconscious. He afterwards lay in a heavy sleep for half an hour, and has since complained much of headache. Answers questions slowly; articulation indistinct. Appears to be quite blind, but hearing is natural. Frequent slight twitching of both eyes to left. Strabismus increased. Both eyelids droop a little, and equally so. When he laughs the left side of his mouth is drawn up, whilst the right scarcely moves. Head still increasing in size. No heat of the head. Pulse 124 regular. Has power over arms, but they are weak. He keeps them for some time in any position in which they are placed by an observer. No rigidity in arms; well marked in legs, most so in right. Cannot stand or sit up. Appetite good. Bowels open twice yesterday. No note was taken for seven months, during which there was but little change in the symptoms.

Jan. 27th, 1862. Is very much in the same condition as at the time of last note. General nutrition moderately maintained, but has probably lost flesh within the last two months. He is usually kept in bed; when he is up is not so well. When asked how he is, says, "Very well, thank you." Has no pain. Articulation impaired. No marked drawing of the

DIAGNOSIS OF THE CASE.

features in smiling. He lies on his back with eyes open ; pupils equal, of full size, rather sluggish under light of candle. Tongue protruded without deviation. Right hand is weaker than left. There is still a tendency for limbs to remain in whatever position they are placed. Legs extended ; can draw either thigh up on abdomen, but has some difficulty in replacing it. Cries when legs are moved about, in fact cries whenever his position is much interfered with. There is no loss of sensation, rather hyperæsthesia than the contrary. Quite blind. Hearing good. Two or three months ago he used frequently to sing whole songs, he now seldom gets beyond the first line of a verse, appearing to have forgotten the rest. His spirits are not so good. No complaints of headache. Has occasional spontaneous pains in lower limbs ; the thighs are not unfrequently rigidly drawn up on abdomen. In this condition he was sent out of hospital as incurable. I heard of him nearly two years later (*Dec.* 1863) as very much in the same state. He died in 1866, but no autopsy was made.

Remarks.—This case may be safely diagnosed as one of tubercle in the brain. The symptoms are just such as frequently accompany this disease. The recurrence of vomiting without other signs of gastric disturbance, with paroxysms of severe headache, unsteadiness, and gradual loss of power in the limbs at an early period, the slow progress of the case, the maintenance of intellectual power for a considerable period, are symptoms indicating organic disease of the brain, which was exerting an increasing pressure on its structure. The enlargement of the head indicated effusion into the ventricles. A tumour in the brain, not tubercular, might have caused just such symptoms, but such tumours are so rare in children, that for diagnostic purposes they may be almost left out of consideration. Cancerous tumours, when they occur, run a much more rapid course than this. The very gradual accession was incompatible with softening or hæmorrhage, affections very rare in childhood. The absence of convulsions until other symptoms had existed six months is worthy of notice ; they often occur at an early period of cerebral tubercle. The treatment consisted chiefly in giving cod-liver oil and iodide of potassium.

I am sorry that no ophthalmoscopic examination was made. In all probability the same condition of the optic disc would have been observed as in the previous case. The great enlargement of the head after the closure of the sutures was a very striking symptom ; the tendency to hemiplegia, great dislike to being moved, with signs of hyperæsthesia, the semi-cataleptic condition of the limbs, the gradual and comparatively slight impairment of intellect, and the maintenance of nutrition, were also interesting. Several cases have come under my care with almost identical symptoms.

In the following case tubercle involved both lobes of the *cerebellum* to a very great extent ; the first noted symptom was a fit of convulsions, after which the patient lived about seven months ; he was never able to stand after the fit. No doubt tubercular disease existed for some time before the fit.

M. S., æt. 5 years, admitted to Hospital for Children the 22nd September. Always a delicate child. As an infant he was precocious, cut his teeth early, and walked at the age of 9 months. At the age of 2½ years had a very severe attack of hooping-cough, which lasted nine months. Five months before admission to hospital he suddenly fell off a chair, and struck the back of his head, he was insensible for an hour, occasionally screaming. The same evening he had another fit, in which his limbs became rigid ; this lasted half an hour. Has never been able to stand since, though until that day he had been running about as usual. Since then has had at intervals pain in the occipital region. No more convulsions. Has slept well ; taken food well, but has been steadily losing flesh. For the last three weeks there has been no pain in his head. He always had a large head, but his mother thinks it has become larger than it was. On admission, extremely emaciated. Lies very listlessly, and when spoken to answers in a very drawling and almost unintelligible voice. Head large, forehead high and narrow ; greatest circumference 20 inches. Pupils equal. No deviation of features. Tongue red, moist. Abdomen retracted. Signs of consolidation in lungs, especially at upper part. Pulse 126 weak and small. Is quite unable to move ; cannot lift arms or legs from the bed. Has no power to support his head.

Sept. 29th. Tendency to diarrhœa. Has frequently screamed at night, and on several occasions the nurse has found him with teeth clenched, eyes fixed and insensible ; this condition lasts for about five minutes. He is usually very drowsy and lies in bed, takes no notice of what is taking place about him. Passes motions without giving any warning. Has vomited several times since admission.

Oct. 1st. Seems a little better. More observant.

10th. Bowels not relaxed now. Occasional sickness. Lies in the same apathetic state.

20th. Emaciation increasing. Sleeps quietly at night. Bowels regular.

30th. Nothing remarkable since last note, but an increase of emaciation.

Nov. 5th. Diarrhœa troublesome. Appetite good. Emaciation extreme. Intellectual condition and motor power as before.

14th. To-day patient seems more lively than usual. No new symptoms.

15th. Unconscious all day. Diarrhœa increased. He gradually sank and died the next morning at 5 o'clock.

Autopsy.—Weight of body 19 lbs.

Pupils very small, equal, about the size of pins'-heads. Calvaria very thin. Sutures closed. Dura mater adherent near longitudinal sinus. Some opacity and granulation of arachnoid near the longitudinal sinus. Arachnoid moist. Convolutions slightly flattened; sulci contain a small quantity of fluid. Substance of hemispheres rather firmer than natural. About 9 oz. of clear fluid in lateral ventricles. No thickening or opacity of lining membrane. No softening of the walls of ventricles or of central parts of brain. Veins of Galen normal.

No trace of lymph or granules at the base of brain. On removing the encephalon, the right lobe of cerebellum was found firmly adherent to the right cerebellar fossa, so that the substance was torn in its removal. The left lobe of the cerebellum was infiltrated throughout, or replaced by tubercle, very little cerebral matter being found to the left of the middle line; the tubercular matter extended over into the right hemisphere in the form of a round mass the size of a large walnut. This part was quite separable from brain substance.

Both *lungs* contained much tubercle. Also the bronchial and mesenteric glands.

The intestines were also the seat of tuberculous ulcers, especially the ileum.

Remarks.— In this case the notable symptoms were at the onset two epileptiform seizures, followed by complete loss of power in the limbs, occasional vomiting, at the earlier periods fever, intermitting occipital headache, intellectual powers gradually rendered dull; the sight unimpaired throughout.

In other respects the case was one of slowly advancing tuberculisation, such as is often met with in children.

The great degree to which the cerebellum was destroyed was remarkable; the part left being not more than about one-third of the whole organ. The absence of tubercles of the meninges with so much tubercular deposit in the encephalon, and elsewhere, is quite in accordance with what is constantly met with.

In another case, a boy 10 years old was ill and kept his bed about two months; the prominent symptoms were occasional vomiting and pain at the back of his head. He lost his appetite and became thin; had no convulsions. His pupils were large and sluggish, and his tongue furred. He died quietly in his bed without a struggle.

In this case a firm mass of tubercle 2 inches by 1¼, with a small cavity in the centre, was found on the under surface of the cerebellum comprising the medulla oblongata. There were 8 ounces of fluid in the lateral ventricles, no softening of central parts of brain, and no tubercle in the meninges.

There were a few old tubercles in the apex of the right lung.

Diagnosis of Cerebral Tubercle.—The signs of tubercular diathesis are the same as in tubercular meningitis. Failing health exists before the cerebral symptoms appear. To distinguish it from meningeal tubercle is not always easy at the onset, nor is this to be wondered at, seeing that cerebral tubercle begins usually in the meninges. In a very large proportion of cases of tubercular meningitis small masses of tubercle are found in the brain substance immediately under the pia mater. When, however, there is a *large mass* of tubercle in the brain it is more usual to find little or no tubercle in the meninges.

Convulsions often occur as the first distinct symptom when there is cerebral tubercle. After a short time there are very often signs of pressure affecting one or more nerves irregularly, such as ptosis of one eyelid, blindness of one eye, double vision, with decided paralysis of the external rectus of one eye, or partial hemiplegia. Headache is a more marked symptom in tubercle of the brain than in tubercle of meninges.

It is not easy to distinguish tubercle in the brain from *other growths;* the signs of pressure will be the same. The points to guide you in the diagnosis are the absence or presence of a tubercular constitution, and the rarity of other tumours; there may be hydatids, and there may be cancer or fibro-plastic growth. I have seen encephaloid cancer attached to the cerebellum in a child 3 years old, in which the prominent symptom was obstinate vomiting for about 2 months, there were no headache, no loss of sight, and no signs of paralysis. Convulsions set in about 14 days before death, and continued to recur very frequently till the last. There was no complaint of headache until the day before death, vomiting continued till the end. The case is published in the "Pathological Society's Transactions," vol. x. p. 26.

Enlargement of Skull from Rickets.—Amongst the public any enlargement of the head is regarded as an alarming sign, and people generally think it must be due to water on the brain. Rickety enlargement is accompanied with other signs of rickets in the limbs, thorax, and elsewhere. Ossification and dentition are retarded. There is often hyperæsthesia in rickets, and profuse sweating of the head. The shape of the head is different, there is not the rounded outline of hydrocephalus, the bones are thickened, especially at the margins of sutures, which are consequently indicated by deep furrows; the anteroposterior diameter is increased; the forehead is high, square, and projecting. The face is small, there is no alteration in the direction of the supra-orbital plates, causing prominence of the eyes; the fontanelles are open but not pulsating, and commonly depressed; there is generally a bruit heard on auscultation synchronous with the systole of the heart.

Hypertrophy of Brain comes on very slowly at an early age, attended with loss of health, dulness, and apathy, with some restlessness and general uneasiness. The head seems too heavy for the child, and it frequently bores in the pillow with the back of its head, and the head sweats much. There are threatenings of convulsions before they really set in, and when they occur it is at a late period. Patients affected with hypertrophy of brain often die of other diseases.

Other causes of Hydrocephalus besides Tubercle and Tumours.—1. Congenital malformation. 2. Intra-arachnoid hæmorrhage becoming chronic. It is much rarer, chiefly met with under 2 or 3 years of age. Convulsions more rarely attend it, and paralysis is less marked as a symptom. In a doubtful case an exploratory puncture might be admissible. 3. Inflammation of the lining membrane of the ventricles. 4. True dropsy of the ventricles. This may occur from obstructed veins, either from spontaneous coagulation or from pyæmia. There are no symptoms distinctive of this rarer form of disease. It is said that this is not an uncommon cause of the intra-arachnoid hæmorrhage and dropsy.

The following case of congenital hydrocephalus may be taken as an illustration of this condition. Tapping was resorted to, but without benefit.

W. C. was brought into the hospital on the 3rd May, at the age of 10 months, with great enlargement of the head. His mother and father are living and in good health. He is their first child. His mother's father died of phthisis. His head measured 20½ inches when he was born; he has had no fit, and has appeared to thrive pretty well; but his head has continued to grow. His mother is still suckling him. He is pale, but not emaciated, seems very intelligent. His head measures 26 inches in circumference. The sagittal and lambdoidal sutures are separated for about ¾ of an inch; the fontanelles are very large. The eyes are much protruded and look downwards; no squint; pupils equal. Iodine has been freely used to the head before admission.

May 7th. Strapping applied firmly to the head in the manner formerly recommended by Trousseau.

14th. The head is quite as large and very tense. It was determined to perform paracentesis. It was done by my colleague, Mr. Smith, with a fine trocar and canula, passed through the anterior fontanelles a little behind and to the left of the centre. About 4 oz. of clear colourless fluid were drawn off, of specific gravity 1007, of neutral reaction. When boiled it did not become turbid; boiled with potash and sulphate of copper it gave an abundant orange-coloured sediment. An elastic bandage was then applied.

15th. During the night the child cried a good deal. He is cold and has a slight cough, looks pale and rather low. Takes food pretty well, and does not seem very different from what he was before the operation. The head cannot be measured on account of the bandage. The next evening the child became very low; skin moist, pulse very weak, cough troublesome. He had a mixture given him, containing a little carbonate of ammonia and citric acid, with three minims of ipecacuanha wine in each dose. The next day he had rallied. The scalp between the folds of the bandage was œdematous. The head was again enlarging; the mother was obliged to go into the country with her child.

In November I heard from the mother, who wrote as follows:—

"Since leaving London his health has been so good that I have not had medical advice for him. He grows a fine child, tall and stout, has only six teeth, but for the last week has been unusually fractious. His head has not increased in size, as it did before the operation, being now 27¼ inches round; for the last week or two it has been very tense. He has no power to walk."

On 24th January I heard again :—"His health continues good; he grows tall and stout, but cannot stand yet. He has now ten teeth. His head continues very open and still increases. It measures now 28 inches."

In the following November he died, aged 2 years and 4 months. He was ill for two months, during which he became blind with one eye, which was extremely painful and much protruded. He was convulsed for several days before his death. A fortnight before this, his head measured 32¼ inches in circumference. These particulars were obtained from his father, who was very accurate in his observations.

No post-mortem examination was made.

The paracentesis does not seem to have been of any use, and as the child appeared to suffer from pressure applied to the scalp, and to some degree from the operation, and as the size of the head was not obviously reduced by the operation, it did not seem advisable to repeat it.

Other affections possibly leading to mistakes of diagnosis :—

Epilepsy.—The first fit may lead to the suspicion of tubercle; it may be accompanied with vomiting and headache, and may be preceded by failure of health. There may be some impairment of motor power, which usually passes off. There will, however, be commonly complete recovery in the intervals of the attacks, unless they are extremely frequent.

Spasmodic Contractions of one limb or more, from reflex irritation set up by worms, gastric disorder, dentition, &c. The only contraction which is at all common as an early symptom in cerebral tubercle is that of the nape of the neck. This may occur as one of the first symptoms. With this exception rigidity of limbs is only met with after the disease has lasted some time and convulsions have occurred.

Essential or Atrophic Paralysis.—This often attacks the limbs of young children between the ages of 6 months and 2 or 3 years. It often occurs in children perfectly healthy after a short feverish attack, without any other signs of disease. It is sometimes ushered in by convulsions. It is followed by recovery of some muscles and rapid wasting of parts which remain paralysed. The paralysis is more absolute, and after a time is usually accompanied by depression of temperature (see chapter on Paralysis).

Neuralgic Headaches in anæmic children from 7 to 10 years of age. These may be accompanied with photophobia and inability to work, with loss of appetite, dyspepsia, and change of natural disposition. There will be commonly anæmia, and often intercostal neuralgia or neuralgia of other parts.

The following cases illustrate the difficulty of diagnosis in some of the cerebral affections of children. Even after postmortem examination, the true pathology of the cases is not obvious. During life it seemed probable that both of the patients were the subjects of cerebral tubercle; nothing of this nature was discovered after death, although in the second patient there was tubercle in the lungs, liver, and spleen.

Intense paroxysmal Headaches. Chronic Hydrocephalus without Tubercle. Calculus in Kidney.

Margaret Spence, æt. 11 years, had been ill for 2 years, with occasional headache. Came under observation 2nd August, 1858. In the previous April she had vomiting and severe headache, followed by drowsiness. Bowels costive. In the end of June more violent headache, with intolerance of light and sound. Pains generally worse at night.

State on August 2nd.—Strabismus and rolling of eye-balls. Eyeballs prominent. Pupils large, equal. Double vision. External rectus of left eye seems paralysed. Tolerably well nourished. Appetite rather excessive. Intellect unimpaired. Spirits good when not suffering pain. During the paroxysms lies on belly with head drawn back on shoulder and inclined to left. Intolerance of light and sound; pain chiefly in occiput; also in vertex, cervical and upper dorsal regions; face flushes and sight is dim. Attacks of pain occur more than once a day, or may be absent for 2 or 3 days. Pulse 90, small and feeble.

8th. After a severe attack of pain, whilst sitting up in bed, died in the nurse's absence.

Autopsy.—Calvaria thin. Pacchionian bodies numerous. Layers of arachnoid adherent over left side of brain at junction of anterior two-thirds with posterior third. Visceral arachnoid roughened, with fine sandy granulations; some opaque spots and lines on it. Brain on section tough and rather pale. Lateral ventricles distended with clear fluid;

lining membrane not altered. No softening of central portions of brain. Membranes at base of brain tough, but not opaque. No trace of tubercle any where. Pons varolii small. On removing pia mater from this part the cerebral substance is torn away with it in small pieces. A calculus in pelvis of left kidney large as horse-bean and a larger one in the ureter, composed of uric acid with phosphatic coating. The edges of mitral valves thickened. Muscular substance of heart rather pale.

Intense Headaches. Suspicion of Cerebral Tubercle. Death. Autopsy. Fluid in Ventricles. Flabbiness of Cerebellum, Pons, and Medulla. Tubercle in Lungs, Liver, and Spleen.

Emily Reeves, æt. 9, a healthy child, without hereditary tendency to tubercle. She was seized in November rather suddenly with severe pain over left eyebrow, accompanied by constant nausea. Her skin was hot. Two days later "scarlatinal rash" appeared, with sore throat. Headaches continued through the fever and more or less afterwards until 12th December, when she was admitted into the hospital. Has slept but little. On admission she is found to be well nourished. Her intellect is unimpaired. Complains of violent frontal headache. Cries out : "What shall I do? Oh, my head!" When told to open her eyes, opens but one at a time, keeping the other firmly closed ; says the candle-light hurts her eyes. Has some trembling at intervals, apparently from pain. Cheeks vividly red, much flushed. Pulse very rapid. Temperature 103·6°. In the night was so cold that she had eight blankets heaped on her. Bowels have acted freely with castor oil. Blister to nape. Ice to head.

Dec. 14*th.* Slept well till 3 a.m. ; after which became restless. Has been hot since 8 a.m. Temperature, 10.40 a.m. 102·6° ; 11.15 a.m. 102·4°, less flushed ; 12.30 p.m. 102·5° ; 2.40 p.m. 101·5°, less headache, less flushed, and less restless ; 5.15 p.m. 100·6°, much less flushed ; 7.30 p.m. 98·5°, headache but slight; 11.30 p.m. 98·5°. There is no albumen in her urine.

15*th.* Headache came on at 5 a.m. ; reached its height about 8 a.m.

16*th.* Temperature varies irregularly at different hours of the day. Headaches persist, but at longer intervals. There is hyperæsthesia of skin generally.

17*th.* Pulse generally rapid. Convergent squint of right eye. She usually keeps one or both eyes shut, says it makes her squint to look with both eyes : probably sees double.

22*nd.* Headaches not quite so violent. Vomiting after everything swallowed. Complains of not sleeping. Often has a dry hacking cough.

24*th.* Is to take 2 grains of quinine every 4 hours.

Jan. 5*th.* The headaches have certainly been less severe since she began the quinine. Signs of tubercular consolidation at apex of left lung were thought to be discovered.

18th. No severe headache for past week. Pulse 136. Double vision. Occasional nausea and vomiting.

22nd. Much improved. To take Oleum morrhuæ. Hyperæsthesia gone. Squint persists.

March 12th. 'Much fatter and of good colour. No remains of head-symptoms except squint, and this is only noticed when she looks suddenly at a thing. Physical examination of chest : Flattening under right clavicle.. Notable dulness under both clavicles. Left supraspinous region duller than right. Respiration everywhere almost normal.

June 27th. Has been an out-patient for 3 months ; one month at Brighton. Has been better but is now falling off again, and headaches, which have never quite left her, are becoming more severe. No double vision. Slight squint. Walks very unsteadily when her eyes are shut.

July 8th. Was readmitted. Three days ago had severe pain between shoulders ; paroxysms of pain in this part have recurred frequently, lasting from half a minute to half an hour. Sickness has returned, and double vision. Very tender in spine, hence lies half over on face. Has lost flesh. Says she is giddy when on her back. When asked to sit up flushes very much ; headache is much increased for a time ; pulse quickened to 158 ; keeps head turned to right. Intellect perfect. Quinine mixture ordered, and linimentum belladonnæ to spine.

July 14th. Yesterday vomited ; had not much headache. To-day has severe occipital headache. Pulse 108, regular.

15th. Headache persisted all day yesterday. Two leeches were applied behind left ear. After this she passed an hour or two quietly ; but headache then returned. At 5 a.m. headache most violent ; doubled up in bed, pressing forehead with both hands and crying bitterly. At 7.30 she complained of feeling numb all over, and then went off apparently to sleep ; was found dead at eight o'clock.

Autopsy eight hours after death. Body kept in prone position. Spinal canal opened. Much blood in vessels in and outside canal. After removing calvaria, pia mater moderately injected, more so posteriorly and near middle line than elsewhere. No excess of red points on section of brain substance. Lateral ventricles distended with clear fluid ; no central softening ; commissures perfect.

Cerebellum unusually flabby, collapsed ; left lobe more so than right. Very soft to touch. Membranes not extra-vascular, and very thin. Grey matter of cerebellum rather pale. Arachnoid generally rather drier than usual. No trace of tubercle in encephalon or membranes.

Pons varolii, medulla oblongata, and upper part of spinal cord (about half an inch) *not nearly so firm* as usual and very pale.

Left Lung adherent behind by old adhesions. In middle part of lung there is some collapsed tissue studded with grey granulations.

Right lung normal.
Liver, weight 22¼ ounces, small; on it a few superficial grey granulations. No other tubercle found in body.

Remarks.—The cases of Reeves and Spence are both of unusual interest. Neither of them was diagnosed during life. There was a suspicion of cerebral tubercle in both. They present many points of resemblance and several important points of difference. They were of the same age, the main feature in each was intense paroxysmal headache, in each there was intolerance of light, vomiting, tonic contraction of muscles of neck, and occasional double vision. In neither was there any impairment of intelligence or amaurosis, or paralysis or rigidity of the limbs. Both patients died rather suddenly after a violent paroxysm of pain.

After death fluid was found in the lateral ventricles, but no cerebral softening or change in the lining membrane. There was no sign of tubercle within the cranium of either. In Spence there were signs of chronic changes in the arachnoid of the convexity, slight softening of the external portion of the pons varolii, and renal calculi. In Reeves there was loss of consistence with pallor, of the cerebellum, pons, and medulla; with a few grey granulations in one lung and on the surface of the liver. The changes in the brain appear to have been in both cases of an atrophic character. In Reeves's case it appeared from the history that the symptoms commenced about 48 hours before the appearance of the eruption of scarlatina, and at the time it occurred to my mind that this specific blood poison was the origin of the entire mischief. On further consideration I am inclined to think that the patient did not have scarlet fever at all, but that the rash was a concomitant of the cerebral disease, more like the flushing which was subsequently observed in her case during paroxysms of pain, dependent on a sort of paralysis of the small vessels.

In Reeves's case the question arose whether the symptoms could be entirely due to neuralgia? Severe paroxysmal headaches, with nausea, photophobia, and hyperæsthesia, are occasionally met with in girls from six years old and upwards, without any organic disease. Irregularity of pulse and retraction of the head may accompany these attacks. The paroxysmal character of the pain, with absolute intermissions, looked like neuralgia, but was not incompatible with organic disease. The elevation of temperature and quickened pulse which were observed during the attacks were noteworthy. It is not clear whether the pain was due to the fever, or the fever caused by the pain, or the two symptoms due to some common cause.

Quinine in full doses appeared to relieve the pain for some time, as it generally does where there is anything like an approach to periodicity in the occurrence of the symptoms, even though there is organic disease.

In Reeves emaciation occurred, a hacking cough, and some dulness with prolonged expiratory murmur was suspected; these all appeared to indicate tubercular disease.

In the following case there was also hydrocephalus without tubercle; there was here a great deficiency of symptoms, especially towards the end of life. Emaciation was great, and there was some loss of consistence in the cerebral substance with a very little thickening of the arachnoid on the convex aspect of the brain. The first symptoms were headache and vomiting five months before death. There was closure of the communication between the subarachnoid space in the ventricles and that of the cord, as is usually the case when fluid accumulates in the ventricles.

Chronic Hydrocephalus. No tubercle or any other discoverable cause.

Alfred Holmes was brought to the hospital moribund; he was 2 years and 4 months of age. The history given was, that he had been attacked with vomiting and pain in his head five months previously. Blisters were applied, but he was not much better for nine weeks, when convulsions occurred. Three or four weeks ago his legs swelled, and since that time have been rather rigid. This swelling still existed; there was much emaciation, but nothing to point in any way to cerebral disease.

Autopsy. Greatly emaciated; body weighs but 23 lbs. Anterior fontanelle about 1 inch square, not elevated. Sutures closed. Dura mater adherent along longitudinal suture. Convolutions much flattened; arachnoid thickened in one or two places. Brain substance, soft and pale. The lateral ventricles, also the third and fourth, distended with clear fluid. Foramen of Monro about the size of a fourpenny piece. Septum lucidum not softened. Under part of fornix adherent to velum interpositum. Velum itself somewhat thickened, yet vessels apparently quite pervious to blood. The cerebro-spinal foramen was quite obliterated by fine filaments. Base of brain quite healthy. All the other viscera healthy but very small. The weights were—heart 1 oz. 3 dr. ; lungs, right, 2 oz. 4 dr. ; left, 2 oz. 6 dr. ; liver, 6 oz. 12 dr. ; spleen 1 oz. ; kidneys, 1 oz. 8 dr. Stomach and intestines normal ; also lymphatic glands. Veins of lower limbs normal.

Remarks.—This case appears to have been one of atrophy of the brain and other organs, without any morbid deposit. The history was very incomplete, but it is evident that the child was treated with lowering measures, and too little attention was given to its obtaining nutritious diet.

In the following case we have severe neuralgic headaches, which were cured with great rapidity by quinine.

Violent headaches of five months' standing, cured by quinine in 24 hours.

Salome Eames, æt. 8 years, a delicate girl, the only surviving child of a scrofulous woman, who died soon after her birth. She has had measles and hooping cough. In the beginning of November, 1863, she began to complain of occasional headaches, shooting up behind the ears, worse on the right than left side. On four or five occasions the headache was accompanied with sickness. The fits of headache were paroxysmal and occurred at first every two or three hours. From November till May the headaches became more intense ; a day never passed without an attack. The pain involved the top and back of the head and the left shoulder ; was ushered in by closure of the eyelids, and a darkness of the upper part of her face. After the pain in head has ceased she often has pain in the left middle finger. She has tenderness in the left iliac fossa. Bowels regular ; tongue clean ; mucous membranes pale. On admission to the hospital on April 28th, she was ordered 2 grains of quinine three times a day. After three doses of this she had one attack of headache ; but subsequently there was no recurrence during the fortnight that she remained in the hospital.

Remarks.—Girls from eight to twelve years of age are not unfrequently the subjects of violent neuralgic headaches, with sickness and intolerance of light and sound.

This case may be compared advantageously with the cases of

Spence and Reeves. In the latter case there was nothing at the outset to distinguish it positively from a mere neuralgic headache. After a time the occurrence of double vision and an elevation of temperature during the paroxysms gave the case a more serious aspect. In her case quinine gave temporary relief, but after a time ceased to have any effect. In neuralgic headache the pain is often more intense, and the remission usually more absolute, than in headache from organic disease. It also continues to recur frequently for a long period without progressive increase of symptoms, or other signs of cerebral lesion. In all cases of doubt, it is well to try quinine, change of air, and good diet.

The next case was one of advanced renal and arterial degeneration, such as is seldom met with in children. The renal disease took its origin in scarlatina, two years and a half previously, and had been at first accompanied with dropsy and convulsions. During the later periods of the disease there was no dropsy. She passed pale, highly albuminous urine, and suffered from intense paroxysmal headaches, amaurosis, and some impairment of intellect; there was also great emaciation.

After death the brain substance was pale and flabby, there was fluid in the ventricles, atheroma of the arteries of the meninges, and hæmorrhage into the arachnoid. The kidneys were atrophied and much degenerated.

Anne Read, æt. 10½ years. Had scarlet fever two years and a half ago; this was followed by dropsy and occasional convulsions. The dropsy was at times much worse than at others. When admitted to hospital there was no œdema. She had a wild distressed look, dilated pupils and impaired sight. Often vomited both on an empty and on a full stomach. Headaches paroxysmal and violent, accompanied or followed by whining, as if from general malaise. Intellect apparently impaired. Emaciation proceeded to an extreme degree. Urine was scanty in amount, pale, and highly albuminous, with very little deposit; a small translucent renal cast rarely found. For more than a week before death she was alternately dozing and screaming out; did not know her friends. Swallowed but little, and this made her sick. Died of exhaustion.

Autopsy. Kidneys extremely atrophied; cortex pale, sandy looking. Much degeneration of Malpighian bodies and blood-vessels. Calvaria very thick and dense in texture, especially in front. Arteries of brain very atheromatous. Hæmorrhage into arachnoid, lining the anterior and left middle fossa with a layer of coagulum. Clear fluid in meshes of pia mater on convex surface of brain, and a considerable amount of fluid in lateral ventricles.

Heart, especially left ventricle, hypertrophied. Endocardium over mitral valves atheromatous ; also the coronary arteries. Stomach much injected in spots, and of a dark purple colour. A layer of soft yellow lymph on the peritonæum. Liver healthy. Spleen tough. Lungs œdematous. Some turbid fluid and soft yellow lymph in pleuræ.

The amount of degeneration in the cerebral arteries was very remarkable at such an early age. It is very likely that the amaurosis was due to degeneration of the vessels of the retinæ. It is very seldom that one finds in the child such a condition of advanced degeneration of tissues.

The next case is also a rare one, in which there was an abscess formed pressing on the middle lobe of the brain, also causing protrusion of the eyeball, with sloughing of the cornea. The primary disease appeared to be caries of the sphenoidal bone.

Caries of Sphenoidal bone. Protrusion of Eyeball, with sloughing of cornea. Abscess in middle fossa of Skull, causing hemiplegia and convulsions.

Edward Sell, a boy nearly 4 years old, was brought to the hospital on the 22nd April, with the following history. He had been quite healthy till the previous winter, when he had hooping-cough. Six weeks before admission he became feverish, and the following morning had a fit of convulsions which lasted an hour. Since then he has not walked ; his left leg was weaker than the right. Five days ago his left arm was found to be paralysed. Three days after the fit his right eyelid began to swell ; an abcess formed, which was punctured. On admission the right eyeball was much protruded ; there was great swelling of the upper lid, and some œdematous swelling over the zygoma. He could see with left eye, probably not with right. Left eyeball always inverted. Left upper prolabium more arched than right ; the right nostril nearly shut, while the left was open. In crying, the right angle of mouth is moved as much as the left. He could move his left leg very little, his left arm rather more ; right limbs naturally. The left thumb inclined stiffly towards the palm. The swollen lid was punctured, and some pus escaped. No bare bone could be felt with a probe. Pulse 124, weak, not quite regular. Respiration 22.

May 23rd. Slowly wasting. Paralysis as before, except that in crying his mouth is much drawn to the left. Left arm and hand more stiffly bent. The last day or two very restless, frequently rubbing his chest.

June 9th. Had a fit in which he lay stretched out straight, his hands clenched. His left eye and left side of his face "worked" a good deal. This lasted two minutes, and was rapidly succeeded by two other similar attacks.

10th. Lies in a drowsy state, speaking a word occasionally, and occasionally crying. The right eye is moveable, everted, protruded, but covered by the lid and much bloodshot. From the line of the eye to the ear and above the meatus there is a bulging as of the bones of the skull.

24th. Has become much thinner. Is very tremulous. The right cornea has sloughed. Desquamation of cuticle over the surface of body generally.

26th. After a slight fit he passed into a comatose condition in which he died.

Autopsy.—Brain substance very pale. Under a thin layer of brain substance of the right hemisphere, near the orbital plates, was found a large abscess, measuring from before backwards 4 inches, from side to side 3½ inches, and from above downwards, 2½ inches. The walls of the abscess were thick and pale. On the inner surface of the right orbit, from the external angular process along the external superior angle of the orbit back to the sphenoidal fissure, the bone was rough and carious. The articulation between the frontal and malar bones and the ala major of the sphenoid is loosened and rough, also carious. The external surface of the ala major is eaten away on the surface as low as the root of the pterygoid plate. Through this wing, on a level with the zygoma, two holes perforate the skull, one ¼ of an inch, the other ⅛ of an inch in diameter. The abscess above named has a pedicle which is attached to the larger of these two holes, but the abscess does not go through the hole. From this point the abscess reaches backwards, lifting up the brain, as it were, from the middle fossa of the skull. A good deal of clear serum in the left lateral ventricle; none in the right, the sides of this one being approximated by the pressure of the abscess. No pus outside the skull, but much swelling of the soft parts above the zygoma. The bone was not at all protruded, as was supposed during life. All the other organs of the body were healthy.

Remarks.—This is a case which was supposed during life to be a malignant tumour, causing pressure upon the brain and protruding the eyeball and the bones of the skull.

The origin of the disease was found to be caries of the sphenoidal bone, a very rare affection. This gave rise to an abscess displacing the right middle lobe of the brain. The occurrence of convulsions as the earliest symptom (soon followed by paralysis in the limbs of the other side and without complaint of pain) is noteworthy. The caries, no doubt, had existed for some time before this, and an abscess was forming. The brain will often bear a considerable amount of slowly increasing pressure without any symptom, and on reaching a certain point, from some exciting cause, convulsions and signs of paralysis occur. We have seen that this is the course often followed by the symptoms in tubercle of the brain.

The paralysis of the left arm and leg was due to pressure on the right middle lobe of the brain, the paralysis of right side of face suggested a lesion near the left side of the pons varolii, involving the origin of the portio dura of the seventh nerve. Nothing of this kind was found post mortem, so that the paralysis was probably due to interference with some of the extra-cranial twigs of the facial nerve by the inflammatory exudation near the zygoma.

The sloughing of the cornea was not due to pressure but to irritation of the nerves regulating the nutrition of the eyeball.

Caries of bone in children is probably more frequently due to external violence than to any other causes. In some cases it results from tubercle, but very often not, and it often arises by extension from the mucous membrane in strumous subjects, or after scarlet fever or other exanthematous disease, especially in the temporal bone by extension from the Eustachian tube or the external auditory meatus.

In this case there was neither external violence (as the part was protected by its situation), tubercle, struma, nor previous acute disease to account for the caries. It is impossible to say on what the caries of sphenoidal bone depended, or what was the determining cause.

ABSTRACT OF CASES DIAGNOSED AS CEREBRAL TUBERCLE.

Name.	Sex.	Age.	Antecedents and Family History.	Course of Symptoms.	Result.
1. G. Smith.	Male.	5 years.	Parents healthy. The first child. Measles and scarlatina. No premonitory symptoms.	Morning sickness; pain in head; unsteady walk; weakness of right hand; slight facial paralysis; intermission of vomiting; occasional paroxysms of sickness, with headache; occasional fits of convulsions. Progressive loss of muscular power in limbs and sphincters; head considerably enlarging; intellect getting weaker, answering questions very slowly. Amaurosis, six months from first symptoms.	Death after 6 years. No post mortem.
2. Jessie Trimm.	Female.	12 years.	Empyema 12 months before, tapping. Recovery.	Headache, listlessness, anorexia; sickness after food; screaming fits at night; double vision. Slow pulse; at times irregular. Strabismus. Loss of sight in 2 months.	Recovery, with loss of sight.
3. M. Sullivan.	Male.	5 years.	Generally delicate. No immediate premonitory symptoms. Hooping-cough 2½ years ago.	Fall on the back of his head. Convulsions, followed by loss of power in legs; occipital headaches. Apathy; occasional screaming at night. Pupils medium sized. Weakness of intellect. No loss of sight. Slow drawling speech. Diarrhœa. Extreme emaciation and exhaustion.	Death in 7 months. Fluid in ventricles. A very large mass of tubercle in cerebellum, involving both lobes. Tubercles in lungs, bronchial and mesenteric glands. Ulcers of ileum.
4. Charles Oakely.	Male.	4½ years.	Consumptive family. Small-pox 3 years ago. Hooping-cough 1½ years.	Loss of appetite; great weakness; headache, and drowsiness at the end of 14 days; frequent vomiting; screaming out for mother in the night. Afterwards could walk about. Bowels rather loose. Then, restlessness in bed; dull in intellect; pupils large. Pulse very	Death in 7 weeks. Tubercle, 3 masses in brain. Fluid in lateral ventricles. Tubercle in liver, peritoneum and intestines; none in lungs.

CASES OF CEREBRAL TUBERCLE.

5. William Lawrence.	Male.	6½ years.	Mother's family, brother and father, consumptive. Fourth child. Healthy till 5 years. Mother has had six births since; 4 at 7 months, and 2 miscarriages.	Severe headaches; sleepiness; sickness after food. Nine months later, his sight began to fail; in 9 months more, nearly gone. Partial loss of power in left arm and leg. Incontinence of urine. Emaciation. Diverging squint, left eye. Pupils large; equal; very inactive to light. Three months before death had five attacks of convulsions. Paralysis increased, with rigidity.	Death, Ætatis 8 years. Extreme emaciation.
6. — Green.	Male.	17 months.	From earliest infancy had been subject to violent fits of screaming, and pain when handled.	*April.*—Much emaciated. Otorrhœa right side. Never convulsed. *May 2.*—Right eye lost sight, and turned inwards; pupil of medium size. Almost constant screaming. *May 30.*—Much weaker.	*June 3.*—Death. *Autopsy.* In pons varolii tubercle size of large almond. Not much fluid in ventricles. No tubercle of meninges; none in lungs.
7. Winifred Martin.	Female.	10½ years.	Father died of consumption. Brother and sister tubercular. Not very strong; bow-legged.	First symptom in *February*, dribbling of saliva. Severe headache and nausea. Right arm became colder, and much weaker than left. Occasionally the power over arm returned for a few seconds. Gradually right leg became weak. Headache continued 4 months. In *July* came to hospital and improved. In *November* sight began to fail; memory failing; articulation less distinct. *December.*—Mouth drawn to left when smiling. Tongue deviates to right. Screams out in sleep. Can only distinguish light from darkness. Pupils medium-size. Spirits good. Pulse 84. Slight resistance to extension of right arm and flexion of right thigh.	Unknown.
8. Anne Fitzgerald.	Female.	4 years and 8 months.	Father died of phthisis. Delicate child. Had a fit 12 months before.	For 5 months has had occasional severe occipital headache, with vomiting; losing flesh; bowels costive; rolling of eyes. Has gradually lost sight, and has not walked for 3 months. Forehead prominent. Lower limbs powerless; rigidity in left knee. Quite blind; pupils large, inactive. Great dislike to being moved in bed. Head enlarged, within 3 weeks, ¼ of an inch; and again, in 10 days, another ¼ of an inch. Stools and water passed in bed.	Unknown.

ABSTRACT OF CASES DIAGNOSED AS CEREBRAL TUBERCLE (*continued*).

Name.	Sex.	Age.	Antecedents and Family History.	Course of Symptoms.	Result.
9. James Hargrave. Chronic hydrocephalus.	Male.	4 years.	No evidence of hereditary predisposition. In good health till 6 months of age.	When 7 months old, head drooped; feverishness; vomiting. A few days later, had fits of convulsions; in the course of 3 weeks had 6 attacks. For 3 months rigidity of limbs. From 8 months of age head began to enlarge. On admission, head very large, especially posteriorly; circumference 24 inches. Body and limbs thin. Talks indistinctly; intellect weak. Will never feed himself, though appetite craving. Bowels regular. Sleeps well. Occasional strabismus. Limbs not rigid; fingers at times stiffly spread out, and great toes often flexed. Anterior fontanelle not closed.	Six years later. Head still larger; cannot lift it up. Talks moderately. Sight deficient. No control over sphincters.
10. Thomas Cooper. Chronic hydrocephalus.	Male.	5¼ years.	General health good. Family not known to be tubercular.	Faintness in street after hot sun; loss of power in right leg; partial recovery; in a few days left leg weak; in 3 months could not walk at all. For five months, occipital headache. After 4 months, some difficulty in micturition. After 6 months, can move legs in bed; cannot stand, though he can move legs when supported. Clumsy and tremulous with hands. Intelligence good; sight good. Headaches, and vomiting occasionally, with pains in limbs. He was then given extract of belladonna, which appeared to relieve him. Slight general hyperæsthesia. Progressive loss of power. After 10 months, head slightly increased; veins of forehead enlarged; pupils smaller. Legs extended; cries when they are bent, but keeps them so for a time, and dislikes them being quickly extended. Slight talipes varus. Has spontaneous pains in legs, also when lifted. Feet always cold. Imperfect control over sphincters and bladder. One month later, pupils dilated; mouth drawn to left.	Not known, after he left the hospital. *Treatment.*—At first iodide of potassium, with tincture of iodine. Then strychnia, gradually increased to 1/12th of a grain three times a day. Next, extract of belladonna, ¼ of a grain, three times a day, with relief. Subsequently pil. hydrargyri, 1 grain, three times daily for a month.

CASES OF CEREBRAL TUBERCLE.

11. Thomas Hughes.	Male.	5½ years.	No tubercular history. Had scarlatina 4 months before.	Severe frontal headache, especially at night, for 6 months. After 4 months, stooping at shoulders; 14 days later, weakness in right ankle, and clumsy with right hand. On admission, right arm paralysed, rigid; also right leg, partially. Left zygomatic muscle paralysed, and right orbicularis paralysed. After 14 days, thinner and paler. Paroxysmal headaches, and pains in hypochondrium. Less deviation of mouth. Internal strabismus right eye. Died one week later, without much change in symptoms till last few hours.	Death. Autopsy. In right hemisphere, near longitudinal fissure, tubercle size of bean; in right side of pons varolii, tubercle size of nutmeg. In left hemisphere, pressing on corpus striatum, was another tubercle, of same size. Lateral ventricles full of fluid. Brain tissue, around tubercles, softened. Tough exudation at base of brain.
12. William Musgrove.	Male.	3½ years.	Delicate child. Family not consumptive. One child older, healthy.	Began to limp on left leg 3 months ago; soon afterwards could not use his left hand. Squinting, yawning, and grinding teeth. Pale and flabby countenance. Pain on right side of head. Passes motions under him. Left leg and arm partly paralysed, and arm stiff. Mouth drawn to right. Soon became insensible. Fits continuously for 14 days before death.	Death 6 weeks later.

Remarks on No. 11 *Case.*—The paralysis of right arm and leg with rigidity was explained by the tubercle in the left hemisphere of brain pressing on the corpus striatum, with softening of cerebral matter around it. There was a large mass of tubercle in the right half of the pons varolii; there was no paralysis on the left side except that of the zygomatic. It is possible that this mass may have interfered with the right portio dura, and would thus cause paralysis of the right orbicularis. The masseters however acted equally. (Brown-Séquard, "Physiology of Central Nervous System," page 201, 1860.) The softening of the central parts of brain probably came on during the last few hours of life.

PYÆMIA AND OTORRHŒA.

PYÆMIA is now well known to surgeons as occurring after operations, especially those implicating bones. It is also frequent after wounds of various kinds, after parturition, phlebitis, and diseases of bone, especially acute necrosis. It may be produced by the introduction into the blood, not only of pus, but of putrid fluids without solid particles, or of minute particles of disintegrated fibrin. Healthy pus has been often injected into the circulation without any secondary abscesses having occurred. To cause pyæmia, either the pus must be morbid or the patient must be in an unhealthy condition.

An attempt has been made to distinguish affections caused by the absorption of putrid matter from those due to the presence of pus, the former being called septicæmia, and the latter pyæmia or pyohæmia. Mr. Savory has shown that such a distinction cannot be maintained (St. Bartholomew's Hospital Reports, vol. i.). At one time all cases of pyæmia were believed to be due to phlebitis; this is certainly one cause, but by no means the most common one.

Mr. Henry Lee maintains that the introduction of pus into the system through a vein can rarely be the first step towards pyæmia or purulent infection. "Some change," he says, "must previously have passed in the blood by which its coagulating power is impaired, or some unusual mechanical means employed before pus can find its way into the course of the circulation." Sédillot, on the other hand, never found any difficulty in producing pyæmia by the injection of pus into the circulation even in healthy animals. He maintained that all cases of pyæmia were thus caused. There is no doubt that he was in error even as regards surgical pyæmia.

But independently of this, there is, in the opinion of many, such a disease as idiopathic pyæmia, a distinct acute specific disease, "pyogenic fever." It is either primary or secondary to some other fever, such as typhoid or scarlet fever.

The parts affected are the subcutaneous tissues and the muscles; much more rarely the joints, the lungs, and other viscera. In this respect the cases differ from surgical pyæmia. They are less frequently accompanied by rigors and delirium, and they are less fatal.

Pyæmia on a small scale is very commonly seen after acute specific diseases; one or two small abscesses are found in the subcutaneous areolar tissue, especially in the scalp; slight fever precedes their formation. Instead of one or two abscesses, there may be many, with considerable febrile disturbance. Locally there are, in the majority of cases, but few signs of inflammation before pus is formed. In other cases there is inflammation which becomes resolved without the formation of pus. Instead of small subcutaneous abscesses, there are in other cases large purulent collections seated deeply amongst the muscles, more rarely in the joints, and still more rarely in the lungs, liver, or spleen.

In some patients, there are purulent discharges from several of the mucous membranes, and every slight abrasion "festers." This condition is probably closely allied to those just mentioned.

Good illustrations of the so-called idiopathic pyæmia are afforded by the two following cases. There was no reason to suspect that there had been any absorption of pus, disintegrated clot, or any putrid fluid. The disease was evidently a general one; the great elevation of temperature which preceded the appearance of the abscesses, and the great benefit derived from the administration of large doses of quinine are especially worthy of note.

Pyogenic Fever following Rheumatism; cured by large doses of Quinine.

C. Sadd, a boy, 8 years of age, always delicate, subject to colds and pains in his limbs, was admitted to the Children's Hospital on the 11th March. His mother's family are phthisical, and his father's rheumatic. Three weeks before admission he had "rheumatic fever;" his joints, large and small, were very painful and swollen, some pale and some reddened. There was also a good deal of sweating. This condition lasted a fortnight, at the end of which a swelling was noticed in the calf of his leg; a few days later a swelling formed over the right scapula; this was punctured, and pus escaped. When admitted he was very pale and thin, with a rather earthy tint of skin. His pulse was 156, quick and sharp in character. Respiration 32. Occasional short cough. There was a copious brownish-coloured discharge from the abscess over the right scapula. At the base of the left lung there was a dull percussion note as high as the angle of the scapula, and on auscultation tubular breathing and bronchophony. He was ordered dilute nitromuriatic acid and cinchona, with 4 oz. of wine and nourishing diet.

March 17th. He seemed better; the pulse had fallen to 128, and the hepatized lung showed signs of resolution. The next fortnight he made no progress; the abscess over the scapula closed, but several fresh ones formed. The temperature observed in the axilla night and morning varied from 99° to 103°.

28th. It rose to 104°, and on 30th to 105°.

30th. He was ordered ½ gr. of quinine with chlorate of potash 4 times daily. Until the 10th April there was no improvement; more abscesses formed; he sweated freely at night, and was rapidly losing strength. He was then ordered 1½ gr. of quinine every 2 hours night and day. This effected a very marked improvement. His temperature, which was ranging from 99·5° to 104·5°, fell to 97° in the morning, and only reached 100·5° in the evening. On the 14th April the quinine was omitted for two days; the evening temperature at once rose to 103°. On the 16th the quinine was resumed, when the temperature fell to 101° of an evening, and 97·5° in the morning. On the 19th a red tender indurated patch appeared on the right calf; it was fomented well, and disappeared without the formation of pus.

May 1st. He was improving. There was a remission of temperature every morning below 99°, and in the evening it seldom rose above 100°. He was gaining strength and flesh. The quinine was now to be taken 8 times a day instead of 12 times.

15th. He had improved so much that he was allowed to be dressed and walk about the ward; in the evening the temperature rose to 102°; he was then kept in bed until the 20th, during which time the temperature never rose above 100·5°, falling every morning to 98°.

20th. He again walked about a little, and there was an elevation of temperature in the evening to nearly 103°. From this time he continued steadily to improve until the 27th June, when he left the hospital in good health. The quinine was given every four hours, with the addition of 1 gr. of ammonio-citrate of iron to each dose. He took altogether more than 2 ounces of quinine with no other apparent effect than to reduce the fever, prevent the formation of abscesses, and improve his general health.

Another boy, named Cumming, had been suffering, before his admission to the hospital, with a low fever, which lasted about three weeks. His brother, who lived with him, had been affected with a similar disease, but suffered from diarrhœa at the same time. The illness was probably typhoid fever.

On admission he was emaciated, his cheeks were flushed, his skin was dry. His abdomen was retracted and slightly tender. Pulse 112, a little irregular. Respiration 24. Appetite voracious. In left acromial angle was an abscess the size of a walnut. Between spine of right scapula and the spinal column was an elongated, hardish, deep-seated swelling ; and just below the left trochanter was a hard tender swelling. Both his legs were much contracted at the hip joints, pushing femurs upwards ; no tenderness. In the urine there were albumen, blood discs, renal epithelium, and granular casts of tubes. This condition of urine disappeared the day after a large abscess near the left hip was opened. His blood, when examined, was found to contain an excess of white corpuscles. At the apex of his right lung there were signs of slight consolidation, probably tubercular. He improved on the use of quinine, but not so obviously as Sadd. There was in several places reddening induration and tenderness, which disappeared without the formation of pus ; abscesses formed were opened and speedily healed ; the temperature ceased to be elevated. The last abscess which was opened, appeared on the thigh with scarcely any warning, whilst he was omitting to take quinine for a few days. He left the hospital, after being under treatment about 11 weeks, in greatly improved health with a sinus leading from the abscess in thigh. This sinus remained open for 2 months, but he had no fresh abscesses.

These cases may be called *chronic* pyæmia, such as are described by Mr. Paget in the first volume of St. Bartholomew's Hospital Reports. He states that they are more rare among the instances of pyæmia following wounds, than among those occurring in diseases ; and that the local evidences of chronic pyæmia are (more often than those of acute) seated exclusively or chiefly in parts of the same tissues; they are more frequent in the trunk and limbs than in internal organs.

In the first case there was exudation in the left lung, giving all the signs of pneumonia.

In the second case there was albuminuria, which disappeared quite suddenly after the opening of an abscess. There was probably blood poisoning in both of these cases. How the poison was introduced in the first case is not at all obvious. Is it not possible that the blood may in itself generate such a poison without its being directly introduced?

In the second case, where typhoid fever preceded the symptoms, it is easy to suppose that the poison was introduced at the surface of some of the ulcers of the intestines which were induced by that disease.

Similar symptoms not uncommonly follow ozœna and other diseases attended with suppuration on mucous surfaces, or slight operations on the urethra.

There are two ways of explaining the coincidence of suppuration on mucous surfaces and multiple abscesses in other tissues. One is to suppose that pus globules, or some of its constituents, are introduced from the suppurating mucous surface which is primarily diseased, and the other is to suppose that there is in the system a condition which leads to the formation of pus both on the mucous membranes and elsewhere in the body. There is one set of cases of pyæmia in which the connection between suppuration on a mucous surface and the blood poisoning is very readily traced. I refer to those cases which depend on otorrhœa and the introduction of poison from the tympanum to the petrosal and lateral sinuses to the jugular vein, and the circulation generally. The lungs are generally the first organs to suffer. The following case is a good example of this. The true nature of the disease was not recognised during life in consequence of the history being imperfect, no mention having been made of otorrhœa as a symptom. The discharge had ceased at the time she came under observation, as is very generally the case when systemic infection takes place.

Kate White, a girl, æt. 6 years, was admitted under my care into the Children's Hospital on the 21st January, with the following history :—

She was believed to be quite well until the evening of 18th January, when she complained of headache. She had been subject to otorrhœa of the right ear since teething. (This was not ascertained till after death.) She had had scarlet fever, measles, hooping-cough and chickenpox. Her parents were poor, and she had been living badly.

Jan. 19*th.* She seemed ill ; her headache continued.

20*th.* She seemed cold and vomited a little ; she had some pains in abdomen. On admission she was pale, and her complexion of a leaden tint ; her skin was covered with numerous petechial spots, larger than ordinary flea-bites. Her temperature was 101·2°. Pulse 136, weak. Respiration not unduly accelerated. Slight strabismus. Pupils small and equal. Surface universally tender, most so on the right side of neck. No discharge from the ears. Abdomen rather retracted. Skin dry. Hands and feet rather cold. Tongue covered with moist brown fur in the middle, and narrow red margins. Slight dulness over lower part of right lung. Respiration at this part rather weak, with slight sonorous rhonchus and a sound simulating fine friction. A mustard plaster was to be applied to right side of chest, and a saline mixture given.

22*nd.* Lips pale, bluish, not turgid ; cheeks faintly purple. Nares dry, coated with sordes. She is peevish, sleepy, lying coiled up as if to keep herself warm. Last night at 6 o'clock she had a short shivering fit, not followed by sweating. Squint continues. Pulse 160, very weak. Respiration 40, easy. A frequent dry cough. Has been sick three times in the night, vomiting mucus. The bowels have acted three times during the night, motions loose, of a dark-brown colour. There is decided slight puffy swelling of the right side of neck. Respiration less weak over right lung ; nothing abnormal heard except a little sonorous rhonchus here and there. Takes food badly, wants cold water ; milk makes her sick. Urine turbid with phosphates, free from albumen.

23*rd.* Shivered again last evening. Passed a restless night without delirium. Looks extremely ill ; pale, muddy complexion. Dislikes being touched anywhere. Keeps eyes closely shut. Breath has a faint sickly smell. Pulse 180. Respiration 60, regular. Percussion note dull over posterior aspect of right lung. Over right lung, front and back, is heard a fine sub-crepitant sound somewhat resembling friction. A little fine sonorous rhonchus over left lung.

She died the same evening.

Autopsy.—Numerous petechiæ on the surface, and some large ecchymoses. Slight discharge from right ear. On removing brain, there is observed a little blackish discoloration of the membranes on the anterior surface of cerebellum where in contact with the pars petrosa of right temporal bone; membrana tympani destroyed; ossicula discharged. Surface of tympanum exposed and rough; no necrosis of bone. Right lateral sinus full of puriform bloody matter. Internal jugular contained a narrow thrombus near the skull, and the walls of the vein and the neighbouring tissues were discoloured green all down the neck.

Right Lung weighed 5½ oz. Some moderately firm adhesions on anterior and posterior aspects. Lobes bound together by adhesions which are broken down tolerably easily. Lower lobe extensively infiltrated with well-defined masses of irregular shape, varying in size from a hemp-seed to a large walnut; in colour dull red, dirty yellow, or almost black. Most of these masses are surrounded by a sort of capsule of irregularly curved outline. The more advanced of these infractions have a very dark colour and a fœtid odour. The surface of the lung corresponding to them varies in colour from red to yellow, and mottled black, red, and yellow. The largest of these products is at the upper part of the lower lobe, near its junction with the upper lobe, which is itself the seat of exudation, more resembling ordinary pneumonia. The middle lobe is free.

Left Lung weighed 3¾ oz. There are similar changes in this lung to those described in the right lung, but less advanced.

Heart contained firmly-coagulated blood and weighed 2¾ oz. The liver, spleen, and kidneys were healthy.

In this case there was little or no pus formed; but there was just the condition described by Mr. Savory in the article referred to; spots and patches of congestion of a dark red and livid colour, slightly raised and indurated. Some of the patches in the lung were softening in the centre, and inclining to suppuration; others were inclining to gangrene. It appears that the more healthy the patient the less disposition is there to the formation of pus. This little girl is said to have been in good health until five days before her death.

The previous existence of otorrhœa (if it had been ascertained), the successive rigors, the sallow complexion and petechiæ, the great prostration, the tenderness on the right side of the neck, the sickly odour of the breath, and scattered rhonchi in the lungs, with slight friction sounds, ought certainly to have enabled one to make a shrewd guess as to the true nature of the disease.

Several cases of this nature are described by Dr. Gull in the "Medico-Chirurgical Society's Transactions," vol. xxxviii.

Reference should be made to Case VIII. of Pleurisy, page 68, where circumscribed empyema was traced to chronic disease of the ear.

Treatment of Pyæmia.—If the source of infection can be detected, it should be removed, if possible. It is very seldom that this is practicable. A case of this kind was under the care of my colleague, Mr. Holmes. A few days after excision of the hip rigors set in, with an anxious countenance, and much depression. There was reason to believe that pyæmia was beginning. Mr. Holmes, believing that the system was undergoing infection from inflammation of the femur, amputated the limb at the hip joint, and the patient did well. Abscesses should be opened early, and wounds should be kept scrupulously clean. The linen should be frequently changed, and the air of the ward be kept as pure as possible by thorough ventilation.* In most cases, however, the source of disease cannot be removed. The indications for treatment are to support the patient well with nourishment, concentrated, but at the same time in a digestible form, such as strong beef-tea and eggs; wine may also be given pretty freely, in quantities regulated by the amount of depression. In the way of drugs, I know of nothing that will do good except quinine in large doses. This has been at times of the greatest service, as in the case of Sadd. I should recommend its being given in full doses until tinnitus aurium and flushing of the face are produced. If there is pain and restlessness, opium should be administered.

Liquor potassæ, or the bicarbonate of potash, has been recommended on account of the power which the alkalies have of promoting oxidation. They appear to have been of service in some cases of chronic pyæmia. Mr. Savory recommends the combination of bicarbonate of potash with carbonate of ammonia. From its power of arresting fermentation and putrefactive changes, sulphurous acid, and its salts of potash and soda, have been recommended by Dr. Polli and others. They have not been found so successful as was anticipated.

* The use of carbolic acid is advisable as a disinfectant.

Nitro-muriatic acid, chlorine and chlorate of potash, have also been severally recommended on account of their oxidating properties; but I am not aware that they have done any good.

Otorrhœa, instead of leading to pyæmia, sometimes gives rise to abscess of the brain, as in the following case :—

Disease of Mastoid Process. Abscess in Middle Lobe of Brain. Thrombus in Lateral Sinus.

Wm. Wareham, æt. 9 years, son of a carman. His mother has had fourteen children, of which six are living. One of them is subject to abscesses; none of the others are subject to discharges from the ears or eyes. He has had measles and hooping-cough; and two years ago was cut for stone. About the end of November, 1863, he complained of violent pains in the back and front of his head; the pain has been almost constant, sometimes more severe than at others. About six weeks later the pain was referred chiefly to the back of the right ear; and he was occasionally delirious. A week later there was noticed for the first time a discharge from the right ear; it was at first thin, afterwards becoming thick; the discharge became very copious. He was admitted to the hospital on the 29th January. The following notes were taken on that day :—Lying asleep. Pulse 76, rather irregular. Complains when disturbed. No rigidity of his limbs. Pale, not very thin. Swallows well. Appetite good. Abdomen rather retracted. Chest rather flat; percussion note over apices of lung dull, equally on the two sides. Sounds on auscultation normal. A pause at the end of each inspiration; expiration attended with a groan. Pupils large, equal; no strabismus. Buries head under the clothes from dislike to light. Hearing with right ear impaired. Intellect is perfect; he is very irritable. Complains of frontal headache. At times he is very restless and wants to get out of bed.

Jan. 30th. He had a very bad night, being kept awake by paroxysms of headache. Pulse 68, regular. Skin cool. He took in the night 5 minims of laudanum without benefit. Is now to take ¼ gr. of extract of cannabis indica every three hours.

31st. Another bad night. Complained of not seeing this morning at 8 o'clock. Seems not able to see so well with right as left eye. Yesterday the right pupil was larger than the left; to-day they are of equal size. Pulse 76, irregular. Tongue clean. Appetite good. He still screams out very frequently from pain in his forehead.

Feb. 1st. Another restless night. Still seems to suffer much, but is not so noisy. At times the right pupil is larger than left. Pulse 80, weak, irregular. Urine pale, with a flocculent sediment of pus and pyoid cells.

2nd. A very noisy night; towards morning he appeard exhausted, and

CEREBRAL ABSCESS FROM DISEASE OF TEMPORAL BONE. 217

dozed between 10 a.m. and 1½ p.m. At this hour he was seized with opisthotonos ; the back was much arched, all the muscles of abdomen rigid ; arms stiff, close to sides, and shoulders carried forwards. Feet and legs extended in a straight line. Muscles of throat very slightly affected. Mouth a little open, left angle depressed. Right pupil dilated, left contracted. Pulse 160, soft, full and irregular. Temperature, 103·75°.
At 2·15 p.m. He could not be roused ; respiration stertorous.
At 3·45. Still comatose. Pulse 100, very strong. Face dark, lips purple, mouth foaming.
He died at 4 o'clock.

Autopsy.—On removing calvaria, dura mater separated easily ; convolutions appear rather flattened ; arachnoid sticky. On slicing the brain from above, the right optic thalamus is found to be higher than the left ; the greater portion of the right middle lobe is found softer and yellow, like rich custard. The fornix, septum lucidum, and walls of descending cornu of right lateral ventricle were disintegrated into shreds. On deeper section, the brain substance is replaced by a dirty greenish-yellow fluid, with shreds of brain substance in it. When the fluid portion is washed away by water, it is found to have been contained in a cavity about as large as a pigeon's egg, with tough walls mottled black and yellow. The walls of the cavity are with difficulty separated from softened brain substance around. On removing the brain, some delicate adhesions are found between the layers of arachnoid over the orbital plates. At the posterior part of the right temporal bone, there is a small point of adhesion, and the brain corresponding to this is black, like clotted blood, with a yellow tinge around ; this is continuous with the softened brain substance observed on the upper aspect. There is a small aperture in the dura mater covering the pars petrosa about the size of a small split-pea. There is a small collection of pus under the dura mater in the neighbourhood of this aperture ; matter also burrows down under the dura mater to the posterior fossa, and the membrane is thick and reddened. In the lateral sinus opposite this part of the pars petrosa was a pale coagulum filling up the sinus. The membrana tympani was intact, but the cavity was filled with a white substance like inspissated sebaceous secretion. There was a cavity in the mastoid cells the size of a pea, which opened externally and communicated with the canal occupied by the lateral sinus opposite the thrombus.

Heart, lungs, and liver healthy.
Kidneys flabby ; cortical portion wider than usual, cloudy.

Otorrhœa extending to base of Brain and Lungs.

A. H., a delicate girl, æt. 5 years, the second child of a phthisical mother. Since the age of 12 months she has had a discharge from the right ear, at times copious, often very fœtid. She was in moderate health until after Christmas, when, on returning from the country where she had visited her grand-parents, she was suffering from gastric disturbance. She vomited the skins of grapes, dried fruits, and oranges. Shortly after this an abscess formed over the right mastoid process. This pointed and was opened. There was a free discharge of fœtid matter from this wound and from the external meatus. From this time she had severe frontal headache for nearly three weeks, frequent attacks of vomiting, and occasional rigors.

Feb. 7*th*, 1865. She had a rather severe fit of shivering ; before and since that her skin had perspired profusely. She had intolerance of light and excessive irritability of temper.

8*th*. I saw her in consultation with Mr. Cooper Rose, of Hampstead. She was very drowsy, but excessively irritable when disturbed. Pulse quick. Was very sick in the morning, not so since. She had a short dry cough, and had several times complained of pain in the chest. Sometimes she eagerly took anything offered to her; at others she refused everything. It was impossible to examine her chest from her great irritability. There was little doubt that disease had extended to the brain from the temporal bone, and it was probable, from the cough and pain in the side, that secondary mischief had been set up in the lungs. I did not see the patient again.

14*th*. I had a report that the child seemed better. Her sickness was less frequent, but occasionally induced by paroxysms of screaming, which seemed partly due to temper and partly to pain. There was less shivering. Her skin moist, not perspiring much. She was very thirsty. There was less intolerance of light. No drowsiness.

She died on the 17th February; her death was not preceded by loss of consciousness, coma, or convulsions. The exact symptoms were not reported to me.

No post-mortem examination was made.

In these two cases there were violent pain and great irritability of temper, with some intolerance of light. In the latter case there had been chronic otorrhœa ; in the former there was disease of the mastoid process of two months' duration only ; it communicated with the external auditory meatus a fortnight before death, but not at all with the tympanic cavity.

Coagulation of blood in the cerebral sinuses may arise from other causes besides caries of the bones following otorrhœa ; it may be caused by caries from any other cause, or by injuries of the bones leading to hæmorrhage from the diploe, or by effusions of blood into the substance of the brain or its membranes. Coagulation or thrombosis may arise primarily in the sinus, as pointed out by Von Dusch (New Sydenham

Society, vol. xi., p. 125), by various influences which retard the current of the blood, or diminish the quantity of the blood, or impede the expansion of the lungs, thus preventing the right side of the heart from emptying itself.

This latter set of cases are distinguished from the former by the thrombus being usually in the superior longitudinal or straight sinus, by the firmness of the clot, and by the walls of the sinus not being diseased. As a result of this form of coagulation, there is often hæmorrhage in the brain and its membranes, but very seldom any secondary deposit in the lungs or other viscera, or any inflammation or abscess of the brain or the membranes.

Discharges from the ears are very common in children; they proceed from a variety of different conditions, but are spoken of under the one name of Otorrhœa. The discharge when chronic proceeds usually from one of the three following states: chronic catarrh of the dermoid layer of the external meatus, catarrh of the membrana tympani, or of the mucous membrane of the tympanic cavity. The first of these may lead on to the other two; or the tympanic mucous membrane may be first affected. In the latter case, instead of a small quantity of thin mucus, there is secreted copiously a viscid matter, which, being too abundant and thick to escape by the Eustachian tube, gradually fills the tympanum and causes rupture of the membrana tympani. So long as there is free exit for the discharge, the disease rarely extends to the cerebrum. But if the discharge is pent up in the tympanum, the bony walls become affected, and the membranes of the brain or the brain itself suffer. Even where there is free egress for the discharge, the dura mater and the bone may become diseased. These discharges from the ear are most common in strumous subjects, and very frequently follow scarlatina, and less commonly measles and small-pox.* Disease sometimes commences in the mastoid cells, and involves the squamous bone immediately

---

* Those who wish a complete account of these diseases should consult "Toynbee on Diseases of the Ear," and a Paper, by the same Author, in the "Med. and Chir. Soc. Transactions," vol. xxxiv. p. 239.

above and behind the external meatus and the posterior part of the wall of the tympanum. From these parts disease is propagated to the cerebrum. In the adult, from the different anatomical relations of the mastoid cells, disease extends to the lateral sinus and cerebellum, as pointed out by Toynbee. In disease of this part discharge often dates from birth, and is usually at first unattended with pain ; so that it is often disregarded. In its earlier stages it seems to be purely sympathetic, and proceeds from the meatus and outer surface of the membrana tympani. As the disease progresses the tympanum becomes filled with mucus or scrofulous matter, and the membrane is ruptured. If a thorough outlet is thus made for the discharge from the mastoid cells, and the health of the patient improves, no further mischief may result; but more frequently the free escape of matter is prevented, and cerebral mischief ensues. In some cases death results from cerebral irritation, without any appearance of disease in the brain, pia-mater, or arachnoid. The dura mater in different cases undergoes different degrees of change ; it is sometimes inflamed and detached from the bone, or thickened and softened, or ulcerated and more or less detached ; in other cases, it is atrophied and adherent to the bone. In a more advanced stage there are adhesions of the arachnoid. A very common further result is suppuration, either diffused or circumscribed, in the cerebrum or cerebellum. Abscess in the cerebrum may result from the disease in the tympanum or the mastoid cells ; in the cerebellum it is not common in young children ; in adults it results from disease in the mastoid cells. There is often healthy brain substance between the abscess and the diseased bone. Suppuration on the surface of the medulla oblongata has been occasionally met with ; it is said to be due to extension of disease from the vestibule or cochlea in the course of the auditory nerve.

Slight diseases of the ear require more attention than is usually given to them, from their liability to induce deafness, or to lead to purulent infection, or to cerebral mischief. The progress of the disease towards the brain is very insidious ; in many cases there is no evidence that the brain has become affected until acute symptoms set in ; the only symptoms are discharge from the ear and some deafness. There is sometimes to be found an unnatural sensibility on percussing the side of the head. Acute symptoms may be excited by various causes, such as a blow on the head, a cold, gastric disorder, stimulating applications, violent exercise, or any depressing influence ; there is sometimes no obvious exciting cause, the symptoms resulting from the course of the chronic disease.

One of the first symptoms of acute disease is cessation of the discharge, which is a result of inflammation ; it is frequently regarded as a cause of the cerebral symptoms instead of an effect of a common cause.

It is very important that attacks of ear-ache frequently recurring in infancy should not be slighted ; the exciting causes should, if possible, be ascertained and avoided. Careful attention to the general health, and avoidance of exposure to cold drafts, are especially to be enjoined. Warm fomentations, and occasionally a leech may be required. Counter-irritation may also be indicated. But each case must be treated, according to its peculiarities, by the surgeon. In chronic otorrhœa the same remark applies ; but there are certain general rules which may be given for all cases. The patients are generally strumous, and consequently require a most nutritious diet, plenty of fresh air, cod-liver oil, and steel. The ears should be frequently syringed with warm water ; mild astringent applications may be injected from time to time. If there is pain, one or two leeches may be applied ; and in obstinate cases iodine liniment (Ph. Br.) may be applied from time to time behind the mastoid process.

The symptoms of extension of disease to the brain or membranes vary very much. There is sometimes severe pain, and a great dislike to being touched. Convulsions do not often occur at the onset, but sometimes at a later period. Vomiting is occasionally present; intolerance of light in some cases. Delirium is not often present. There may be strabismus, unequal dilatation of the pupils, or these symptoms may be absent. There are sometimes rigors when abscess is forming, and often great prostration. As the cerebral mischief advances, stupor and coma may set in. From the inconstancy of the symptoms a diagnosis is often very difficult. The occurrence of any symptoms indicating cerebral irritation in a patient suffering from chronic discharge of the ear, especially if the discharge is arrested, is a circumstance which must always be regarded as very probably due to extension of disease from the ear to the brain. When acute symptoms have appeared, the case is nearly hopeless ; if, however, there be tenderness on percussing that side of the head, with slight fretfulness and pain, it may be advisable to establish and keep open a blister behind the mastoid process, and keep the patient very quiet, with his head raised, and on a simple diet, such as bread and milk and beef-tea ; the bowels should be gently acted on, and the ear fomented and syringed with warm water.

CHOREA.

THIS disease is characterised by irregular clonic spasms of the voluntary muscles, and by a partial loss of control over voluntary muscular movements. It is especially a disease of childhood, affecting twice as many girls as boys, and preferring the age of from 6 to 12 years for its first onset; but it is not limited to childhood. Of 422 cases treated as out-patients at the Children's Hospital, the numbers of each sex at different ages were as follows :—

		Males.	Females.	Total.
From 3 to 6 months		1	2	3
,, 6 ,, 12 ,,		1	4	5
,, 12 ,, 18 ,,		1	1	2
,, 18 ,, 2 years		1	3	4
,, 2 ,, 3 ,,		3	3	6
,, 3 ,, 4 ,,		6	5	11
,, 4 ,, 5 ,,		4	16	20
,, 5 ,, 6 ,,		7	23	30
,, 6 ,, 7 ,,		18	30	48
,, 7 ,, 8 ,,		17	34	51
,, 8 ,, 9 ,,		17	41	58
,, 9 ,, 10 ,,		23	57	80
,, 10 ,, 12 ,,		23	81	104
Totals		122	300	422

The number of cases under 2 years of age is higher than would be expected from statistics derived from other sources.

Of 37 cases, the ages at which the first attack of chorea occurred was noted, and the results were as follows :—At 5 years, two ; at 6 years, two ; at 7 years, ten ; at 8 years, six : at 9 years, nine ; at 10 years, five ; and at 11 years, three.

Amongst 174 in-patients, some of whose cases are first registered with the out-patients, the ages at which the cases of each sex came under treatment were as follows:—Between 2 and 3 years, one girl; between 4 and 5, one girl; between 5 and 6 years, five boys and ten girls; between 6 and 7 years of age, were six boys and eight girls; between 7 and 8 years, nine boys and fifteen girls; between 8 and 9 years, seven boys and seventeen girls; between 9 and 10 years, fourteen boys and twenty-nine girls; between 10 and 12 years of age, thirteen boys and thirty-nine girls.

Of the 174 cases, 54 were boys and 120 girls; considerably more than half came under treatment between 6 and 10 years of age, namely, 36 boys and 69 girls; between 10 and 12 years of age the proportion of females to males was higher than at earlier ages, 39 to 13, or 3 to 1. These numbers give results that agree very closely with what may be gathered from statistics on a larger scale; namely, that chorea is rare under 5 or 6 years of age, that it is very frequent between 6 and 10 years of age, attacking more girls than boys even at this age; that between 10 and 12 it is also very common, at this age attacking three times as many females as males. At the Children's Hospital patients are not admitted above the age of 12 years, so that we have no statistics as to the occurrence of chorea from 12 to 15 years of age, that is, about the age of puberty.

Of 100 cases occurring at all ages, tabulated by Hughes in the "Guy's Hospital Reports," (2nd series, vol. iv., 1846), 29 were between 12 and 15 years of age, 9 being males and 20 females; at 15 years of age there were 5 females and 1 male. One cause of the greater frequency of chorea in girls than in boys is their greater susceptibility to emotional excitement, especially fright. The excitability of their system, and the tendency to irregular local determinations also play a part in the production of the disease. The sudden suspension or the delay of the catamenia is sometimes coincident with the accession of chorea. It does not appear, however, that the period of puberty is more prone to the disease than the period between the second dentition and puberty.

From 6 to 10 years of age more than half the cases occur. The children attacked are very commonly characterised by vivacity and restlessness of disposition. They are, many of them, subject to headaches or flying pains in the limbs or stomach. It is not very often hereditary. M. Sée found records of 18 cases in which a choreic patient had either father or mother who was subject to chorea. A number of others had parents with tendencies to nervous diseases, and were subject to neuralgia, hysteria, paralysis, or madness.

Rheumatism and chorea are often found in the same family, as pointed out by Dr. Begbie and M. Sée. Of 37 cases I found 15 who had themselves been rheumatic, and 7 others one of whose parents was said to be rheumatic. In the earlier editions of their works, Barthez and Rilliet and Dr. West expressed doubts as to the frequency of the connection between rheumatism and chorea. In their later editions they admit that the two diseases often occur in the same patients, and Dr. West now believes that the rheumatic diathesis is a powerful predisposing cause of chorea.

I do not go so far as M. Botrel, who says that "it ought to be considered as a rheumatic affection, and finds its physiological cause in rheumatism of the nervous centres." But I believe in a very close connection between these diseases. In reference to this subject must be mentioned the frequency of cardiac disease in chorea; of 37 cases in my note-books there was probably organic disease of the heart in 25, and in 4 others there was evidence of functional derangement; whilst in 8 only was there no sign of cardiac disturbance. Now it is very well known that the commonest cause of heart disease is rheumatism. The younger the patient the more frequently is rheumatism accompanied by endocarditis. Articular rheumatism is often latent in childhood, its only symptoms being feverishness, with but slight pain in joints; whilst there are physical signs of endocarditis or pericarditis, with scarcely any subjective cardiac symptoms.

Endocarditis is commoner in children than is usually supposed. It is not always dependent on rheumatism; it may follow scarlatina or measles, or be independent of any other disease. It is sometimes taught that there is frequently to be met with in chorea a systolic murmur at the left apex of the heart, of *dynamic* origin, which cannot be referred to inflammation or organic change of the mitral valve or the inner surface of the ventricle. This murmur has been ascribed to disordered action of the musculi papillares. Another explanation has been suggested by Dr. Andrew (" Bartholomew Hospital Reports," vol. i., p. 28) for transient mitral systolic murmurs. He says that regurgitation may be produced by altering the shape of the mitral orifice from oval to round, and that irregular contractions of the heart may effect this change.

Murmurs of dynamic origin, not due to organic change, should disappear on the recovery of the patient from chorea. Such a disappearance of the choreic murmur is, in my experience, are. It is not uncommon, however, for the murmurs in chorea to vary in character with different beats of the heart, and at times completely to intermit.

Inasmuch, however, as the murmur does not usually disappear on the recovery of the patient, I think the cause is organic, but that the effect is modified in degree by irregularity of the heart's action. Such abnormal contraction of the musculi papillares as would produce a murmur with every contraction of the heart must be rhythmical, and would be altogether unlike other choreic movements, which are essentially without rhythm.

My own experience leads me to doubt the existence of dynamic apex murmurs in chorea; that is to say, murmurs produced in hearts entirely free from organic change. If such murmurs ever occur, they are certainly rare. Organic murmurs of the heart, on the other hand, are common in chorea, and I am inclined to believe that organic disease of the heart often exists in chorea when there is no murmur.

Vegetations on the auricular surface of the mitral valve frequently cause no murmur, and have been several times found in fatal cases of chorea. (See Dr. Ogle on this subject, "British and Foreign Medico-Chirurgical Review," January, 1868.) In 10 cases, out of 16 analysed by Dr. Ogle, there was fibrinous deposit or granulations upon some part of the heart's valves or lining membrane. In 6 cases there was congestion of the brain or spinal cord, in one softening of the cord, in one there was a peculiar dull yellow colour in portions of the grey matter of the cord, and in one there was softening in the central parts of the brain.

A systolic murmur heard at the second and third left costal cartilages, dependent on an anæmic condition of the blood, is by no means uncommon in chorea, and differs in no respect from the murmur of uncomplicated anæmia. This murmur disappears as the condition of the blood is improved.

Some of the most refractory cases of chorea are those in which there is heart disease. In fatal cases there is not unfrequently endocarditis or pericarditis. In 84 cases collected by M. Sée there were 16 of endocarditis or pericarditis; in 10 cases reported by Dr. Hughes there were 6 with endocarditis or pericarditis; in 2 fatal cases observed by myself there was endocarditis with mitral disease.

What is the link of connection between chorea and heart disease? Is it that rheumatism, a frequent cause of heart disease, is also a cause of chorea? Or is it that chorea depends on heart disease, which may be excited by rheumatism or in other ways?

The late Dr. Kirkes, in one of his last papers ("Medical Times and Gazette," June, 1863), proposed a very ingenious and plausible theory to explain this connection. He dwells upon the frequent presence of a systolic murmur in chorea, and states that valvular disease often exists without the presence of a murmur to indicate its existence, especially when granulations occur on the auricular surface of the flaps of the mitral or tricuspid valve. He also calls attention to the fact that a condition, which has been not uncommonly found post-mortem in fatal cases of chorea, is softening of parts of the cord or of the cerebrum, and occasionally intense capillary congestion of other parts, more or less simulating hæmorrhage. The patients who suffer from chorea are very impressible and emotional, and very liable to derangement of the nervous system. Putting these facts together, he asks, may not endocarditis be the first link in the chain, causing fibrinous exudation on the cardiac valves? Particles of this exudation are carried forward into the smaller vessels of the cord and brain, giving rise to deranged function in these parts, and sometimes leading to local softening of nerve tissue. A forcible objection to this theory is that we do not find symptoms like those of chorea produced by embolism, when it is proved to exist, or when experimentally induced in animals.

There are difficulties in the way of adopting any theory yet proposed as to its pathology. One main difficulty exists in the circumstance that there is no constancy in the post-mortem appearances which have presented themselves in patients who have died from chorea. The most frequent conditions have been changes of the membranes of the brain and cord, consisting either of recent or old exudation, or of simple serous transudation, the result often of pulmonary or cardiac disease, which was the immediate cause of death. Less frequently the remains of inflammatory products have been seen in the spinal meninges. Softening of the spinal cord or of the central portions of the brain have also been seen. Tubercle of the brain has been found four times. Of 84 post-mortem examinations recorded by Sée, in 32 only was it ascertained that any change was present in the central organs of the nervous system. In 16 of the 84, it is expressly stated that the nervous system was unchanged. It must be remembered, however, that changes inappreciable by the naked eye often occur in the nerve substance; and we may hope that with the advance of microscopic research in morbid anatomy, many nervous diseases (amongst others, chorea) now regarded as simply functional, will be shown to depend on organic changes. Dr. Lockhart Clarke has already made observations in tetanus which confirm this notion.

Romberg classes chorea amongst affections of the spinal

cord. Dr. Reynolds, on the other hand, thinks it is the brain and not the spinal cord which is at fault, for the following reasons :—1. Clonic spasm of the incessantly repeated character met with in chorea is not a phenomenon of spinal irritation, whilst tonic spasm is. 2. The movements in chorea are, in some measure, under the influence of the will. 3. The spasms cease during sleep, whilst excito-motor phenomena are increased by the withdrawal of the will ; fixing the attention on other objects also diminishes the intensity of choreic movements. 4. The special occasions of increase or induction of choreic movements are attempts at volitional action and emotional changes. 5. The phenomena of chorea appear to point to changes in the central ganglia of the brain. In health, changes in the optic thalami and corpora striata are expressed, through volition or emotion, in the balancing and co-ordinating of movements.

Amongst the exciting causes of chorea, *fear* is certainly one of the most frequent. It was assigned as a cause in 31 out of 56 cases collected by Dufossé and Bird, in 34 out of 100 cases reported by Hughes, in 25 out of 128 by M. Sée, in 9 out of 31 by Dr. Peacock, in 9 out of 38 of my own cases. The symptoms of chorea do not generally follow the fright immediately ; there is a period intervening, which varies from a few hours to a week or more. When the interval is a long one, it may fairly be doubted whether there is any real connection between the alleged cause and the disease. On the other hand, it must be borne in mind that chorea often sets in so insidiously that it may be present for some days before its existence is recognised. In some cases the occurrence which frightened the patients was of so trivial a character, that one must suppose that undue susceptibility previously existed ; occasionally we meet with cases in which the patient, already suffering from the earlier symptoms of chorea, has them much aggravated by exposure to fright.

There are, however, cases in which patients in apparently good health have been frightened by something really of an alarming character, and within a very short period the symptoms of chorea have shown themselves. A child recently died from chorea brought on by fright at the Fenian explosion in Clerkenwell. Amongst my own cases one patient was much frightened by seeing a person in a fit, another by a fall from a ladder, and a third by seeing her father beating her mother. Extreme joy is said to be sometimes the exciting cause.

It might have been expected that sympathy and imitation would be an occasional cause, as is the case in hysteria. I have never seen anything to confirm this anticipation. The disease is never induced by the assemblage of several choreic patients in a ward of children, nor does it appear that the symptoms are in any way aggravated by mutual association.

It has been said that chorea is more frequent in the autumn and winter than at other seasons. Of my cases, 13 occurred in 6 winter months and 14 in 6 summer months: 2 in January, 5 in February, none in March, 3 in April, 2 in May, 4 in June, 2 in July, none in August, 2 in September, 1 in October, 3 in November, 2 in December. Season in this country does not seem to have much to do with it. It is, however, much more common in damp and cold climates than in warm and dry ones; in this respect resembling rheumatism. It is said, however, that rheumatism is very common in Geneva, whilst chorea is very rare.

It is not easy to ascertain whether it is more prevalent in Paris or in London. Of 84,968 cases admitted to the Hôpital des Enfans Malades in 21 years, only 531 were cases of chorea, which gives the proportion of 1 to 161.

Amongst 122,621 out-patients treated at the Hospital for Sick Children in London, only 406, or 1 in 322 had chorea; of 5585 in-patients, 174, or 1 in 32 were cases of chorea. The mortality amongst the 174 cases, was 1 from chorea and 1 from diphtheria supervening on chorea.

Symptoms.—Some authors state that before the pathognomonic symptoms of chorea show themselves, there are certain premonitory symptoms; namely, loss of spirits, an exaggeration of the patient's natural timidity, increased irascibility, capricious appetite, impaired digestion, constipation, and tympanitic abdomen, with headache and flying pains in the limbs. These symptoms last usually a week or more before any of the distinctive symptoms are observed.

The peculiar movements of chorea very commonly begin in a gradual manner. In some cases their accession is quite sudden. The child is observed to have awkward, fidgety movements, recurring at frequent intervals; the parents often think the child has formed a silly habit. The movements are very often at first limited to one side; the upper limb is usually affected before the lower. In a few days the leg also becomes the seat of similar movements, and the muscles of the face soon become contorted into a variety of grimaces. Shortly afterwards the limbs on both sides are involved, and the muscles of the trunk. The characteristic movements in chorea are clonic spasms unattended with pain, something like the restless movements of a child put out of temper. They may be independent of any voluntary movements, but are usually aggravated by voluntary efforts, especially such as involve the co-ordination of a number of muscular movements. When such efforts are made there is a combination of the choreic spasms, with imperfectly executed voluntary movements. Emotion increases the irregularity and spasmodic character of the movements, the patient can generally do better when he is not conscious of being observed. The course of the disease and its intensity vary extremely in different cases. In aggravated cases he is quite unable to hold any object in his hand, or to walk. When asked to put out the tongue, he does so after a pause with a sudden jerk, and retracts it with equal suddenness. His articulation is impaired or rendered totally unintelligible; deglutition is difficult. The child has a vacant look, semi-idiotic; he is made to cry with scarcely any obvious cause. It is not easy, in severe cases, to say to what extent there is impairment of intellectual power. The grimaces of the patient and his inability to talk or to exercise his will in executing the simplest movement give him an imbecile and silly appearance, which inclines an observer to consider him idiotic. There is scarcely any way in which he can give evidence of such intellectual power as he retains. In less severe cases, however, it would seem that the intellectual power is impaired, the memory is weakened, and there is difficulty in fixing the attention to any subject. This impairment of intellectual power generally disappears with the abatement of the other symptoms, but in some cases there has been left permanent dulness of intellect.

A symptom not unfrequently met with in chorea is more or less complete *paralysis*, affecting those limbs which are most under the influence of the choreic spasms. The paralysis is sometimes the first symptom, as in Case 24; but more commonly it comes on later, and sometimes remains for several weeks after the choreic movements have subsided.

Dr. Todd considered the paralysis similar in character to epileptic paralysis. Trousseau says there are usually present pricking sensations, and more or less loss of sensibility on that side most affected by the chorea. Impairment of sight has been observed by Trousseau and Sée. There is nearly always more or less constipation, probably from impaired peristaltic action of the intestines, and very rarely there has been seen loss of power over the sphincter ani, as in Case 5.

The movements of chorea usually cease completely during sleep, though in aggravated cases the sleep is disturbed or almost entirely wanting; this loss of sleep may lead to delirium, coma, and fatal exhaustion.

In some cases the movements are so violent that there is the utmost difficulty in preventing patients from inflicting injury upon themselves by blows against the bedstead or throwing themselves on to the floor. In these cases bruises and sores are often formed from the blows inflicted, and these occasionally give rise to erysipelas, abscesses, and sloughing of the skin over projecting bones.

An intercurrent febrile affection will often put an end to chorea which is previously subsiding, whilst it will aggravate chorea which is at the time on the increase. In some cases the chorea will subside as the fever gets better.

Diagnosis.—From the greater chorea of the Germans or the

DIAGNOSIS OF CHOREA. 233

great dance of St. Guy, it is easily distinguished. That disease is more allied to hysteria ; it is manifested at intervals of several days. Between the attacks the patients have entire control of their muscles. The spasms are not induced by attempts at voluntary action ; there are both clonic and tonic spasms, and they are often attended with catalepsy. There is often mental aberration, ecstasy, and during the attacks insensibility to external impressions. The attacks occur in violent paroxysms, and are induced by irritation or by acting on the imagination in any way. Mesmerism induces such attacks at times. There are other nervous conditions which are characterised by rhythmic movements, and called "chorée" in France ;—a tendency to run forwards, to rapid rotation of the head, one of the limbs, or the trunk ; a tendency to irregular oscillation of some part ; or to repeat the same word an indefinite number of times. These conditions are quite different from ordinary chorea.

From hysteria and paralysis agitans, also, it is easily distinguished.

There is a peculiar nervous twitching, having some resemblance to chorea, and often called *partial* chorea, which differs entirely from the disease under consideration. It is called *muscular tic*. It consists in a rapid contraction, uniform and involuntary, attacking constantly one muscle or set of muscles, especially those of the face. It is quite beyond the control of the will, and not increased by voluntary effort. It is generally incurable ; it is not attended with ill health, and is of no consequence except from the peculiarity of appearance caused by it.

The average duration of cases of chorea is differently stated by different authors. Sée gives it as 69 days. This is very near what was observed in 30 attacks under my own observation. The mean was about 10 weeks, the longest was 28 weeks, and the shortest 2 weeks. As a general rule, primary attacks last longer than relapses ; the average duration of primary attacks in my cases, including a few which were very protracted, was $12\frac{1}{2}$ weeks. It would seem that it is more difficult to cure chorea in cold, damp weather, than in fine weather.

Occasionally chorea is acute, and proves rapidly fatal. I have notes of two such cases. Sée reports 9 deaths in 158 cases, or about 6 per cent., which represents a very high mortality. Occasionally chorea becomes very chronic; it is then usually limited to one set of muscles, and is not severe.

Treatment.—It is very difficult to judge of the relative efficacy of different modes of treatment in a disease which, if left to itself, usually gets well in the course of a few months, and which does not very suddenly subside under any plan of treatment. It is necessary to consider at what period of the illness the treatment is commenced; whether the patient is suffering for the first time from the disease, and what length of time the attack lasts altogether. There is no doubt that amongst the poor, hospital régime, without medicine, is of use. This is due probably to regularity of living, better food, the absence of undue excitement, and a regulated temperature. A nutritious diet alone will often go far towards curing the patient. This must be always borne in mind. On this account I generally wait a few days to see whether the patient is improving or getting worse before I administer drugs. A great number of remedies have been vaunted in chorea, but when tried by others than those who introduced them they have failed to do much obvious good.

Occasional purgatives and tonics, especially steel, were attended with as much success as any plan of treatment which has been tried in a number of cases. I have also seen good results from the employment of baths of sulphuret of potassium—4 ounces to each bath. I have tried them in 8 cases, and in 5 with decided benefit. Sée states that of 57 cases in which these baths were used, 50 were cured in an average space of 22 days.

Gymnastic exercises are strongly recommended in France as a means of treatment. Sée states that of 22 cases thus treated 18 were cured in an average space of 29 days. M. Blache gives more striking results ("Mémoires de l'Académie de Médecine," vol. xix. pp. 598—608): 108 patients were treated by gymnastics alone, without drugs. In 34 cases the chorea was of moderate severity, in 74 of extreme violence. All of the former class recovered in an average period of 26 days after 18 series of exercises. Of the 74 severer cases, 68 recovered in 45 days after 31 series of exercises, and the other six in 122 days, after 73 series of exercises.

The mode of proceeding adopted in these cases is thus described. The child lying in bed on its back is held perfectly still for 10 or 15 minutes by the teacher of gymnastics and several of his pupils. A series of kneading or shampooing movements are then commenced with the open hand upon the limbs and over the chest, afterwards along the back, and especially the back of the neck and the muscles in the intervertebral grooves; these movements are followed by rapid frictions of the same parts. About an hour is thus employed, and the process is repeated every day for three or four days in succession. The shampooing process is still partially continued, and the limbs of the patient are made by the operator to perform a series of regular and perfectly rhythmical movements for some time. Thus, whilst the patient is still lying on a mattrass, with his arms extended by his sides, the operator takes hold of his wrists, and bends the forearm on the arm, raises the arm, extends the forearm, and by reversing the movements again, replaces the arm by the side of the child. In the same way the leg is bent on the thigh, and the thigh on the trunk and again replaced, and these movements repeated for a number of times in succession. At the end of 10 or 12 days the child is allowed to go into the gymnasium, and is then exercised by the professor of gymnastics with other children.

I have frequently prescribed gymnastics, and with benefit; but I have not resorted to the preliminary shampooing and passive movements which are such an important part of the French method of treatment. The children at the Hospital in Great Ormond Street are trained in gymnastics set to music on the system followed by Dr. Dio Lewis of Boston. The great points to be considered are to exercise as great a number of the muscles as possible without fatiguing any of them.

Shower baths are useful in the later stages of the disease, when the patient is not too timid and too much excited by them.

Arsenic in full doses has been strongly recommended by Begbie and others. I have given it an extended trial, and am satisfied that it is a valuable remedy in a fair proportion of cases ; whilst in some cases it entirely fails.

Of 19 children to whom I gave it, and noted the results, it appeared to be of service to 12 ; and the mean duration of treatment in these cases was 14·7 days.

Narcotics, such as opium, belladonna, cannabis indica, or conium, are of little or no use. Antispasmodics, such as assafœtida, or valerian, are also useless.

Strychnia, strongly recommended by Trousseau, appears to me to be injurious in the acuter forms of the disease ; in the more chronic form, and when there is a tendency to paralysis, it has been of service.

Zinc in rapidly increasing doses is much praised by some, but I have not found it answer my expectations.

Iodide of potassium, when the patient is subject to chronic rheumatism, may be useful.

In all cases the clothing should be warm, the diet nutritious and easily digestible. When the pulse is weak, and appetite not good, wine is indicated. If the patient be of tuberculous family, or of spare habit, cod-liver oil is useful, with or without camphor dissolved in it.

When the weather is fine, open air exercise should be enjoined. Excitement of any kind must be avoided.

Dr. West speaks highly of antimony in full doses in acute cases of chorea. It is sometimes tolerated in very large doses without doing any good.

MM. Gillette and Roger gave 3 grains of tartar emetic on the first day of treatment, 6 the second, 9 the third ; then allowing a pause of from three to five days, and recommencing with 4 grains the first day, 8 the second, 12 the third, and so on, after a pause of three or four days for another period of three days. It is a plan of treatment to which I have strong *à priori* objections, which have been rather confirmed by a slight experience. Chorea is certainly a disease of weakly ill-nourished children; and it does not seem likely that such a depressing agent as antimony can be useful.

CASE I.—*Chorea, preceded by Rheumatic Fever. Decided improvement in 9 days from arsenic.*

G. Wigger, a boy, æt. 9 years, tolerably healthy. In December, about the 11th, he had a cold, attended with very severe pain in his knees, which were swollen, but not reddened. His skin was hot, and he was thirsty. He kept his bed three days, and at the end of this time he seemed helpless; could scarcely move his limbs. There was some involuntary twitching of them. This symptom became worse for a fortnight, implicating his upper and lower limbs. He came under treatment on the 11th of January, and was ordered liquoris sodæ arseniatis ♏x. in aquæ ℥ ss. ter die; with a powder of jalap and calomel at first. On the 16th the dose of arsenical liquid was increased to 12 minims.

Jan. 20th. Much better. Can now walk. Tongue clean and tolerably steady. Skin dry and scurfy. Pulse 108, hardly regular. At the apex-beat of the heart there is heard a blowing systolic murmur, which is also faintly audible at the inferior angle of the left scapula.

26th. Discharged well. The murmur, however, remains.

CASE II.—*Chorea, rather severe, not traceable to Rheumatism. Treated with arsenic; improved in 16 days. Cured in 30 days.*

E. C., a thin, sickly-looking girl, came to the hospital on the 31st of October with severe chorea, affecting the right side more than the left. She had been suffering from it for six weeks. No cause could be assigned for the attack. She could not hold her tongue out of her mouth. Her pupils dilated equally. Pulse 92, slightly irregular. Nothing abnormal observed about heart. She was ordered 4 minims of Fowler's solution three times a day, which was increased after 5 days to 6 minims. By the 16th November her symptoms had much abated. The dose of Fowler's solution was reduced to 3 minims, and on the 20th it was omitted and followed up by steel wine. On the 29th November she was discharged quite well.

CASE III.—*Chorea in a Child subject to Epileptic Fits. Cured in a month by arsenic.*

Emily Usher, æt. 8 years, has been subject to fits since the age of 2 years. Her mother states that she has four other children living, all of whom except the baby have been subject to violent convulsive attacks between the ages of 2 and 5 years. The last fit which Emily had was about 8 months ago ; in it she was strongly convulsed for 4 hours, was livid and foamed at the mouth. It was followed by stupor and profuse sweating, which lasted several hours. She has never been strong ; her appetite has always been bad. The week before Christmas last she was rather suddenly seized with great unsteadiness of the left arm ; her left leg soon became similarly affected, and afterwards the limbs of the right side. She became gradually worse, until the 4th February she could hardly walk ; her speech was quite indistinct. She had pain in her shoulders and hips. On the 8th February she came under my care. The choreic movements were not very violent. Pulse 96, not quite regular. The first sound of the heart at the apex is somewhat murmurish. Her pupils are not quite equal, left rather larger than right. She was ordered 10 minims of the liquor sodæ arseniatis three times a day.

Feb. 20th. Her chorea was decidedly better. She was ordered 2 grains of compound aloes pill every night.

March 8th. Her chorea was all but cured. Heart's sounds as before. She was discharged to-day.

In this case, although the patient is said not to have been rheumatic, there was pain in the joints during the course of the disease.

CASE IV.

The case of E. W. is an illustration of a more obstinate form of the disease. The patient had suffered from the disease three times. Her first attack had lasted many months, and the third, for which she was treated by me, lasted nearly four months ; arsenic, sulphur baths, gymnastics, steel, shower baths, and strychnia were tried in succession without any good result. Strychnia seemed to aggravate the symptoms. She was ultimately cured by country air.

CASE V.—*Aggravated Chorea ; severe pains in the limbs, much increased by movement. Numerous periosteal swellings rapidly formed on the head and rapidly absorbed; enlargement of lower end of femur. Recovery.*

F. W., the son of a surgeon, aged 5 years. He had always been a delicate child ; naturally fidgety. He has one elder brother who is strong and healthy, and two children younger than Frederick are in good health. I first saw him on the 17th November. His mother stated that six weeks ago he was frightened by a dog. Three weeks later he was walking in the street, when he said he should be blown away (there was not much wind at the time). His body was drawn to one side, and he became red in the face. His mother thought he was going to have a fit. From that time he has been noticed to be very awkward in all he does ; this has much increased till the present time. Now he has general choreic movements. His speech is much impaired. At times he can speak distinctly, but more frequently cannot make himself intelligible. His appetite is good, but sometimes he can scarcely swallow anything ; the food is jerked out of his mouth by his tongue. He has a very silly, idiotic look. Has complained often of pain in his teeth ; I can see nothing amiss with them. I ordered calomel and scammony powders, and steel wine.

Nov. 23rd. Not any better. I gave him camphorated cod-liver oil ʒij. three times a day, and hydrarg. chloridi gr. iij. alt. mane.

Dec. 3rd. Worse. Cannot sit up at all. If taken up by his arms, without very carefully supporting his legs, he has severe pain. Is constantly grinding his teeth. Has lost flesh. His legs are constantly in motion. Says no words but "yes" and "no." His appetite is not good. Can swallow if his food is put well to the back of his mouth. His bowels are very costive. He is often very sick. The pulse is very quick ; his heart's sounds are free from murmur *Under his scalp are a number of hard round lumps*, about the size of peas and horse beans ; the skin moves over them, but they are firmly attached to the bone. They are neither red nor tender on pressure. I gave him a purge of calomel and jalap ; to go on with his camphorated oil.

7th. Chorea better ; but he is very feverish at night, and the swellings on the head have much increased in number. There are now numerous smaller elevations, smaller than peas, as well as the larger ones. They are quite hard ; are not tender on pressure. There is a swelling on the inner condyle of his femur. Dr. Jenner saw him with me, and feared that these growths might be malignant. The other suppositions were that they might be due to syphilis (which is negatived by the history) or to rheumatism. In favour of this theory is the circumstance that there is now a loud systolic murmur at the heart's apex. He is now to take potassii iodidi gr. ij., potassæ bicarbonatis gr. iij., ter die ; and hydrarg. chloridi gr. ij., jalapin gr. ij., pulveris jalapæ gr. v. as an occasional aperient.

11th. On the 9th he had a severe convulsive fit, followed by sleep and stertor. The swellings on the head are larger. Chorea is better ; he sleeps better. His bowels are very costive.

21st. Some of the lumps on his head have disappeared. His knee is still enlarged, and the lower end of his right ulna is enlarged. He last night complained of pain in his thumb. He is losing flesh ; is sometimes feverish at night, but not always. Appetite very capricious ; for two or three days he was very sick, rejecting almost everything he swallowed. His bowels are still very costive. Chorea is sometimes better and sometimes worse. Syrupi ferri iodidi, ℞xx.; potassii iodidi, gr. j.; aquæ, ʒij. Ter die.

27th. Many of the smaller swellings have disappeared; there is one on the occiput which has much increased in size, and is now tender, but still hard. There are only five distinct lumps on his head now. There is also tenderness near the clavicles and in his fingers. Knee and wrist less swollen. Sickness has ceased. Bowels confined. Chorea less. Urine turbid. He has some pain in passing his water; this is an old symptom with him.

Jan. 7th. A good deal better. Chorea nearly gone; sleeps better. He can now stand, and bears to be lifted without screaming. He can talk. He cannot retain his motions when the desire for relief comes. Still has pain in micturition. All the swellings are gone except on the occiput, where the larger one is split up into several about the size of peas. There is no tenderness now. Murmur audible at apex-beat as before.

22nd. His mother wrote to me. " My little boy is much better. The lumps on his head are, I fancy, quite gone. They at one time increased very much. We gave him the prescription advised on the 7th December, and it has proved most beneficial. He can manage to walk now, but the chorea has increased. Any exposure to cold brings the pains back in his arms and legs." I prescribed liquor potassæ arsenitis, ℞ ij.; potassæ bicarbonatis, gr. iij.; potassii iodidi, gr. ij; aquæ camphoræ, ʒ ss. Ter die.

Feb. 18th. Better. Some chorea and want of power in left arm. A loud systolic murmur at apex, which beats a little outside nipple line; also heard at angle of scapula. Temper still irritable.

This case is a remarkable one from the occurrence of the swellings on the head, and near the knee and wrist. They must be regarded as due to rheumatic periostitis. It could not be ascribed to syphilis, for the patient never exhibited in infancy any signs of the disease; his father denies ever having had syphilis in any form, and the first-born in the family is a strong healthy child; his mother has had no still-births or miscarriages, and there is not the shadow of suspicion that she has suffered from syphilis.

Short Abstract of Cases of Chorea.

VI. E. P., a girl, æt. 12. The fourth attack in six years. (Her mother has had rheumatic fever.) Her first attack was in January; the second, a slight one, in the month of May; the third began in summer, lasted five months, and was most severe. The present attack commenced in the right arm. There was a systolic murmur at the base of the heart and an occasional murmur audible at the apex. Treatment: steel and aloes.

VII.—S. E. L., a pale girl, æt. 10, had rheumatism seven weeks before. Choreic movements began at the end of February; at the same time the patient complained of pain as if her hair was being pulled out. (*This symptom was also noted in Case* XVI.) Systolic murmur at apex of heart, which remained. Treated with steel. Recovery in 8 weeks.

VIII.—G. H., a boy, æt. 11. Had acute rheumatism seven months before. Chorea began six weeks before admission in April. No heart disease. Treated with baths of sulphuret of potassium for nine nights, followed by steel. Recovery in 14 days.

IX.—E. W., a girl, æt. 8. A delicate child of tubercular aspect, with a consumptive father. Eleven weeks ago frightened by a man who nearly fell on her in the street. In four days later choreic movements were first observed. They became worse and worse for six weeks. Speech impaired; could not walk; sleep disturbed; lost flesh. Systolic murmur at apex; left side more involved than right. Treated before admission with cod-liver oil and steel wine for 24 days without benefit. Then steel wine and tincture of hemlock. When admitted, sulphuret of potassium baths with occasional aperients. Improvement in a few days; baths continued 13 days, followed by steel wine. Recovery in 14 days after admission. Heart murmur persisted.

X.—E. D., a girl, æt. 7. Mother rheumatic. Chorea came on gradually four weeks before. Treated as an out-patient with a dose of antimony followed by warm bath every night; and spirits of nitrous ether ten minims in camphor julep, three times a day. Cured in 14 days by Vinum ferri and hospital régime.

XI.—A. K., a girl æt. 11. Family very phthisical. Sudden attack without known cause. Admitted four days later. Heart's action irregular; apex beat outside nipple; systolic murmur. Baths of sulphuret of potassium for nine days with great benefit. In three days afterwards, well. Total duration 18 days.

XII.—R. S., a boy, æt. 9. Had "brain fever" three years ago. Fourteen days has had pains in bones with choreic movements; sleep disturbed. Heart's action irregular, an intermitting, musical, systolic murmur. Sulphuret of potassium baths for 18 days; then Vinum ferri and decoction of aloes, 1 month; Liquor arsenicalis, 18 days; sulphate of zinc, 1 week; then iodide of potassium and bicarbonate of potash for 14 days. Not much improvement. Sent to the country for a month. Came back well. Heart murmur remained.

XIII.—A. K., a delicate girl, æt. 11, with tubercular tendencies. Six weeks before admission chorea began, much increased 10 days ago by a fright. Sulphuret of potassium baths every other day for 12 days with much benefit; a purge every third day at the same time. Steel wine for a week. Recovery in 19 days.

XIV.—*An attack of Chorea, proving fatal by intercurrent diphtheria.*—M. P., a delicate girl, æt. 10. Mother rheumatic. Nine weeks ago the chorea began; right arm worse than left; partial paralysis. Speech impaired, and then quite lost for some time. Difficulty in deglutition. Face rather livid, mouth open, rolling of eyes; convulsive working of corners of mouth. Heart's apex an inch outside and half an inch below nipple; no murmur. Action irregular. Tonsils pale and swollen. Slow improvement on the use of arsenic. Recovered power of speech slowly. In hospital 2 months; sent to the country for a month. Came home well in December. Soon after broke her right arm. When arm was nearly well chorea returned. Readmitted on 9th February, almost as bad as ever. Now distinct murmur at apex with systole. Treated with belladonna and sulphate of zinc; then with hot air baths, and antimony in small doses. 6th March: Tonsils large, pale, and painful. 12th: Arsenic and quinine. 22nd: Not any better. Vin. ferri, ʒij.; strychniæ, gr.$\frac{1}{32}$. Ter. die. 27th: Increased dose of strychnia to gr. $\frac{1}{24}$. 31st: Diphtheritic patch seen on tonsils. Died on 4th April from exhaustion and extension of diphtheritic membrane to air passages. *Autopsy.*—Chronic adhesion in pericardium, chiefly on posterior and outer aspects. Tricuspid and mitral valves diseased, studded with vegetations on their auricular aspects. Echymoses under peri- and endo-cardium. Weight of heart 5½ oz. Sinuses of dura mater contained coagula. Vessels of pia mater congested. No softening or other change in brain. Diphtheritic membrane extending from larynx into the bronchi.

This case is interesting in many aspects. The co-existence of paralysis of right arm, with loss of speech, was noteworthy in connection with Broca's observations, and those of other recent observers which show the frequent connexion between aphasia and right hemiplegia. The vegetations on the cardiac valves correspond with Dr. Kirke's and Dr. Ogle's observations. One cannot help thinking that the child's death was caused by the strong tendency in the blood to deposit fibrin, as seen on the valves of the heart, interlaced with the meshes of tricuspid valve, in the cerebral sinuses, and in the air passages. The lungs exhibited a condition suggestive of an early stage of embolism, very similar to what is seen at any rate when small foreign particles have been introduced into the veins, as insisted on by Mr. Savory in his articles on pyæmia. The section of pulmonary substance was of a bright red colour studded with a number of darker spots, looking almost like an early stage of apoplexy, but found to be due to dark fluid blood which can be removed by pressure. There were several spots of subpleural ecchymoses, varying in size from pins' heads to split peas.

CASE XV.—*Another fatal case of Chorea, not in the Children's Hospital.*
C. C. R., a boy, æt. 14. Father rheumatic; mother excitable and nervous; a cousin had chorea. Seven years before had chorea lasting 3 months. Three or four months ago had scarlatina followed by rheumatism. Was frightened on a ladder; two days later chorea began, chiefly in left arm.
Nov. 28th. Very violent choreic movements; not ceasing completely in sleep. Heart's action irregular; slight murmur at apex. Treatment at first, strychnia, which appeared to aggravate symptoms. Then tincture of conium, followed by large doses of assafœtida. With this belladonna was combined, internally and as a plaster to the spine. He became worse; had no sleep; skin of nates and elbows became inflamed from friction. Lips became dry, tongue brown. Fancied he saw frightful objects. Died Dec. 2, after an attack of dyspnœa.
Autopsy.—Small red spots on right arm and thigh. Slough on back. Sinuses of dura mater full of blood; some of the veins inflamed. Old lymph in pericardium near the origin of the great vessels, with ½ oz. of red serosity. Left ventricle hypertrophied. Mitral valve studded with vegetations. Spinal cord softened for 2 inches in the middle of the dorsal region. Pia mater injected. *Brain:* grey matter generally too red. Red points on section too numerous. Grey matter of pons varolii very vascular, also of corpus striatum; dark in tint, becoming redder on exposure. Cerebellum also of an unusually dark tint.

XVI.—R. D., a delicate girl, æt. 10. Had been subject to a barking cough, and to becoming blue when cold. Two months ago, two days after being frightened, choreic movements began; and has since had frequent palpitation of the heart and panting. A loud systolic murmur audible both at base and apex. Some hypertrophy. Complains of pain as if her head was tied with a cord and her hair being pulled. Improved at first with cod-liver oil; at the end of 6 weeks almost well; remained 3 months longer without being quite free from chorea.

XVII.—L. W., a girl, æt. 5. Mother's family consumptive. A delicate child; intellect not very bright. A month ago choreic movements began soon after being frightened by her father beating her mother. No heart affection or rheumatic history. Treatment: sulphate of zinc, arsenic, and strychnia, in succession. Very slow recovery, lasting 4 months. Relapse in 2 months. Treated with baths of sulphuret of potassium. Recovery in a month.

XVIII.—W. P., a boy, æt. 7. Mother and grandmother rheumatic. Slight chorea, lasting 6 months before I saw him. Whilst recovering from chorea subacute rheumatism set in, during which aloud systolic murmur audible at apex, and chorea was in abeyance. On recovering from rheumatism the murmur became inaudible, and slight choreic movements were observed.

XIX.—E. B., a boy, æt. 11. Third attack of chorea. The first, 2 years ago, lasted 4 or 5 months, coming on suddenly without apparent cause. Second attack, 10 months ago, lasted 2 months, during which had pain in left leg and arm. A loud systolic murmur at apex of heart. Treatment and result not noted.

XX.—E. N., a delicate girl, æt. 9. Her second attack. Is occasionally subject to painful swellings of knees and ankles. Heart's action irregular, without murmur.

XXI.—W. S., a boy, æt. 9, who had acute rheumatism last winter. Six weeks ago had pain in joints for one week, which was immediately followed by chorea. No heart murmur. Well in a month.

XXII.—E. D., a delicate girl, æt. 9. Father a "rheumatic" subject. Has had frequent headaches. One night in September, 1862, was suddenly attacked with "twitching and snatching," and in the morning could not dress herself. Her right side was more affected than the left. She was under homœopathic treatment, and continued ill 5 or 6 months. In November, 1863, a second attack of chorea, which came on gradually; the left side worse than right. Speech impaired for a short time. Admitted January 23. Heart's sounds normal. Treated with purgatives, cod-liver oil, and steel wine. Well in 14 days.

XXIII.—M. A. T., a delicate girl, æt. 10½. Her father has "rheumatic gout." All her brothers and sisters are delicate. About 4 weeks ago, without any obvious cause, she began to fidget and to talk oddly. For one day legs seemed stiff, and when they were touched, she screamed and said she was hurt. Left side more affected than right. Heart's first sound at apex, murmurish. Second sound reduplicate over pulmonary cartilage. Treated with liq. arsenicalis. Well in 14 days.

XXIV.—E. C., æt. 9. A miserable-looking child, of weak intellectual power. Was well until Nov. 1st, when she had pains in her limbs and chest, with cough and feverishness. On Dec. 17th, her mother noticed she could not use her left arm; afterwards could not use left leg. There was no unsteadiness or jerking observed. When admitted on Jan. 1st, used left arm very imperfectly; walked very unsteadily, protruded tongue and held it between the teeth in a way not unusual in chorea. The heart's impulse was very heaving and visible in the 2nd, 3rd, and 4th costal interspaces; and there was a distinct systolic murmur audible at the apex. Under the use of quinine and a liberal diet, she was nearly well on the 2nd February. She remained in the hospital another fortnight, at the end of which the chorea and paralysis had disappeared, but the cardiac murmur remained.

PARALYSIS.

CHILDREN are liable to suffer from nearly every variety of paralysis which occurs in adults. The varieties, however, which are most frequent in adults are rarest in children, and there is one variety which is very common in infancy but almost unknown in adult life. Paralysis is a term which includes loss of sensation as well as loss of motion. I propose to treat especially of paralysis of motion.

Motor paralysis is caused by a number of distinct conditions:—1. An organic change in the muscles, as in inflammation of the muscular fibre and possibly in progressive muscular atrophy. 2. An organic change in the conductors of motor impulse, the nerves supplying muscles, part of the cord or brain; either of these may be affected by inflammation, hæmorrhage, or degeneration. 3. An organic change in that part of the brain which is the organ of the will; this condition and the preceding usually go together. 4. A morbid change in the blood such as may be induced by typhoid fever, scarlet fever, measles, diphtheria, or anæmia. 5. By a reflex influence transmitted from afferent to efferent nerves and thence to muscles, or from some part of the brain or cord which is the seat of irritation. Dr. Brown-Séquard has especially insisted on the fact that paralyses of cerebral origin are sometimes reflex in their nature, paralysis not being due to that portion of brain which is the seat of lesion having lost its function, but to the transmission of irritation from the morbid part to some other portion whose function it is to control muscular movement. The fourth and fifth named conditions are often combined.

The cases of paralysis to which I wish especially to refer

(because very frequent in infancy and almost unknown in adult life) are probably due to disease in the spinal cord. The names applied to these cases have been "essential or idiopathic paralysis," "infantile or spinal paralysis," "atrophic paralysis," or "paralysis of children combined with fatty degeneration of the muscles." Cases have been described by Kennedy under the name of "temporary infantile paralysis," presenting almost identical symptoms at the outset, but followed by complete restoration within a month or six weeks. The two following cases may be taken as specimens of this disease.

A boy, F. P., son of a tailor, was in very good health until 12 months of age. He began to walk at the age of 10 months; suddenly, whilst walking, his left leg gave way under him. He was brought to me at the Hospital for Sick Children when 16 months old. The left lower limb was completely powerless, falling about in the most lax manner possible. The muscles felt flabby, but there was no distinct difference in size between this limb and the right one. It was usually colder than the unaffected leg. He only remained under my care for 1 month, during which no alteration was observed.

Anna M'C., aged 16 months, was brought to me with the following history :—

She was born a strong child and thrived well until 11 weeks old. Whilst at her mother's breast, the child seemed to the mother to get heavier, having less spring. She looked, however, brighter than usual. After this the child could not sit up as before. Two days later she was taken to a hospital, when it was found that her legs had entirely lost their power. The child since then has grown, and, with the exception of her lower limbs, is healthy. The legs are still both of them paralysed, flabby, and not so well nourished as the arms. The left calf and left nates are somewhat more atrophied than the right. She cannot bear the least weight on her legs. She can straighten the right leg on the thigh when the thigh is held firm by another person. When the upper part of this leg is held she can move the toes. She can throw about the left lower limb, moving at the hip-joint, nursing it as a doll under the arm of the other side. The left foot is extended on the leg, and cannot be flexed; when this foot is held she can bend the toes. She has only been able to do this lately. Her legs do not suffer much from cold.

These two cases illustrate very well some of the features of this affection, the suddenness of the attack in a healthy child, the complete paralysis of motion, involving at first all the muscles of the limb, in the course of time becoming localised to some muscles only, and leading to atrophy, more or less loss of temperature, and in some cases deformity, from the action of muscles which recover being unopposed by their antagonists, which still remain paralysed. In these cases no fever was observed at the commencement; it is present in the majority of cases, but is often overlooked by the mother or nurse, especially in the case of the poor, who have not much time to devote to their children.

The children, when attacked by the paralysis, are usually in excellent health and robust. The loss of power is sometimes ushered in by convulsions or delirium, but more frequently by a simple feverish attack, occurring quite suddenly and lasting only a few hours or from 2 to 14 days, with, or more frequently without, vomiting or diarrhœa. Occasionally paralysis is observed after a night's rest without any premonitory symptom, or even during the day without any warning. The age at which the attacks occur is most frequently between 6 months and 3 years, the period of primary dentition. The feverishness, when present, does not always subside immediately on the appearance of the paralysis, it may continue for several days later.

Occasionally paralysis appears in the course or as a sequela, of an eruptive fever, scarlatina, measles, or typhoid, for instance. Severe pain is sometimes present in the paralysed parts at the time of the attack and for a varying period afterwards. This symptom often leads to the suspicion of an injury having been inflicted by a fall, a blow, or a strain, a suspicion which is usually quite unfounded, as in the following case.

William N., æt. 18 months, son of a labourer. His father and mother were healthy, and he appeared quite robust and well until he was 6 months of age. He was left in charge of an aunt, and on the mother's return he was found screaming. His aunt stated that he had not met with any accident, and that nothing had happened which could explain his screaming. On moving his legs his pain was increased; from that time until now he has not been able to move his left lower limb. The limb has lain quite helpless, and could not be excited to movement by tickling the foot or by voluntary effort. For three weeks the child cried when this limb was moved. There was no swelling or redness ever to be seen. Since that time the limb has wasted; it is usually colder than the other. The right thigh now measures 8¼ inches, the left only 6¾; the right leg 6, the left leg 5 inches. The child's general health is good. He cannot use the left lower limb at all, it is quite flaccid.

The pain which attended this case at its commencement was unusual in its intensity and its continuance. Some may be inclined to think that the child met with an injury during his mother's absence, but it would be difficult to account for the symptoms even on this supposition. Other cases are recorded by authors in which there has been tenderness or pain on movement of the paralysed parts where no accident could have occurred to the patient. Laborde mentions a case where this was combined with tenderness near the spines of the lumbar vertebræ, which lasted two months, and rigidity of the lumbar muscles. In one case that came under my care the paralysis was preceded six weeks before by a feverish attack, apparently of a rheumatic character, which lasted a fortnight.

Loss of power is complete *at first;* usually in one limb, in both legs, or in one arm and one leg, or in the four extremities, and sometimes in the trunk and neck.

There is very rarely paralysis of sphincters; incontinence of urine, and some difficulty of defœcation have, however, been met with as rare phenomena.

The sensibility of the part is not destroyed, but is usually blunted for a time only. There is no disturbance in the functions of the special senses.

Reflex movements are usually abolished, but M. Laborde has shown that this is not always the case, even in the earliest period of the paralysis. In the later periods of the disease, if the atrophy is very partial, reflex paralysis is not observed.

There is no progressive increase of paralysis; but, on the contrary, some of the muscles first paralysed very often subsequently recover their power, whilst others remain powerless. If the four extremities are involved, the lower limbs are usually paralysed most completely, and remain so after the recovery of the arms. Partial or complete relapses may, however, occur; paralysed parts which have recovered again losing power.

There is no rigidity of muscles but absolute relaxation, so that the paralysed limbs fall in the most lifeless manner when left unsupported. During the first week, or at any rate for 3 or 4 days after paralysis, the electro-muscular contractility remains intact (according to M. Duchenne, jun.); at the end of this period, in the cases which are on the point of rapidly recovering, all the muscles still retain their sensibility, whilst in those cases which are about to persist and be followed by atrophy, faradisation * of the muscles gradually loses its power of exciting muscular contractions in some or all of the affected muscles. The cases in which complete recovery takes place rapidly have been distinguished from the others in which more or less paralysis is persistent and followed by wasting of muscular tissue. They are described together by Barthez and Rilliet under the name of essential paralysis. M. Duchenne says they are two distinct affections, although he admits that during the earlier period their symptoms are identical, and that it is impossible to distinguish one from the other. The one, he says, always terminates by a rapid recovery, and is properly called " the temporary paralysis of infancy." The other has a long duration and ends by atrophy or the fatty degeneration of a greater or smaller number of muscles, and later by deformity due to disturbance in the equilibrium of the muscular forces.

It may be convenient for purposes of description to make this distinction, but it must at present remain doubtful whether there is any essential difference in the pathology of the cases. The difference would seem to be rather one of degree than of nature. Some may be disposed to maintain that the temporary paralysis is due to mere reflex irritation whilst the more persistent form is dependent on organic changes in the cord. On this point I shall speak further when discussing the pathology of the disease.

* Faradisation is the term used for the application of electricity, induced by a voltaic or a magnetic current passing through a coil of wire, and frequently interrupted.

In reference to the frequency and duration of premonitory fever, M. Duchenne, jun., has given the following statistics:— Of 70 cases, in 7 there was no perceptible fever, in 1 fever lasted 1 hour, in 1 it lasted 2 hours, in 11 cases it lasted an entire night, in 2 cases a day and a night, in 6 from 2 to 3 days, in 7 cases 4 days, in 5 cases 5 days, in 3 cases 8 days, in 1 case 10 days, in 1 case 15 days, and in 25 cases no definite information could be obtained.

Of 23 cases under my own care no fever was observed in 3, of 5 others no accurate history of the attack could be obtained, and the rest suffered from fever of varying intensity, lasting from a few hours to 14 days. There seems to be no relation between the severity and duration of the premonitory fever, and the persistence or otherwise of paralysis. The older the child at the time of the attack, as a general rule, the more marked is the fever.*

More than half of the cases occur between the ages of 6 months and 2 years; in Duchenne's cases, 37 out of 56; in Dr. West's cases, which, however, include some paralyses of cerebral origin, 27 out of 43; of my own, 8 out of 12; the other cases ranged from a few days after birth to the age of 10 years. Paralysis is sometimes congenital, but is then either due to cerebral defect or some intra-uterine lesion of the paralysed limb, especially dislocation of the humerus.

The parts affected are, in the order of frequency, one leg, one arm, both legs, all four limbs, the arm and leg of one side, the arm of one side and the leg of another, and most rarely the two arms. When all the limbs are affected, one or more of them usually soon recover power, whilst the other or others remain paralysed and become more or less atrophied. Hemiplegia is not often present at *the outset*, nor is paralysis of the leg and arm on opposite sides, but they are consequent on general paralysis and the restoration of some limbs, whilst others remain paralysed.

After the paralysis has existed some time, unless recovery

* M. Laborde says, that in cases where paralysis is extensive, the febrile state is scarcely ever absent.

takes place, the affected muscles become atrophied; in some cases this occurs very rapidly, the temperature of the part is lowered, and the subcutaneous veins are smaller. The nutrition of the whole limb is affected, the bones do not grow as on the other side, so that the paralysed limb may be an inch or two shorter than the sound one. The period which elapses in different cases before atrophy commences, and the rapidity with which it proceeds, differ extremely without any obvious reason. Wasting may commence within a month, but more frequently it is not apparent for several months. Some muscles are more prone to rapid wasting than others; for instance, the deltoid and the tibialis anticus waste much more frequently to a marked degree than the other muscles of the upper or lower limbs respectively.

According to Mr. Adams the groups of muscles affected in this disease may be thus arranged in the order of frequency: 1. The muscles of the front of the leg, forming the extensors of the toes and the flexors of the foot; 2. The extensors and supinators of the hand, these muscles being always affected together; and, 3. The extensors of the leg, and with them generally the muscles of the foot, as in the first group. When single muscles are affected, the order of frequency is said to be this: 1. The extensor longus digitorum of the foot; 2. The tibialis anticus; 3. The deltoid; and, 4. The sterno-mastoid.

At the end of six months or more, if the muscles are not recovering their power or their sensibility to faradisation, the muscular fibre undergoes degeneration; the transverse striæ disappear first, then the longitudinal markings, instead of the transverse striæ amorphous granules appear, which are soon replaced by distinct fat globules. Whilst these changes go on in the paralysed parts, the general health and nutrition of the patient are usually quite unimpaired.

The causes of this disease are not understood, the only thing known is that it occurs more frequently between the ages of 6 months and 2 years, the period of dentition, than at any other period of life. It cannot be traced directly to the irritation of teeth pressing on the gums or nerves, but it probably occurs at that age on account of the general susceptibility of the spinal nervous system at that period of life.

The disease does not appear to be much more frequent amongst ill-fed and poor children than amongst the well-fed and the children of those in easy circumstances. Exposure to cold and the occurrence of measles, scarlet fever, and typhoid, have been observed as exciting causes. It is doubtful whether it occurs more frequently in families subject to other forms of paralysis.

Diagnosis.—In the premonitory symptoms which usher in this affection there is nothing pathognomonic, when fever exists there is nothing distinctive in its character, nor is there anything peculiar in the convulsions by which it is, as a rare occurrence, preceded. When feverish symptoms, whose cause is not understood, occur in children between the ages of 6 months and 3 years, it is desirable to notice carefully whether the patient has lost power over either of its limbs. Loss of power in one of the legs or both is sometimes overlooked for some time in children who have not begun to walk.

The points of distinction between this kind of paralysis and that due to cerebral causes are the following : hemiplegia is rare in the former, and if present it does not affect the face and usually has been preceded by paralysis of the limbs on both sides ; and the leg usually (but not always) recovers more slowly than the arm, whilst the converse is true in cerebral paralysis. Cerebral paralysis is, more frequently than spinal, congenital. Convulsions of a more severe and persistent character frequently precede those cases of cerebral paralysis which are not congenital. Muscular rigidity occurs usually instead of complete relaxation of muscles. There is not the tendency to wasting of the limb nor to depression of temperature in cerebral paralysis. Reflex irritation generally causes movement in cerebral but not in spinal paralysis.

In spinal paralysis there is an entire independence of action between the non-paralysed and paralysed muscles, no effort to use the former will in the least degree induce contractions of the latter.

Galvanism affords great assistance in this diagnosis; in cerebral paralysis the muscles retain their complete electric contractibility; whilst in "essential" paralysis of the graver type this is lost partially or wholly soon after the commencement of paralysis.

General paralysis of cerebral origin is generally combined with signs of hydrocephalus, a want of symmetry in the head, amaurosis, intellectual disturbance, or some other symptoms clearly referable to the cerebrum.

Paralysis from myelitis involving the whole thickness of the cord is attended with paralysis of the sphincters, loss of sensation, and alkaline urine, with a tendency to an increase of paralysis instead of a decrease.

Paralysis from Caries of the Spine is usually readily recognised by some prominence or tenderness of the spines of vertebræ which are diseased, and by pains in the spinal region, with paralysis of the bladder and rectum. Occasionally cases occur in which these symptoms are not present. The history of the attack and its sudden occurrence will usually distinguish the essential paralysis.

Reflex Paralysis is usually preceded by some visceral affection or dental irritation; it may, however, be impossible to detect any external irritation. Some of the cases described by Kennedy as temporary paralysis are probably of this nature.

Progressive Muscular Atrophy is rare in children but not unknown. Eight cases are described by Dr. Meryon, four being brothers in the same family, and the other four belonging to only two families. These cases, though classed with Cruveilhier's atrophy by Dr. Roberts and others, present peculiar features, which seem to bring them more nearly into alliance with cases of progressive paralysis, with enlargement of some muscles and without obvious localised atrophy,—cases to which attention has lately been called on the Continent by Duchenne, by Eulenberg, and others, and by Lockhart Clarke and others in England. This disease was under consideration at the Pathological Society during the present session. A case was brought forward by Mr. Adams and one by myself;

and they will be published in the "Transactions" for 1867-68. My case was briefly this :—

A boy, aged 11 years, of healthy parentage. He began to walk at 21 months of age ; could never walk well, waddling very much with his legs apart. He has never been able to run or jump ; frequently tumbling when attempting to go quickly. His calves were always large and out of proportion to his other muscles. His power of walking, since he was 3 years old, instead of improving has gradually declined, till for six months past he has been quite unable to walk, and cannot now raise himself from his bed or stand unless supported. His heels are drawn up and the lower part of spine arched forwards when he attempts to stand. Lately his upper limbs are failing, so that he cannot cut his food or grasp anything firmly. He has complete control of his sphincters. His intellect is not very bright, but not very deficient. He is reported to be very slow in learning. The upper limbs and his muscles generally are very thin. His calves are still of fair size, but they have a peculiar doughy feel. There is no localized atrophy of any part.

This case seems to belong to a class of cases of which a considerable number have now been observed, which present certain features in common. Their pathology is not understood. The first symptoms usually appear in childhood in an increasing difficulty in walking, and a tendency to arch the lower dorsal and lumbar spine forwards, and to separate the legs when attempting to walk ; gradually this power is lost and weakness extends to the upper limbs, and still later to the respiratory muscles ; the sphincters and other involuntary muscles are not implicated. The patients usually die of inflammatory disease in the lungs or bronchi or some other intercurrent disease, before attaining adult age. There is usually no sign of cerebral or spinal disorder, and with the exception of the progressive loss of muscular power, the patients appear to be in good health. In some cases, whilst there is loss of muscular power, the muscles are enlarged either generally or those of the calves especially, so that the disease has been sometimes called paralysis with hypertrophy. It is not, however, true hypertrophy, but the enlargement is due to deposit of areolar tissue and fat between the muscular fibres. After a time the muscular tissue wastes and the interstitial deposit is absorbed.

In Meryon's cases there was granular degeneration of the muscular fibrillæ, and no interstitial change is described.

In the case I have just related the calves are stated to have been much enlarged at one period, and even when under observation they were decidedly large considering how long the boy had been unable to walk.

No disease of the nervous system has been detected in any of these cases. It remains for future observation to discover whether there is any change in the spinal cord, the sympathetic nerves, or in the blood to explain this disease of nutrition.

Progressive muscular atrophy may be recognised by the very different manner of its progress; atrophy exists before paralysis and gradually leads to weakness; it is accompanied often by quivering of the affected muscles, with a tendency to increase and involve other muscles. Electro-muscular contractility is only lost when atrophy is far advanced, there is not the loss of temperature which is met with in the atrophic paralysis of children. The following cases appear to have been cases of progressive muscular atrophy.

Harriet Guy, æt. 3½ years, a daughter of healthy parents, was in good health until 11 months ago, when a watery bleb appeared upon the back of her right thumb. This bleb broke on the 3rd day and got well in a short time. Soon after this her mother observed that she had difficulty in opening the right hand; she could close it as well as usual. This symptom has increased until the present date. Her mother thinks that the little finger of the left hand is becoming weak as the thumb did at first. She has been taking nux vomica lately, and it is thought with some benefit. She has a slight discharge from her ears. The right index finger assumes a peculiar position, the first phalanx being drawn back towards the dorsum of the metacarpal bone, the second phalanx extended, and the third slightly flexed. There is no resistance to altering this position, but it is again speedily resumed by the patient. The muscles on the radial side of second metacarpal bone are wasted, so also is the ball of the thumb. She cannot well control her right thumb; when placed away from index finger she cannot draw it back. The muscles on the back of right fore-arm are less developed than on the left side, especially on the radial side. The atrophied muscles were galvanized for a few weeks without any obvious effect, and the patient discontinued to come to the hospital.

Margaret Dolan, æt. 8 years, admitted into Children's Hospital in

September, 1867. She had typhoid fever in the hospital 14 months before, and had not been strong since. Her mother first noticed flattening of the left buttock 8 months ago, she soon after observed that there was wasting in the right fore-arm and left shoulder. On admission the child looked not very well nourished. Complained of no pain. Her left shoulder has lost its rounded form from wasting of the deltoid; the left supra-spinatus muscle is also wasted; she can, however, raise her arm to the top of her head, and move it about in every direction. The right fore-arm is also wasted. The left gluteal region is much flatter than the right. She can, however, stand on left leg and abduct the leg perfectly. There is very conspicuous, in the left gluteal region, a condition which is less marked, though present, over all the atrophied parts, namely, a *puckering* and *alteration* of the colour of the skin, which is *pale* and *mottled*. There is a constant quivering motion observed, as she stands, in the fibres of the left glutæus maximus. In the left lumbar region the pigment of the skin is irregularly distributed; there is a white patch surrounded by abnormally brown skin. On the hands, knees and feet, are thin crusts, the remains of recent blebs. A few days later there was noted decided atrophy on the outer side of the left thigh, about in the position of the tensor vaginæ femoris. I gave her liberal diet with a mixture containing steel and arsenic, and the atrophied parts were to be faradised regularly. It was found that they were quite as much stimulated to contraction by the induced current as the healthy parts. A little later there was observed decided atrophy on the back of the right arm above the elbow for about two inches; the skin looks contracted and bluish-white. There is slight wasting on the outer side of the leg for about an inch and a half below the knee. Under faradisation continued for six months (the mixture being omitted at the end of one month), some of the muscles regained their form and size, the skin resumed a less mottled appearance. The left gluteal region still remains (February, 1868) much flattened, especially towards the outer part; the skin of this part now looks well nourished. The child's health appears excellent.

Remarks.—This case differs from any case that I have been able to find recorded. It differs from most of the cases of progressive muscular atrophy in the age of the patient, in the course of the symptoms, and the localisation of the atrophy, also in the condition of the skin, which seemed to share in the atrophy. It is unusual for atrophy to start from so many different centres, and for one set of muscles to recover whilst others are becoming implicated.

The occurrence of blebs on the skin in this case and the

preceding is noteworthy, the nutrition of the skin appears to have suffered with that of the muscles beneath. The atrophy is no doubt due to some condition of the nerves controlling the vascular supply of the affected parts.

The muscular quivering seen in Dolan's case has been frequently observed in Cruveilhier's atrophy, and has been seen also in animals where the anterior columns of the cord have been injured.

When deformity has occurred in consequence of atrophic paralysis, it is important not to confound this with *congenital deformity*, or with deformity from *spasm* of muscles. In the case of paralytic talipes equinus, the deformity arises from the gastrocnemii acting whilst the tibialis and the flexors are paralysed or acting very feebly; these latter muscles will be found not to be stimulated to contraction by galvanism. The most frequent deformities are talipes equinus and equinovarus; there is one deformity which occasionally occurs, and is said by Laborde to be caused in no other way, it is an elevation of the dorsum of the foot from paralysis of the gastrocnemii, with an exaggeration of the plantar arch from the action of the peroneus longus or the flexors of the toes and the flexors of the foot. The history of the case will assist the diagnosis.

Mr. Adams states that the following deformities also occur from this disease: equino-valgus, calcaneus or calcaneo-valgus, and talipes varus. "When both feet are affected equino-varus of one foot is generally found with equino-valgus of the other."

In congenital club-foot there is usually bony deformity as well as displacement of the foot.

There is sometimes in *rickets* a great want of muscular power, simulating paralysis, this symptom may be quite disproportioned to the signs of rickets in the bones or elsewhere. The gradual accession of the muscular weakness, the other marks of rickety disease, especially in the bones, and the diffused and at the same time the incomplete character of the paralysis should prevent one from mistaking such a case for atrophic paralysis.

Pathological Anatomy of Atrophic Paralysis of Infancy.

The affected muscles undergo the following changes in different cases, or in various muscles of the same case :—1. The transverse striæ become less apparent and separated by wider spaces, which are filled with opaque granules, which are not dissolved by ether, but are sensibly acted on by dilute acetic acid. 2. The transverse striæ disappear, and there is an abundant appearance of granular substance. 3. There remain but slight traces of longitudinal fibres, filled with granules with a larger quantity of connective tissue between the bundles. 4. The granules have disappeared, and empty transparent tubes of myolemma, with a few scanty granules on their walls, remain, with more connective tissue and some elastic fibres. 5. In some cases fat globules take the place of the granular matter in the muscular fibre and in the cellular tissue between the bundles of muscular fibre. This change is not universally present in cases even where atrophy has proceeded to an extreme degree.

A child under the care of M. Roger was paralysed from the age of 8 months, and died with cerebral symptoms, vomiting and strabismus, followed by coma, when 2 years and a few months old. The paralysis was at first general, and subsequently localised in the lower limbs. The meninges of the brain were considerably congested. The anterior columns of the cord were translucent and greyish red; the lateral columns were slightly affected in the same manner. The change in appearance involved the whole thickness of the anterior columns, and was accompanied by a consistence less than normal; the lateral columns were only affected in their cortical portion. The change extended from the cervical to the lumbar region of the cord. Examined microscopically the affected parts presented a remarkable proliferation of the elements of connective tissue, both cells and nuclei; the latter measuring from $\frac{1}{8333}$ to $\frac{1}{5555}$ of an inch, with a nucleolus scarcely visible; the former less abundant, with a nucleus and nucleolus; these elements were dispersed in a finely granular substance composed of fibrillæ of extreme tenuity. In the parts most affected scarcely any nervous tubules could be recognised; when present they were varicose. No other morbid change was discovered. The muscular fibres were atrophied, but had not undergone fatty change.

Another case is recorded by MM. Laborde and Duchenne, which was under the care of M. Bouvier. The child was two years old; he had suffered when 12 months old from fever, with convulsions, during which he lost consciousness and became blue; he also seemed to suffer very acute pain. From this time he could not walk; it is doubtful whether he lost the use of his upper limbs; if he did the power was soon regained. The lower limbs became rapidly atrophied. He died of pneumonia following measles. Shortly before his death each leg was forcibly flexed upon the thigh with the knee turned outwards; the feet, especially the left, were strongly extended on the leg, the sole turned inwards and upwards and the toes flexed (in a state of "Talipes equino-varus"). Any attempt to restore the feet to their natural position gave severe pain. The hands (especially the right) were turned back and rotated outwards; the fingers slightly flexed, the elbow extended. Attempts to restore the natural position were futile and caused much pain. The muscles of the legs, except the gastrocnemii, had lost their excitability to galvanism, and in the upper limbs the electro-muscular contractility was much diminished, especially in the deltoid and the muscles of the back of the arm. The cutaneous sensibility was increased as evinced by pinching; the intelligence was unimpaired; there was neither rigidity of the spine nor spinal tenderness.

On *post mortem* examination, the cerebral meninges were congested, and there was a little reddish serum in the lateral ventricles. A thin layer of fluid, slightly turbid, was found in the spinal canal outside the membranes; in the arachnoid was found a little more than 3 ounces of a yellowish turbid fluid thick and almost gelatinous. The pia mater of the cord was of an intense red colour throughout; on its surface were numerous vascular networks, and whitish milky-looking streaks.

The cord was of tolerably firm consistence; its external appearance was normal; on section the cord had a gelatinous quasi-transparent appearance. Microscopically, in the anterior columns the longitudinal nervous tubules appeared to be fewer in number than in the normal state; those which remained were swollen and varicose; this was especially the case in the cortical parts. In these parts there were some granular corpuscles; the capillary vessels, both of the pia mater and the peripheral nervous substance, presented an increase of their nuclei, and their walls were studded with a great number of exudation corpuscles. This condition of the vessels and the nervous tissue was especially marked in the lumbar region, and it extended to the fibrous prolongations enveloping the roots of the nerves. The nerves of the limbs appeared unaltered, except that the left sciatic nerve presented a comparative scarcity of primitive nerve tubes.

This case cannot be regarded as a simple case of infantile paralysis. According to M. Laborde there were two distinct sets of morbid changes—the one recent, accompanied with congestion and exudation, the other, of older date, indicated by appreciable changes in the intimate texture of the nerve substance. I confess, however, that the evidence of anything more than a comparatively recent inflammatory change of the meninges, and the cortical portion of the cord, is very scanty.

These are the only cases with which I am acquainted in which changes in the cord have been discovered in connection with infantile paralysis.

Rilliet and Barthez state that they examined the cords of two children who died of this disease, and could find no lesion. It does not appear that they examined them microscopically. M. Fleiss records a case of paralysis of one arm, in which he discovered congestion of the spinal meninges on the level of the brachial plexus of the paralysed side, but no change in the cord.

At present the evidence of change in the cord from actual observation rests mainly on the one case under M. Roger; the second case lends additional support in the same direction, but it is not so conclusive as could be wished. *Further post mortem* evidence is certainly required to settle the matter.

In coming to a conclusion on this point, it must be remembered that negative observations are of no value, unless the cord has been examined microscopically with the improvements introduced by Mr. Lockhart Clarke and Dr. Bastian. It has been conclusively shown that a spinal cord whose colour, form, and consistence are quite normal may be found to have undergone considerable degeneration. An interesting case bearing on this point is reported by Dr. Bastian in the "Medico-Chirurgical Transactions," vol. 50, page 499 : " Case of Concussion Lesion, with extensive secondary degenerations of the Spinal Cord, followed by General Muscular Atrophy."

In this case there were positively no naked-eye appearances of disease ; yet there were found, after hardening the cord in chromic acid, very extensive areas of degeneration, consisting of atrophied nerve-fibres, new connective-tissue elements and granulation corpuscles. The changes appear to have been very similar to what were found in M. Roger's case of infantile paralysis.

M. Cornil (" Gazette Médicale," 1864, p. 290) has published a case of paralysis in a woman aged 49 years, which had dated from the age of 2 years. In this case there was complete fatty degeneration of the muscles with atrophy of the primitive fibres ; fatty degeneration of the nerves, with atrophy of the nervous tubuli, and atrophy of the antero-posterior bundles of the cord, with production of amyloid corpuscles throughout its entire extent. The real nature of this case is somewhat uncertain, though it may very probably have been a case of the disease now under consideration.

Heine is of opinion that this disease consists in a congestion of the cord followed by watery effusion, and subsequently by atrophy of the cord. This theory is so far rather supported by fact. Dr. Radcliffe and M. Brown-Séquard seem to have come to the same conclusion. M. Duchenne argued that this disease must have its seat in the spinal cord on the ground of analogy; he says: "In nearly all traumatic lesions of the cord, or its membranes, the muscular symptoms consequent thereupon are *exactly the same* as those observed in the atrophic paralysis of infancy. In both, paralysis marks the outset of the disease, then after a shorter or longer period the muscles supplied from those parts of the cord least affected recover their voluntary movements and nutrition, whilst those which receive their nervous supply from parts more profoundly injured, atrophy and become fatty." There is some truth in this, although the case is stated rather too strongly.

To sum up, the atrophic paralysis of infancy may be regarded as, in all probability, an affection of the spinal cord, especially the anterior and part of the lateral columns. Symptoms of paralysis, not at first to be distinguished from true atrophic paralysis, sometimes occur, and disappear in a week or two without leading to atrophy. These cases must either be regarded as due to reflex irritation, or to a lesion of the cord similar in kind, but less in degree. It is not difficult to suppose that just as, in the graver cases, many of the muscles at first paralysed entirely recover at the end of a short time, so in the slightest cases all the affected muscles after a while recover. Some authors distinguish the cases of temporary paralysis from the cases followed by atrophy, and regard them as totally distinct diseases; but I confess that the reasons for doing so do not appear quite conclusive.

Prognosis.—In regard to life this is favourable; the disease does not prove fatal. As regards restoration of power in the limb, at first no opinion can be given; the character of the initial symptoms affords no clue to what may result. When the fever is intense, paralysis of the four limbs often occurs; but it may shortly afterwards be localised in the legs or in one limb. In cases ushered in by convulsions the progress of the case is not always more unfavourable than in other cases. When paralysis has lasted a few weeks, the most valuable means of prognosis is furnished by faradisation. The muscles that do not act to this stimulus will in all probability become atrophied, and if most of the muscles of a limb are thus affected, the growth of the limb will be arrested. By this means we may learn what kind of deformity is likely to ensue, and we may be guided in directing the orthopœdic treatment. Even when muscles are much atrophied and insensible to galvanism, they may be restored by persevering in suitable treatment for a long period, if the treatment be commenced within twelve or eighteen months of the attack.

Duchenne reports a case in which improvement took place after paralysis of the arm of four years duration, by faradisation and localised gymnastics continued for two years. The only parts of the limb which retained their electro-muscular contractility when the treatment was commenced were the inner part of the anterior third of the deltoid, the middle third of the trapezius, and the short abductor of the thumb. After the treatment the patient could raise the arm outwards and forwards, and bend or extend the fore-arm on the arm, and could oppose the thumb to any of the fingers. The nutrition of the limb had much improved, and its temperature was well maintained.

If the electro-muscular contractility is not destroyed, but diminished, we may anticipate a *complete* restoration of the muscles thus affected by the use of galvanism and other suitable means.

Treatment.—In the first instance attempts should be made to discover any source of eccentric irritation, and if discovered to remove it. If necessary, the gums should be lanced. An aperient should be administered, and the motions examined to determine the presence or absence of worms. Tepid baths are useful to equalize the circulation and to allay irritation. Counter-irritation to the spine by means of mustard or turpentine appears indicated. Some recommend leeches, but I should not be inclined to follow this advice. After the first stage has passed, and the circulation in the paralysed part is feeble, and the temperature lowered, friction with stimulating liniments (such as camphor or turpentine) should be used; the part must be kept wrapped in flannel or wool. Strychnia in small doses may be administered. I have used it hypodermically, but without any obvious benefit.

In the later periods stimulating saline and terebinthinate baths should be used, and frictions afterwards employed. To prevent deformities from the contraction of the muscles antagonistic to those most paralysed, bandages and orthopœdic instruments may be requisite.

To remedy great deformity in the feet tenotomy, especially of the tendo achillis, is sometimes requisite. It must not be adopted until time has been allowed to see the extent of permanent paralysis, and is not indicated where all the muscles of the limb have lost their contractility. It should not be performed until it has been found that orthopœdic apparatus is insufficient to restore the limb to its shape.

The paralysed muscles should be galvanised for five minutes three or four times in the week. A coil apparatus should be used; the most convenient one is that constructed by Scherer of Dresden, which can be obtained of Pratt, 420, Oxford Street. This apparatus is in a very small compass, and is very readily put in action; and even when in constant use, the plates of zinc require to be amalgamated about once only in two months; there are but two cells, so that the amalgamation is a trifling matter. The magneto-galvanic machine cannot be so conveniently adjusted, and is not nearly so available for clinical purposes.

The poles of the battery should be furnished with several handles for sponges of different sizes, and also with two metallic flat circular discs about the size of a shilling piece, which are to be covered with wash leather. These, when used, should be quite wet (not dripping), otherwise the galvanic action is limited to the skin, and does not penetrate to the muscles. The two poles of the battery should be near together, and should be applied consecutively over the various parts of the muscles under treatment. The intensity of the current must be regulated by the effect produced on the corresponding muscles in a sound limb, some muscles being much more irritable under the galvanic stimulus than others.

Duchenne recommends that the intermissions of the current should, in young children, be rather slow, and thus the pain of the operation is much reduced. This can be effected only by a cogged wheel to be turned backwards and forwards by the hand or by the magnetic battery. It is not, however, a precaution that is really needed, the pain is not usually so great as to require it.

In some cases the most powerful induced currents (faradisation) have no effect, whilst the muscles act more energetically under the influence of a direct galvanic current of low tension slowly interrupted, than the corresponding muscles of the sound limb. (See Case XII.) In such cases the direct current of low tension should be employed, "galvanisation" instead of "faradisation."

There is reason to believe from the observations of Mr. Radcliffe that this current is often useful in cases in which the paralysed muscles are not contractile under the influence of faradisation.

It is a curious fact that in proportion to the recovery of susceptibility to an induced current and of voluntary power, the direct galvanic current loses its effect on these muscles.

If the case come under treatment at an early period, counter-irritation to the spine and a course of iodide of potassium should be tried.

At the same time that the paralysed muscles are galvanized, the child should be taught for a few minutes every day to attempt to use the weak muscles by voluntary effort.

If muscles have been paralysed for more than a year or two, are much atrophied, and are not stimulated to contraction by faradisation or galvanisation, they have probably undergone such complete atrophy and degeneration that attempts to restore their power are futile. It will be necessary to use orthopœdic apparatus as a substitute for these muscles; resorting also to galvanism, gymnastic exercises, and other means to strengthen the other muscles of the limb.

When one arm is paralysed to a great extent, the other arm may be tied up so that it cannot be used, and the child tempted in various ways to use the paralysed limb as far as practicable. When the deltoid only is paralysed, by fixing the shoulder the patient can often use the fore-arm and hand. The shoulder may be fixed by the hands of a second person or by a suitable apparatus.

I will now give the notes of some cases of paralysis due to *cerebral* causes, and direct attention to the points which distinguish them from the cases of atrophic (spinal) paralysis which we have been considering.

The following case appears to have been of cerebral origin :—

J. W. P., a moderately healthy boy, was seized, when 2 years and 1 month old, with a succession of convulsive attacks which were followed by drowsiness and delirium. He was kept in bed for a week. On getting up he was found to totter in walking, and at the end of 14 days his right hand was observed to be weak so that he could not hold his food ; on the following day his wrist completely dropped on the fore-arm. When I saw him six weeks later he could raise his hand to the horizontal position, but not higher ; he could not straighten the 2nd and 3rd phalanges of fingers of right hand ; when they were extended by an observer they slowly returned to the flexed position. The electro-motility seemed to be unaffected. There was *no atrophy or loss of temperature* in the limb. In six weeks he had almost quite recovered the healthy condition.

In the next case there was incomplete temporary hemiplegia, following convulsions, in a very young child.

A girl, aged 10 months, of good constitution, had suffered from hooping-cough, during which she had several attacks of convulsions. The cough was now leaving her, and she was progressing favourably, when she was seized with a fit, which lasted two hours. After the fit her right arm and leg were found to be powerless. I saw her in consultation with my friend Mr. Newton of Wimpole-street five days later. She had regained some power in the paralysed limbs, but they were distinctly weaker than the corresponding limbs on the other side. There was no rigidity ; no paralysis of the face. In the course of the following week she had completely recovered the power, and continued well.

In the following case the lesion was no doubt cerebral ; its exact nature is uncertain :—

W. M., a delicate boy, aged 3½ years, was observed to limp on his left leg. About this time he was noticed to yawn a great deal and complain occasionally of pain on the right side of his head. Shortly after this his left hand was noticed to be less useful than the right, so that he could not pick up his toys with his left hand. He came under my notice three months later. There was then incomplete paralysis of left leg and arm, the arm was bent and somewhat *rigid*. When crying or gaping the *mouth was drawn* to the right; sometimes his left eye was not closed so completely as the right. There was *no atrophy* of the paralysed limbs or loss of temperature.

Another case of hemiplegia, with contraction, may be here given. Although the paralysed limbs were colder, they were not atrophied.

An apparently healthy girl, G. B., was observed when 18 months old to become *gradually* more awkward in the movements of her right arm and leg. These limbs appeared colder than those on the other side. She was brought to me when 4 years old. The chief points observable were contraction of the gastrocnemii of the right side, causing slight talipes equinus; the right thumb and fingers flexed rather firmly. The nutrition of the two sides seemed nearly equal.

In the following case there was weakness of one arm and leg following two severe convulsive attacks (epileptiform).

M. A. Bailey, æt. 8 years, had been delicate from her infancy, with a tendency to attacks of lividity. Six months previously she had been seized with a fit, in which she had fallen down quite insensible; her face was pale, eye-balls rolling about, and her limbs convulsed. This lasted nearly half an hour, and was followed by stupor lasting three or four hours, when a second fit, very similar to the first, seized her. After these attacks she appeared almost insensible for two days, at the end of which she was able to get up, but her right arm and leg were incompletely paralysed as at present. Her intellect has been less bright since these fits; she has been subject to frequent headache, which is sometimes violent and lasts 24 hours, at others is slight, and of much shorter duration. She had occasionally suffered from headaches before the fits. When I saw her the only paralysis observed was that she could not grasp so well with the right as the left hand, and dragged her right leg a little in walking. There was *no failure of nutrition* on the right side or any difference of temperature observed. Occasionally she has pain in her right knee at night, making her cry out. Her father and uncle died of cerebral disease, as well as other children of her mother. The father suffered from fits, followed by hemiplegia, and at last leading to insanity.

Facial Paralysis, or paralysis of the portio dura of the 7th cranial nerve, is as frequent in children as in adults. It comes on suddenly after exposure to cold, or without obvious cause; sometimes it arises more gradually from disease of the temporal bone implicating the course of the nerve, or it occurs in new-born infants from injuries received at birth. The angle of the mouth on the other side is drawn up; there is no contraction of the eyebrow on the paralysed side. The eye cannot be completely closed on the affected side; there is an escape of tears over the cheek, and the conjunctiva becomes injected and covered with a film of mucus, because not kept clear by the movements of the lid. When the mouth is closed and the cheeks are blown out, the muscles on the affected side offer less resistance than on the other.

The following case may be taken as an illustration of the ordinary course of the symptoms, and is interesting from the observations made with the galvanic chain :—

George Hill, æt. 5, admitted in November, 1867. A healthy boy three weeks before admission, complained of face-ache, and there was some swelling on the left side of his face. A week later his mother noticed that his mouth was drawn to the right side in laughing. On admission (Nov. 11th) the left side of his face was a little swollen in the parotid region. This side of the face is quite immovable even in laughing; the left eye cannot be closed; the tears run over the cheek on this side. The paralysed muscles tried with.faradisation found quite inactive. But a Pulvermacher's chain of 30 links applied in the course of the levator or depressor anguli oris is found to produce decided contractions of those muscles each time the connection is interrupted; no effect is produced in the non-paralysed side. Gradually this condition disappeared, so that on 1st January there was no effect produced by the Pulvermacher's chain; and the difference in the two sides of the face was disappearing; the left eye could be closed but not quite so firmly as the right. There was at no time any paralysis of the limbs or of the tongue.

The treatment of this disease is simply to apply warmth to the affected part, and to avoid exposure to cold. Whether the cure was hastened in this case by the daily use of the galvanic chain I cannot say; but it seems probable that when such a condition of electro-motility exists, as in this case, its employment may be of use. This condition has been detected in several cases of atrophic paralysis as mentioned before, and it is found to exist in lead palsy, in some cases of muscular atrophy, and in paralysis from traumatic injury of a nerve. As the functions return to the paralysed muscles, the irritability to galvanism of a low tension disappears, whilst the irritability to the induced current (electricity of high tension) is restored.

It is of course very important, especially as regards prognosis, not to mistake mere facial paralysis for hemiplegia of cerebral origin. In the latter case the paralysis is not limited to muscles supplied by the seventh pair; in fact these muscles usually escape, so that the eye can be shut, and the eyebrow contracts. The muscles engaged in mastication (the masseters, temporals and pterygoids) are supplied by the fifth nerve, so that they are not paralysed in facial paralysis, but in cerebral hemiplegia they are sometimes affected; the tongue does not deviate from the middle line in facial paralysis. The limbs of course are entirely free.

Postscript.—The galvanic apparatus for the use of electricity of low tension (referred to on page 265) should consist of fifty or one hundred of Daniel's cells, with an arrangement by which only a few or the whole number of cells can be put in action. Such an apparatus is supplied by Elliot Brothers, Strand.

INFANTILE ATROPHIC PARALYSIS.

Sex.		Antecedents of Patient.	Age at time of attack.	Symptoms of Attack, and subsequent course.	Age and Condition of Patient when coming under notice. Treatment.	Result.
Male.	Female.					
	S. J. White	Previous health good; walked at age of 14 mths.; scalded on chest when 16 mths. old; wound open 6 weeks.	16½ months.	Sudden loss of ability to walk. Some loss of flesh.	18 months. Muscles of leg very flabby; not distinctly atrophied. Can move legs from side to side when lying; but cannot raise them from bed. Steel Wine. Cod Liver Oil. Soap and Cantharides liniment.	Complete recovery in three months.
F. Perkin.		Healthy; walked when 9¼ months old.	12 months.	Whilst walking, left leg suddenly gave way.	16 months. Not a very strong child. Has but four teeth. Left lower limb completely paralysed as regards motor power; colder and flabby; not much atrophy. Strychnia.	At the end of a mnth. in *statu quo*: not seen again.
	A. M'Carty.	A well-nourished child.	11 weeks.	Whilst being nursed, the child seemed to have less spring, and to feel heavier in mother's arms. She looked brighter than usual; her back seemed weaker, and, two days later, could not move her legs.	16 months. General health good. Legs, when raised, drop in a lifeless manner; when right thigh is held by observer, she can straighten leg on thigh; when leg is held, can move foot; can move left lower limb at hip joint, and with her hand put it under, opposite axilla, like a doll; left foot extended on leg; when foot is held firm, can move toes (she can only do this lately); lower limbs flabby, and ill nourished; left nates and calf atrophied.	Unknown.

CASES OF INFANTILE ATROPHIC PARALYSIS.

				...and has regained some power. Walked when 18 months old.	runs about, but, from weakness of left leg, falls and bruises herself often. Very cheerful.	
5. (Vol. IV., p. 101.)	W. Meadows.	One of 6 children; 4 dead; not strong.	16 months.	After an attack of diarrhœa, could not use left leg.	18 months. Has 12 teeth; not rickety; left leg smaller, more lax, and colder than right; can move it from hip joint, but cannot rest on it; rectus femoris acts when faradised; other muscles not perceptibly. To be galvanized, and rubbed with stimulating liniment.	Slightly improved in a month; afterwards not seen.
6. (Vol. IX., p. 90.)	W. Newcome.	Only child of healthy parents; apparently strong and healthy.	6 months.	Left in charge of aunt; found by the mother screaming; which became more violent on moving legs. Since then has not moved left leg properly; pain on movement lasted 3 weeks; no redness or swelling; no evidence of any accident having happened.	18 months. General health good; left lower limb smaller and colder; no motor power; when raised, it falls like a dead limb; no reflex movement when sole of foot is tickled. Left thigh measures 6¾, right 8¼; left leg 5½, right 6 inches. Galvanism.	Six weeks under treatment, with slight improvement.
7.	L— N—.	Previous health good.	18 months.	Whilst cutting canine teeth, sickness and screaming, followed by drowsiness, which lasted 3 days. The right leg was swollen and red for 2 months; quite powerless.	2½ years. General health good; right lower limb atrophied and paralysed. Temperature, right hand, 85°, left, 92°. Right thigh measured 7¾, left 8½; right calf 5¼, left 6¾. Faradisation acts scarcely at all on muscles of right leg and thigh.	Unknown.

INFANTILE ATROPHIC PARALYSIS—(continued).

Number of Case, and Reference.	Sex.		Antecedents of Patient.	Age at time of attack.	Symptoms of Attack, and subsequent course.	Age and Condition of Patient when coming under notice. Treatment.	Result.
	Male.	Female.					
8. 1865.	R. Pimm.		Good.	2½ years.	Had pain in head, and fever, lasting 2 weeks. It was called rheumatism; but there was no pain in his joints. Four weeks after this, his right leg was powerless; the right arm seemed weaker than left, but soon regained power.	3 years. Health good; arms of equal power; right lower limb paralysed and atrophied; the thigh an inch smaller than left; foot extended on leg; when attempting to walk, leg hangs, and foot turns outwards; tickling causes no movement; electromuscular irritability is not quite lost. Galvanism. Strychnia by mouth, and, subcutaneously; an instrument applied.	Slight improvement in 6 weeks; not heard of since.
9. 1865.	C. Green.		Previous health good.	3½ years.	Sudden loss of power in lower limbs, without any previous symptom. The left soon recovered; but right remained powerless, and has lost flesh; is usually cold.	3 years and 9 months. Health good; upper limbs of equal power; right leg weaker than left; can walk, or raise it in bed; it is wasted, and tends to give way in walking; foot colder. An apparatus applied.	Able to walk with apparatus.
10. 1865.	S. Allen.		Health good.	5 years.	Sickness recurring, especially after food. Discomfort and weakness in left leg, which, in 5 days; led to paralysis; then delirium, lasting	10 months after attack. Both legs weak and wasting; unable to walk; can move limbs at hips; tendency to talipes equinus of left, and varus of right foot; legs usually cold.	Unknown.

CASES OF INFANTILE ATROPHIC PARALYSIS.

11. 1866.	Theresa Barrett.	When teething, subject to screaming fits, and intolerance of light.	14 months.	Fever, screaming, and rolling of head; for 3 days insensible, and twitching right arm. Had 2 or 3 other similar attacks. A month before had strained right arm; 3 weeks later it began to waste, and has continued to do so.	22 months. Deltoid, and supra and infra-spinatus, muscles of right side atrophied and paralysed. They are excited by magneto-galvanic current.	Five weeks later decided improvement.
					This case is probably not of the same character as the preceding. The attacks were *cerebral*, and the atrophy may have resulted from an injury.	
12. 1866.	H. M. Brown.	Health good.	11¼ months.	One night fretful, and not sleeping well; in morning, feverish, with red face, and right eyelid drooped; same evening, right arm was observed to be powerless. Remained unwell for 3 days.	1 year. Right arm powerless; within last few days only could move fingers; left angle of mouth drawn up a little; slight wasting of right deltoid and biceps. These muscles do not act to faradisation, but they respond to an interrupted galvanic current, which does not excite muscles of left side. This excitability was lost after a time, as the limb was improving.	Galvanism carried on at intervals. Improvement.
13. 1863.	E. Munns.	Feverish attack a month before; health otherwise good.	2 years and 10 months.	Left lower limb almost suddenly became powerless.	3 years and 11 months. No power in left leg below the knee; not decidedly wasted. Six months later, some wasting of leg; arching of foot: can walk with a slight limp. Galvanism.	
14. 1867.	Jane Wilkinson.	Health good.	5 months.	Feverish and restless in night; on being taken up she cried, and her legs were found to be powerless.	6 years. Both legs atrophied, and fall about, when raised, in a lifeless manner; left cannot move at all; right can move from hip joint a little. Sensation seems impaired in right limb: not in left. No contractility to interrupted galvanic current; induced current causes action of right rectus, and left gastrocnemius and glutæi: no other muscles. Electro-sensibility reduced in both lower limbs.	In *statu quo.*

T

INFANTILE ATROPHIC PARALYSIS—(continued).

Number of Case, and Reference.	Sex.		Antecedents of Patient.	Age at time of attack.	Symptoms of Attack, and subsequent course.	Age and Condition of Patient when coming under notice. Treatment.	Result.
	Male.	Female.					
15. 1867.		Emily M'Gill.	Health good.	18 months.	Feverish for several days.	6½ years. Both limbs paralysed; slight power in right limb during last 6 months.	In statu quo.
16.	James Smith.		Health good.	2 years.	Feverishness, and loss of appetite, for 2 or 3 days; then, pain in legs, especially near hips; legs powerless; in a week the left began to recover.	2 years 3 months. Wasting of muscles of right leg. The induced current has no effect on these muscles; an interrupted direct current causes slight muscular contraction on paralysed side; but not on the other. Right thigh 8⅞, left 9¼; right calf 6¼, left 6⅜. Galvanism, with direct current.	Unknown.
17. 1867.	William Roberts.			2 years.	Said to have had "brain fever" for a fortnight; both legs paralysed for 6 months; the left has remained so.	4 years. Left lower limb partially paralysed; adducted, flexed on abdomen; not rigid. Wasting of nates, thigh, and leg; cannot extend leg on thigh, or flex foot upon legs; no movement on tickling sole. Cremaster acts well on touching the front of thigh.	Unknown.
18. 1867.	Otterwill Brandon, 5, Philips Buildings.			5½ years.	Poorly for a fortnight, with swelling in neck. Paralysis observed one morning, on getting him up, in arms and legs.	5 years and 6 months. Speaks thickly; uses arms and legs clumsily; can stand, and move his arms; cannot walk or feed himself; stands half on tip-toe; no rigidity; sensation natural.	Recovery.

CASES OF INFANTILE ATROPHIC PARALYSIS. 275

Name	Health	Age	History	Condition	Result
	Health good.	1 year and 9 months.	Fever, with swelling of left ankle and knee, lasting 2 days; poorly for a week; on recovery, could not use left leg well. This leg has not grown, and is usually colder than the other.	2 years. Left lower limb, as a whole, a little smaller than right; no localised atrophy. In walking, everts the foot, and throws it round; often falls, when trying to run.	Gradual restoration of power
				Vinum Ferri. Linimentum Camphorae to be rubbed in.	
	Health good.	1 year.	Twelve months ago, lost power in all his limbs one morning: he had been hot and feverish previously. In a month he could use arms, and right leg began to get better.	2 years. On admission both leg and arm on left side were paralysed; general health good; paralysis of left lower limb, with atrophy, which increased somewhat under observation; some want of power to grasp with left hand; wasting of the muscles of thumb, and also of arm and forearm, to less degree. Very bright, intelligent boy. Faradisation.	Arm regained power; leg improving. Interrupted by two or three feverish attacks.
Gertrude Bithny, from Leicester.	Good health.	3 years.	Had "gastric fever" when 3 years old, and has not walked since.	7 years. Healthy. Paralysis of right lower limb, with atrophy, and one inch shortening; slight tendency to talipes equinus; cannot straighten leg; has some power to move thigh on trunk. Electro-muscular contractility nil in legs. Faradisation.	Fitted with an instrument, and sent to country. Walking with a little difficulty.
Frances Helpin.	Good health.	6 months old.	Weakness of left leg; said to have come on gradually. Had diarrhoea at the time. The left foot gradually got stiff, the toes pointing down.	16 months. Left leg smaller than right; foot extended, and turned slightly inwards; resists flexion; anterior tibial seems wasted; gastrocnemius rigid; can move toes of left leg. Left calf, 5¼; right, 5¾ inches.	In statu quo.

INFANTILE ATROPHIC PARALYSIS—(continued).

Number of Case, and Reference.	Sex.		Antecedents of Patient.	Age at time of attack.	Symptoms of Attack, and subsequent course.	Age and Condition of Patient when coming under notice. Treatment.	Result.
	Male.	Female.					
23.		Harriet Hughes.	Good health.	2¾ years.	"Rheumatism," lasting six weeks; afterwards unable to walk; right leg quite powerless; the left weak.	2½ years. Appears in good health. Both feet extended on legs, with slight inversion. Can move left lower limb from hip as one joint; but not the right. Foot resist flexion; cannot be brought to a right angle with leg; the left more nearly than right: right leg usually colder than left. Faradisation has no effect on anterior tibials or extensors of toes of either side. The interrupted direct current, of low tension, acts freely on anterior tibials, extensors of toes, and peronei of both sides.	Under treatment.
24.	Francis Church.*		Good health.	1¾ years.	A sudden feverish attack; no vomiting or convulsions. 2 days after, his legs were weak; but able to walk in six weeks. Left arm quite powerless, and right nearly so.	2¼ years. Healthy looking. Can walk; nothing abnormal in legs. Much wasting in both deltoids; more left than right; also supra- and infra-spinati, and left biceps. Cannot raise left shoulder, or flex fore-arm; can slightly raise right shoulder, and can bend fore-arm well. No action with direct current. The induced current acts on all the atrophied muscles to some extent.	Much improvement under faradisation.

* In this case the prognosis is good, because all the atrophied muscles contract under the stimulus of faradisation.

ASCITES.

AMONGST the less common diseases of children may be classed ascites. Several cases which have come under my care within the last four years may be brought together and their symptoms compared.

The occurrence of ascites as a part of general dropsy from renal or cardiac disease needs no special mention. It is met with also frequently as a result of tubercular or non-tubercular peritonitis. There is one cause of ascites met with in adults which I have reason to believe sometimes occurs in children; namely, adhesive inflammation of the hepatic veins. This usually begins with inflammation of the capsule of the liver, and gives rise to symptoms of obstruction exactly similar to those due to occlusion of the portal vein. This may give rise to a lobulated condition of the liver which has been mistaken for true cirrhosis, which is rare in children but not unknown. (Path. Soc. Trans., xiv., page 175.)

In the following case ascites was attended with great enlargement of the liver, probably fatty, and with probable tubercular disease in the bronchial glands and lungs.

James Ings, æt. 3 years 3 months, one of 5 children, of whom two died in infancy, and whose father and mother enjoyed good health, was brought into the hospital on the 15th July. His mother gave the following account of him :—That he had suffered eight months previously from "bronchitis," which began with a fit and lasted seven weeks. Since that time he had not been strong, and his abdomen had gradually enlarged. Two months ago his legs became swollen. He had had two or three more fits within this period of eight months. The appetite had continued good, there had been no sickness, and his bowels had been regular.

On the 15th July the following notes were taken of his condition :—
Of clear complexion, rather thin, weighing 23 lbs. There is some rickety enlargement of the ends of bones with curvature of the legs. His skin cool and moist; lips slightly livid; cheeks of good colour, not anæmic. Pulse 136, weak. Respiration 50. No movement of nares. Tongue clean and moist.

Abdomen enlarged, symmetrical.

The liver reaches in the vertical line of nipple from 4th interspace to near Poupart's ligament; in the axilla from 5th space to crest of ilium in the midsternal line from 4th cartilage to an inch below umbilicus. The edge is rather sharp, hard and even. The left lobe cannot be very thick because stomach note is heard through it on deep percussion. There is no tenderness, and the sense of resistance is not great; the organ is moveable, and fluid is felt in front of the lower third of the right lobe, and easily displaced. There is a considerable amount of fluid in the peritoneum, causing distinct fluctuation and dulness in the flanks. Fluid escapes into tunica vaginalis and can be readily returned into abdominal cavity. To take Ol. morrhuæ, ʒj.; syrupi ferri iod. ℥xv.; potassii iodidi, gr. j. Ter die.

July 17th. Urine scanty. It contained a little pus and was alkaline two days ago, but is now clear, free from albumen and acid. His bowels are costive.

22nd. Examination of chest. Both apices in front dull and hard on percussion. First bone of sternum almost absolutely dull. Suprascapular regions also dull. At both apices in front and behind; inspiration normal, expiration longer than inspiration, bronchial in character. In other parts the respiration sounds are normal. Veins of neck tortuous and dilated.

29th. Ascites has increased. Veins in abdominal wall distended.

Aug. 7th. He was attacked with pyrexia which lasted two days, during which petechiæ appeared on the hypogastrium. Some small ulcers on his tongue; his bowels were relaxed, and there was ear-ache, with pain and serous discharge. These symptoms passed off in a few days and left him quite lively and in good spirits. The abdomen became somewhat reduced in circumference, varying from day to day between 23 and 24 inches. The thoracic signs remained unchanged, and he left the hospital apparently in tolerable health.

Remarks.—The probable diagnosis in this case is tubercular deposit in the bronchial glands under the upper part of the sternum and in the upper part of both lungs, with enlargement of the liver, probably fatty. The fluid in peritoneum was possibly due to enlarged glands pressing on the vena portæ, but this part of the diagnosis is very uncertain. The attack of "bronchitis" was probably the commencement of tuberculisation.

In the following case there is great ascites with nothing to account for it; the liver is somewhat enlarged. There is no evidence of hepatic disturbance.

Alfred Smith, æt. 2½ years, one of five children, with healthy parents. The first child was born at seven months and died in a month; the 2nd only lived three months, and the 3rd died at the age of five months. None of these children had snuffles or any eruption. There is one child younger than Alfred, aged six months, in good health. Alfred appeared quite well till six weeks before admission, when his abdomen began to swell and rapidly increased. His bowels were rather confined at first, and he was sick several times. The bowels have been rather loose recently; he has had no pain.

Nov. 9th. Rather thin and pale, but not emaciated. Pulse 136, weak. Skin natural. His abdomen measures 25½ inches at the most prominent part, this is just above the umbilicus, which is slightly protruded. Superficial veins not much enlarged. There is perfect fluctuation in every direction. Percussion resonance everywhere deficient except above the level of navel and to left of middle line. No solid mass of any kind can be felt, nor enlargement of liver or spleen. The upper limit of hepatic dulness is at 4th rib. There is considerable œdema at the lower part of back, and slight œdema in the feet. The heart and lungs appear healthy. Urine passed in 24 hours measured 10¼ oz.; specific gravity 1020, of a light yellow tint, free from albumen, with a copious deposit of mucus and epithelium.

14th. The abdomen is smaller, measuring 24¾ inches; the liver can now be felt; its lower margin is about three fingers' breadths below false ribs; it does not feel nodulated on the surface.

Dec. 1st. Much improved in general condition. He about this time contracted hooping-cough, during which his ascites almost disappeared. He was discharged on 29th December and re-admitted on 25th March. His abdomen had been swelling for six weeks, and now measured 26½ inches. The ascites was somewhat reduced in the hospital, and he was discharged on 8th May.

Nov. 12th, 1867. He was again admitted. During the previous four months he had not been so well, frequently suffering from pain in abdomen. Has had a cough lately.

13th. His condition was very much the same as when previously in the hospital, except that his conjunctivæ had a slightly jaundiced tint. Abdomen now measured 28¾ inches. The lower edge of liver could be felt three fingers' breadths below xiphoid cartilage. He had a little general bronchitis; and some general deficiency of resonance on percussion over the back of right lung. After this he had a mild attack of scarlatina, which he passed through without any apparent injury. In fact on the 16th December his abdomen had fallen in circumference to 25 inches ; the edge of the liver could be felt more distinctly, it seemed rather harder than natural. His appetite was good ; he was cheerful and had a good colour. He had no œdema in his feet.

Remarks.—This case is remarkable for the manner in which the ascites came on in an apparently healthy child without any other symptom except moderate constipation and two or three slight attacks of sickness. It was noteworthy that the fluid was almost entirely absorbed during the course of hooping-cough, and was very much less after an attack of scarlatina than it had been before. He has been in University College Hospital, but without benefit. I have treated him with iodide of potassium, with small doses of bichloride of mercury, also with diuretics and diaphoretics ; he also has had quinine and iron. He has not received benefit from any of the medicines administered.

The pathology of the case is very obscure. It can scarcely be supposed to be cirrhosis of the liver : independently of the extreme rarity of the disease in children there is neither the cachexia, the dyspepsia, the pale stools, the contraction and granular condition of the liver, nor the splenic enlargement which one expects to meet with in cirrhosis. There is no sign of tubercular disease nor of renal nor cardiac disease.

A case in many respects similar to that of Alfred Smith has been recently under my care, and formed the subject of a communication to the Clinical Society.

The patient was in this case also a boy, whose abdomen began to enlarge when he was $4\frac{1}{2}$ years old, after a feverish attack accompanied by sore throat which lasted about 3 weeks. He had no pain in the abdomen, no sickness, and no diarrhœa at the time. There was some looseness of the bowels a month later.

When he came under my care there was great distension of the peritoneal cavity, the abdomen measuring 28 inches in circumference. The liver was enlarged, but not the spleen. All the characters of ascites were well marked, but there was no other dropsy. The veins of his neck were enlarged, and pulsated synchronously with the heart. His pulse and heart's impulse were very weak. His lips and face were of a deep livid tint. The urine was scanty, concentrated, and free from albumen. Some of the fluid was drawn off by paracentesis and found to be of a clear lemon colour, with some flakes floating in it. It was of an alkaline reaction, specific gravity 1019, not precipitated by acetic acid, but solidified by boiling. The case will be published in the "Transactions of the Clinical Society." (See *Postscript*, page 284.)

The pathology of these and similar cases of ascites, which come on without the least sign of inflammation, quite independently of tubercle, cancer, or any renal or cardiac disease, requires elucidation. The symptom sets in rather suddenly and remains as a chronic condition without much disturbance to health, beyond what is due to distension and pressure. I am inclined to believe that they are due to *obstruction* in the venous system, probably in the trunk of the vena portæ, or it may be in the larger branches of the hepatic vein near their entry into the vena cava. Amongst conceivable causes of such obstruction may be mentioned, thrombosis in the course of the veins, plastic exudation narrowing them on the outside, or an enlarged gland or tumour pressing upon the veins.

It has been recently shown that there are free openings between the lymphatic vessels and the cavity of the peritonæum. It is possible that obstruction of the thoracic duct or of the lymphatic glands connected with the lymphatics of the peritonæum might cause effusion into this cavity. Dr. Habershon mentions two cases in which the fluid of ascites had a milky appearance; in one of these cases the thoracic duct was obstructed, and in the other the lymphatics of the abdomen generally were much distended.

These facts may be borne in mind in our attempts to explain obscure cases of ascites, though they do not, I think, throw any light on the cases which have come under my notice in children.

In the next case the cause of the ascites is also doubtful.

It would seem to be some morbid condition involving the liver, as it was ushered in by jaundice at first.

Henry Bate, æt. 10 years, a tolerably healthy child until 2½ years ago, when he had an attack of jaundice with considerable enlargement of abdomen. This lasted two months, during which he had oil of turpentine given to him in doses of 6 minims, repeated thrice daily. The jaundice and swelling of stomach gradually subsided together, and he remained well until six weeks before admission, when the abdomen again began to swell, and he has been losing flesh and has appeared languid ; he has also complained constantly of frontal headache. His motions have been of a light drab colour, partly hard and lumpy, and partly relaxed. His appetite has been moderately good, except at breakfast time.

April 8th. On admission he was found to be moderately nourished, pale, with long eye-lashes ; chest rather contracted immediately below the level of nipple, and margins of costal cartilages spread out below. Tongue moist, with a thin white fur. Pulse 94, feeble. Respiration 22. Abdomen full, not very hard. From an inch on each side of umbilicus to the lumbar regions percussion note is quite dull. A sense of fluctuation felt from side to side of abdomen. Bowels open, motions of a pale colour. To take calomel, gr. j. ; pulv. rhei, gr. viij ; and extracti taraxaci, gr. vj. ; potassii iodidi, gr. ij. ; decocti scoparii, ʒss. Ter die. The measurement of abdomen varied from 25 to 26 inches. The superficial veins of abdominal walls were considerably distended. Neither the liver nor the spleen seemed to be enlarged.

21st. In much the same condition. Appetite moderately good. An enema of barley water and tincture of assafœtida was given, which dispelled some flatus ; but left the abdomen full, dull on percussion, and fluctuating as before. He remained in the hospital for a month, during which his symptoms continued much the same ; his motions were abnormally pale ; there was no bile or albumen in his urine ; he had frequent frontal headache. He had a slight attack of erysipelas of the face, which lasted only a few days.

May 22nd. When he left the hospital his general condition was improved and his abdomen was a little smaller. For a short time he was treated with bichloride of mercury in doses of $\frac{1}{24}$ of a gr. three times a day ; under this treatment the abdomen became more swollen and the medicine was discontinued.

Remarks.—The enlargement of the superficial abdominal veins appeared to indicate obstruction in the portal circulation of the liver and complementary increase of the collateral vessels, in order to carry some of the portal blood direct to the heart without passing through the liver by the hypogastric and epigastric veins.

In the seventh volume of the "Pathological Society's Transactions," page 227, I have recorded the case of a child, aged 13 years, who died of hæmorrhage into the stomach and intestines, accompanied with a lobulated and indurated condition of the liver (cirrhosis). This girl had several times suffered from ascites and jaundice.

The peculiarities presented by cirrhosis in children is a subject which requires further investigation. It is not due to spirit-drinking, and runs a more chronic course than in the adult.

The *diagnosis* of ascites in children requires some care. Mere tympanitis may be mistaken for ascites, unless care be taken to ascertain whether there is fluctuation from one lumbar region to the other, when pressure is applied in the middle line, without which a sense of fluctuation may be given when there is no fluid in the peritoneal cavity. The upper line of percussion dulness, which will be altered by raising the child from its back to a sitting posture, should also be noted; if fluid be present in the peritoneal cavity it will, of course, gravitate to the most dependent part. In cases of obscurity the bowels should be freely relieved of fæces and flatus by a good turpentine injection.

A very unusual case of congenital disease simulating ascites has come under my care, and is reported in the "Medico-Chirurgical Transactions," vol. xlviii. It was due to enormous distension of the pelvis of the kidney which caused great abdominal enlargement, the circumference being 27 inches when he was $3\frac{1}{2}$ years old. It was at first mistaken for ascites, but a more careful examination showed that the swelling was not symmetrical, that the right flank was much more resistant and prominent than other parts, and that there was absolute dulness at the umbilicus and on the right side, but not in the left flank. On tapping the tumour five pints of a clear, non-albuminous fluid were withdrawn; it was found on examination to have all the characters of dilute urine. Tapping was several times repeated, but with only temporary relief. There is obviously an obstruction of the ureter due to congenital defect; there is, however, at intervals a communication established between the kidney and the bladder, because on several occasions when the accumulation has reached a certain point there has been an unusual flow of urine from the urethra which has partially reduced the swelling. The case is a very remarkable one; not long ago the boy was living and no new symptoms had occurred.. The swelling is so great as to prevent him walking about, though he is 8 years old, and it interferes much with his respiration and general nutrition.

The prognosis and treatment of ascites must remain in an unsatisfactory position until more is known of the diagnosis and pathology of the disease. If there be tubercular disease, this should be treated in the ordinary way, the general health in every way should be improved, and the diet should be nutritious and of an easily digestible character.

If there be abdominal pain or tenderness, the application of mustard plasters or of iodine liniments is advisable. The abdominal walls may with advantage be supported by a flannel or elastic bandage. The bowels should be regulated by occasional aperients, and the action of the liver encouraged by carbonates of potash and soda with taraxacum and a bitter infusion, such as calumba. Diuretics, such as scoparium, digitalis, spirits of nitrous ether may be used, but do not seem of much service. Turpentine appears to have been of use in Bate's case. Paracentesis should only be resorted to in very extreme cases, when dyspnœa is excessive, and compression of the lungs very serious.

When the case is chronic and no benefit is derived from the use of diuretics, it will be better merely to attend to the general nutrition of the patient, give the syrup of the iodide of iron, and support the abdominal walls with a bandage.

Postscript.—The patient referred to on page 280 died whilst this book was going through the press. He suffered from effusion into the right pleura during the last week of life. The liver was cirrhosed, and this was the cause of the ascites; the heart was universally adherent to the pericardium by old adhesions. There was old cretified tubercle of the bronchial glands; and some recent miliary tubercle on the pleura.

SCARLATINA.

Of all the diseases incident to childhood there is no acute disease which presents greater varieties of aspect than scarlatina, and there is none in which it is more difficult to predict the probable course and result. There is no disease which so often brings, with very short notice, unexpected deaths into a healthy household; it attacks alike the rich and the poor, and selects often for its victims the most robust and healthy.

Amongst acute specific diseases it ranks *first* in regard to mortality. On the registers of the Registrar-General it stands second, being preceded by "typhus;" but this heading, it must be remembered, includes typhus proper, typhoid, and many other cases of disease certified vaguely as "*fever*."

Scarlet fever has, since the beginning of 1848, a period including three epidemics of cholera, numbered more victims in London than cholera. From the latter disease the deaths in three epidemics were 30,452, whilst from scarlatina between January, 1848 and December, 1866, the deaths in London numbered no less than 52,461.

It is peculiarly a disease of childhood, 95 per cent. of the deaths occur under 15 years of age, only 1·75 per cent. of the deaths occur above 25 years of age. The highest mortality occurs during the second, third, fourth, and fifth years of life.

This disease was first distinguished from measles 300 years ago by Philip Ingrassias of Naples. It was then known as *Rossalia* or *Rossania*. It was, however, constantly confounded with measles until the end of last century, when its distinctive characters were more fully pointed out by Withering.

It is a disease whose intensity varies extremely at different periods. In the time of Sydenham, it was so trifling that he speaks of it as "the mere name of a disease much less to be feared than measles." It assumed the form of a very fatal epidemic in 1672-86, when it was described by Morton.

Graves, in his " Clinical Lectures," states that "in the years 1800-1804, scarlet fever committed great ravages in Dublin, the type which it assumed was most virulent, sometimes terminating in death as early as the second day. It thinned many families in the upper and middle classes, and left not a few parents childless. From the year 1804 to 1831 scarlatina epidemics recurred, but always in a mild form, so much so that many were led to believe that the fatality of the former epidemic was chiefly, if not altogether, owing to the erroneous method of cure adopted by the physicians of Dublin, and that the diminished mortality was entirely attributable to a cooling regimen and the timely use of the lancet."

Graves adds : "This was taught in the schools, and scarlet fever was every day quoted as exhibiting one of the most triumphant examples of the efficiency of the new doctrines. This I myself learned, this I taught, how erroneously will appear in the sequel." He goes on to say, " The experience derived from the present epidemic has completely refuted this reasoning, and has proved that, in spite of our boasted improvements, we have not been more successful in 1834-5 than were our predecessors in 1801-2."

The error committed by the Dublin physicians is one into which we are all prone to fall, not only in reference to scarlet fever but also to many other diseases. At one period and in one place the prevailing type of disease is much milder and more tractable than at other periods and in other places ; whatever plans of treatment are adopted under the former conditions, the majority of the patients recover. We must not conclude that the modes of treatment which have been adopted are necessarily the best, and that all other plans of treatment are objectionable, even although the mortality has been higher in the latter case than in the former. The difference in the mortality of scarlatina at different periods in the same place and under similar plans of treatment is very striking. In the course of 11 years the annual mortality from scarlet fever, in the London Fever Hospital, varied from 1 death in 40 cases to 1 in 6; or, in other words, from 2·5 to 16·5 per cent. In the Hospital for Sick Children, the annual mortality has varied in different years from 9 to 31 per cent. A large number of the cases admitted to this hospital, are cases which have been neglected during the early periods of the disease and are suffering from this neglect.

Dr. Richardson has suggested that there is usually almost a constant proportion between the number of mild and of malignant cases at all periods, and that in any epidemic where there are many malignant cases there is always a proportionate number of mild cases, and *vice versâ*. This is certainly contrary to my experience, and is disproved by the statistics of the two hospitals just quoted, as well as by the history of the disease in Dublin, and in this country in Sydenham and Morton's time respectively.

The only known cause of the disease is contagion, we do not know whether it ever arises *de novo* in the present day. I am strongly inclined to the belief that it does sometimes originate without exposure to infection.

The period of *incubation* varies from a few hours to six days. That the disease may show itself within 24 hours of exposure to infection is proved by cases related by Trousseau and Dr. Murchison, and I have myself met with cases which prove the same point. Whether the latent period ever exceeds 6 days, it is more difficult to decide. Persons frequently show the first signs of disease more than six days after the first exposure to contagion, but it is difficult to say when the poison entered the system, if the person have been continuously exposed to its influence, and to exclude the possibility of a more recent exposure in doubtful cases. Three sisters came under my care at the hospital: one of them was attacked with scarlatina on the 14th September, was brought to the hospital on the 15th September, and remained there until the 18th October. Eight days after her return home a sister, who had been with her at first, showed signs of the disease. The second patient remained at home ill from the 26th October. Another sister, who lived in the house and not at all isolated, remained free from all symptoms until the 10th November, when the disease attacked her. These cases are interesting to show the difficulty of deciding the length of the incubating period where there is continuous exposure to infection.

There are, however, some cases on record where a much longer interval has elapsed between the last exposure and the outbreak of the disease. In the "Lancet" (vol. ii., 1864), Dr. W. M. Saunders, R.N., relates two cases which occurred at sea at no less than 51 and 54 days respectively after exposure.

As to the properties of the *virus* the following points are known : it cannot be carried far by the air, probably not many feet; it is readily carried by clothes, by persons who do not contract the disease, or by other fomites, to an indefinite distance; it retains its power of propagation for a very long period, especially if not freely exposed to the air; it is destroyed by heat below the boiling point of water. From these properties we may argue that the virus is not very volatile, and that it is either an organic solid, or is usually attached to an organic solid such as epidermic cells. The infection may be retained in a room for weeks or even months. A very remarkable instance of this is given by Dr. Richardson in "The Asclepiad." He relates it as follows:—

"During my early career I assisted a practitioner at Saffron-Walden. In our union district we had an outbreak of scarlet fever. At a short distance from one of our villages there was situated on a slight eminence a small clump of labourers' cottages, with the thatch peering down on the beds of the sleepers. A man and his wife lived in one of these cottages with four lovely children. .... The poison of scarlet fever entered this poor man's door, and at once struck down one of the flock. It seemed to me that I had saved the remaining children by obtaining their removal to the care of a grand-parent who lived a few miles away. Some weeks elapsed when one of these was allowed to return home. Within twenty-four hours it was seized with

SCARLATINA AFTER SURGICAL OPERATIONS. 289

the disorder, and died with equal rapidity to the first. We were doubly cautious in respect to the return of the other children. Every inch of wall in the cottage was cleansed and lime-washed. Every article of clothing and linen washed, or, if bad, destroyed. Floors were thoroughly scoured, and so long a period as four months was allowed to elapse before any of the living children were brought home. Then one child was allowed to return. He reached his father's cottage early in the morning, he seemed dull the next day, and at midnight in the succeeding twelve hours I was sent for, to find him also the subject of scarlet fever. The disease again assumed the malignant type, and this child died."

He further adds :—" I have always believed that in this instance the thatch was the medium in which the poison was retained." This would seem to be the only way of explaining the phenomena, unless we believe in the possibility of an independent origin.

We do not know on what conditions the receptivity of a subject depends, why some persons contract the disease and others exposed to the same degree of infection escape, or why a person resists free exposure to infection at one period and succumbs to the disease after slight exposure at another period. Surgical operations certainly seem to render persons peculiarly receptive of the disease ; but, as a general rule, cases occurring under these circumstances are not very severe, in fact, are often so mild that doubts arise whether they are really scarlatina or a non-specific rash. That they are really scarlatinal seems to be proved by the following observations, well stated by Dr. Gee in Reynolds' " System of Medicine" (vol. i., p. 350) :—" First, that the disease occurs in epidemics ; secondly, that in a given epidemic a severe case occasionally relieves the monotonous recurrence of a very mild form; thirdly, that a precisely similar scarlatinilla attacks, in the same epidemic, patients who have not been subjected to operation and who have no open sore ; and lastly, by way of a veritable experimentum crucis, that however freely these patients are exposed to ordinary scarlet fever contagion afterwards, they do not contract that disease." In the Hospital for Sick Children, of the children who contract scarlatina, a very large proportion have been the subjects of a surgical operation within a week before the rash appears.

U

The conditions which determine the severity of an attack of scarlatina are not understood. As stated already, during some epidemics there is a preponderance of malignant cases. Family constitution exercises a great influence over the character of the disease. The circumstance that one or more members of a family have suffered from the disease in a very violent form is a sufficient reason for increased anxiety, should any other members of the same family contract the disease. Scarlatina does not seem to be less frequent in well-drained places than in those which are ill-drained, the mortality from the disease is not obviously reduced by ordinary sanitary improvements. Overcrowding, as might be expected, somewhat promotes the spread of the disease. It is stated that defective drainage occasionally renders attacks of scarlet fever more virulent ("Edinburgh Medical Journal," Oct. 1859). I have not myself been able to obtain evidence of this effect, although I have carefully sought for it in cases that have come under my notice.

Social position does not seem to exercise any influence over its mortality. With the view of obtaining information on this point, I compared the social position of 500 children who died in St. Pancras from scarlatina, with a corresponding number of children who died from all causes. The results were as follows :—

		Scarlatina.	All causes.
Children of	gentlemen	7	7
,,	professional men	15	9
,,	merchants, tradesmen	100	79
,,	clerks	31	25
,,	artisans and skilled labourers	205	195
,,	policemen, soldiers, and postmen	12	23
,,	household servants	25	49
,,	common labourers	109	121

The families of the classes on the upper part of the list suffered proportionally more from scarlatina than from other causes, whilst the families of those on the lower part of the list suffered more proportionally from other causes of death than from scarlatina. It must be remembered that the children of the poor suffer a much higher mortality from all causes than those of the rich, so that these figures do not show that scarlet fever is more fatal to the rich than the poor, but merely that it is one of the diseases of which the mortality is not so much reduced by a comfortable social position as is the mortality of many other diseases.

Of several persons equally exposed to the same infection, one escapes entirely, another has a mild attack of the disease, a third has simply a sore throat, and a fourth has the disease in a most virulent form. Such results as these may ensue from exposure either to mild or malignant forms of the disease.

In this, as in other exanthemata, one attack usually constitutes a protection against another in the same person. Recurrence of the disease once or even twice in the same person is, however, not an unheard of event. Dr. Richardson states that he has himself had three distinct attacks of scarlatina. Some persons, whenever they are exposed to the infection, suffer from sore throat more or less scarlatinal in character. Relapse of the disease within a month of its first occurrence is not very rare.*

I have seen a young woman in the Fever Hospital suffering from a second attack of scarlatina, the first attack having occurred five weeks previously. She had quite recovered from the first illness, and was acting as nurse. In both seizures the rash, the throat, and other symptoms were characteristic. The relapse (or recurrence) was less severe than the primary disease.

Attack.—The commencement of scarlatina is usually very sudden, occurring without previous warning; in this respect it differs from measles, and still more strikingly from typhoid fever. Occasionally the patient complains of a sore throat for a day or two before any other symptom arises. More frequently

---

* A student, who was present when this lecture was delivered, informed me that he had himself contracted scarlatina three times, and that a week after his third attack he had a distinct relapse.

without any warning the patient becomes suddenly ill ; after a slight chilliness the skin becomes hot, the pulse is quickened, there is sometimes sickness, and usually more or less soreness of the throat. Rigors or severe headache are rare. Convulsions occur but seldom at the onset of this disease; they do not invariably indicate a malignant form of the disease. They are rarer in this disease than in measles or small-pox. Languor and sleepiness, with disturbed sleep at night, frequently precede the rash. Slight frontal headache and aching of the limbs are usual symptoms.

At the end of 12 or 24 hours the *rash* is usually to be seen ; occasionally no symptoms are observed until the appearance of the rash. Barthez and Rilliet say it was the first symptom in 4 out of 24 cases. I have found it to be so in 3 out of 26 cases, not including any that have followed closely on surgical operations. The rash is sometimes postponed until the third, fourth, seventh, or, Trousseau says, even to the eighth day. In some of these cases of supposed delay, the rash has appeared partially and has receded, again to return at a later date.

The eruption appears first at the root of the neck, upper part of the chest, loins, and arms, soon spreading to the face, abdomen, and legs. It sometimes comes out at once over the whole integument. It consists of minute scarlet points (puncta), which disappear on pressure, but rapidly return on the removal of the pressure. The puncta may be quite discrete and distinctly elevated above the surrounding skin. They are often confluent, and on the forehead and cheeks especially, a uniform red blush is the appearance presented. The colour is sometimes rather purplish or dusky, and in other cases paler than usual. It is sometimes limited to the neighbourhood of the joints and the bottom of the back. The hair follicles are at times elevated and slightly reddened. As the redness of the skin subsides a slight yellowness often remains for a day or two. Petechiæ are not uncommon near the axilla and pubes when the rash is copious ; they are not necessarily of serious import. About the eyelids, hands, and feet, subcutaneous swelling from œdema occasionally arises. The duration of the rash varies from one day to five or ten, the average duration being about four days. When very copious and intense it usually lasts the longer periods. The intensity of the rash begins to subside in a typical case on the fourth or fifth day. As the rash declines miliary vesicles, with semi-transparent or turbid contents, often appear more or less copiously near the top of the sternum and on the abdomen; they are less frequent in children than in adults; their contents dry up in a day or two and desquamation ensues. When the nail is firmly drawn over the skin at the height of the rash, a white line soon follows, lasts for some seconds, and is lost; strong pressure gives rise to a red line with white streaks on each side of it. These appearances have been regarded by Bouchut as diagnostic, but they are to be seen in many forms of erythema.

The *pulse* in scarlet fever usually rises in frequency on the first day and continues to be very rapid until the fifth day, when there is often a fall. It usually falls with the temperature. A pulse from 140 to 160 on the first day of disease is not necessarily of grave omen. I have seen a boy, 3½ years old, whose pulse numbered 180 on the second and third days of scarlet fever, which did not prove fatal. The pulse is usually above 120, except in the mildest cases. An abnormally unfrequent pulse, with some irregularity, is by no means rare between the fifth and tenth day.

Temperature.—This may range from 100° to 105°, very rarely rising to 106° in the axilla. Of 28 cases carefully observed by Dr. Ringer and Dr. Gee at the Hospital for Sick Children, a considerable fall of temperature, nearly to the healthy standard, took place on the 4th day in 4 cases, on the 5th day in 11 cases, on the 6th day once, on the 7th day in 3 cases, on the 10th day in 4 cases, on the 15th day in 2, and on the 20th in 1 case; whilst in 2 cases the fall of temperature was so gradual that it was impossible to say exactly on what day it reached its normal point. Dr. Ringer's observations seem to indicate that there is a tendency to a cycle of five days, so that a crisis occurs more frequently on the fifth or some multiple of five, than on the intervening days. Even when the temperature continues high till the 10th, 15th, or 20th day there is usually some fall on the 5th day, but it may subsequently rise even higher than it had risen at any period before the 5th day. The range of temperature on any one day after the first fall may be very slight or considerable ; when there is a very decided remission of pyrexia each morning, the case is usually about to terminate favourably. The hour of day at which the highest temperature is attained is usually in the afternoon, between 2 and 8 o'clock, but it may be at any hour between 9 a.m. and 11 p.m. If, after a complete fall, the temperature again rises to any considerable extent, this is generally due to some local lesion, either of the kidneys, throat, pleura, or of the endocardium or pericardium. An attack of urticaria occasionally intervenes and causes an elevation of temperature.

The *sore-throat* is as constant a symptom as the scarlet rash. In a typical case there is bright redness of the soft palate, tonsils, and uvula, with slight papular elevations on the soft palate. At the back of the pharynx and tip of the epiglottis there is also sometimes bright injection. There is also more or less swelling of the reddened parts, especially of the tonsils and uvula. In the tonsils, besides hyperæmia and œdema, there is excessive secretion of thick yellow matter, which appears studding the tonsils at the follicular openings, or covering them with a thin uniform layer. Ulceration soon follows in many cases. Except in grave cases of scarlatina anginosa, ulceration does not extend beyond the tonsils until after the fifth day. The lymphatic glands below the angles of the lower jaw swell and are often tender. As the rash subsides, the ulcers of the tonsils often become deeper, and there is suppuration of the lymphatic glands, with points in the side of the neck. In other cases ulcers now appear on the uvula and soft palate, sometimes on the hard palate and back of the pharynx, being preceded by greyish patches surrounded with reddened mucous membrane ; the lymphatic glands are more swollen, and the whole front of the neck is enlarged.

These new ulcerations may heal or spread rapidly to adjoining parts. In the worst cases hard brawny swelling of the front of the neck is rapidly produced; softening may occur in one or two spots, and the skin slough to give exit to thin, shreddy pus. The swelling does not subside until large masses of connective tissue have sloughed away, the skin being undermined and riddled with holes. The brawny swelling may extend up on to the face and down over the clavicles. In cases of very extensive sloughing, death usually results; this may be caused by general exhaustion, hastened in many cases by difficulty in swallowing, vomiting, or diarrhœa. One of the large veins of the neck may be involved, causing fatal hæmorrhage; or a communication may be established between the pharynx and the surface, so that food and drink taken by the mouth escape more or less completely at the side of the neck.

In more favourable cases resolution occurs, and absorption of the effused lymph which causes the hard swelling; in other cases the inflammation is limited, and an abscess formed at one side of the neck. During the height of the fever the lymphatic glands of the groins and axillæ are often swollen.

The mucous membranes of the nose, of the lachrymal sac, and conjunctivæ, sometimes share in the inflammatory congestion; a thin glairy discharge exudes from the nostrils; this soon dries into a crust, which obstructs the entry of air and causes the patient to breathe with open mouth, leaving the tongue dry. The nasal discharge often excoriates the orifice of the nares, and has a very offensive odour.

The *tongue* is usually covered with a thin white fur, except at the tip and edges on the first day; the fur becomes thicker on the second and third days, the papillæ being more or less completely covered, then beginning to clear off from before backwards, leaving the tongue deep red, with prominent papillæ. Occasionally the tongue presents no abnormal appearances throughout. In unfavourable cases the tongue becomes dry and brownish about the fifth day. Vomiting does not often occur during the eruptive period except in very unfavourable cases. The bowels are usually slightly costive;

diarrhœa is an occasional symptom, especially in the graver cases, and it may prove a most troublesome symptom.

During the eruptive period the *urine* presents the following characters: it is diminished in quantity, the urea is not increased, and the chlorides are diminished; the pigment is not necessarily increased. Dr. Gee has carefully noted the phosphoric acid, and finds that it is in normal amount for the first three or four days; on the fourth or fifth day it is notably diminished, remaining for four subsequent days at a half or a third of the normal amount, and then assuming the healthy standard. A similar diminution of phosphoric acid has been found during the climax of pyrexia, or soon after, in ague and measles. Uric acid appears (judging from careful quantitative analysis in one or two cases) to be retained during the second, third, and fourth days, and to be excreted in excess on the fifth day.

Desquamation occurs in the great majority of cases of this disease. Of 23 cases where this symptom has been carefully looked for, two showed no desquamation during three weeks that they were under observation, whilst two others, who left the hospital respectively on the fourteenth and sixteenth days of the disease, showed no signs of desquamation up to the time of their discharge. The amount of desquamation is usually in proportion to the intensity of the rash. The date of commencement has varied in the cases of which I have notes from the 6th to the 25th day; two patients began to desquamate on the 6th, three on the 7th, three on the 8th, four on the 9th, one on the 10th, two on the 11th, two on the 12th, one on the 21st, and one on the 25th day. The desquamation may be so slight as to cause a mere harshness of the skin, and to exhibit on close inspection a few minute, branny scales here and there; or it may be so abundant as to cause large flakes of cuticle, and to cover the sheets with almost handfuls of the same. On the hands and feet where the cuticle is thick, the largest flakes of cuticle are seen. Desquamation often begins near the clavicles, sometimes on the forehead, at others near the elbows, at others in the groins. Desquamation may extend over many weeks, especially on the hands and feet, a succession of exfoliations taking place. When I have noted its termination carefully, it ended once on the 17th day, twice on the 25th, and once on the 29th day. In other cases that left the hospital severally on the 16th day, between the 20th and 25th days, between the 25th and 30th days, between the 35th and 40th days, between the 41st and 45th, and between the 45th and 50th days, desquamation had not ceased upon the hands and feet when the patients were discharged. When desquamation begins early, it is usually due to the rupture of miliary vesicles. Sometimes desquamation begins on the papular elevations of the follicles, and spreads from them to the surrounding skin, or it may not extend much beyond the papules.

Albuminuria is a very usual symptom during the desquamative period, less frequently met with during the eruptive stage. Of 72 cases, albuminuria was present in 47 ; but some of these came under treatment specially for the renal symptoms, so that the proportion of cases attended with this symptom is higher than it would be if all the cases of scarlatina which occurred in the same time had been observed and reported. There is no doubt that in different epidemics the relative number of cases of albuminuria varies. Some observers (Begbie, Holder, &c.,) have found albuminuria in nearly every case ; others in about one-third or half of the cases. I have found it once on the 2nd day, twice on the 5th day, three times on the 9th day ; in two other cases during the 2nd week, twice in the 3rd week, three times in the 4th week. Once it was present from the 11th to the 13th day, and then disappeared until the 20th day. In 8 out of 21 cases observed by Abeille, the urine was albuminous before the 6th day. When albuminuria occurs before the end of the first week it usually lasts only for about one day ; but at later periods it seldom ceases under a week.

When the throat suffers in scarlet fever more than usual, ulceration being very rapid and extensive, the term *anginosa* has been used to designate this form of the disease ; many cases of "cynanche maligna" are of this character.

The term *malignant* is often used for any case of the disease which proves fatal before the end of the second week, whether from the complications, such as the ulcerated throat and its results, or intercurrent inflammations, or from the intensity of the disease manifested more especially by its effect on the nervous system.

Scarlatina Maligna is a term more strictly applied only to the latter class of cases. These may be divided into two sets : those in which there is considerable febrile excitement, followed by exhaustion, and those in which collapse occurs with scarcely any preliminary excitement.

In the former (the *ataxic* variety), after delirium and sleeplessness, the rash comes out well, the throat is sore, and the temperature is very high. Delirium, more or less noisy, continues. The eyes are bloodshot, the pulse frequent and moderately full, but compressible; vomiting is a frequent symptom, and diarrhœa is occasionally present.

In a day or two depression ensues, muttering, with very weak pulse, dusky injection of the cheeks, vomiting, and sometimes tympanites. Stupor succeeds, speedily followed by death. Convulsions and coma occasionally precede death.

In the latter, or *adynamic* variety, the patient is at first pale and exhausted; he vomits, and the bowels are relaxed. There is subsultus tendinum, with great anxiety, and feeble delirium. The pulse is small and weak, sometimes very rapid. The parts exposed to the air feel cold, though the axilla, and still more, the rectum, may reach 105°. There may be an imperfect reaction, a slight eruption more or less dark in tint appears, but the pulse continues weak and rapid, and the extremities are cold; livid palor of the face, with stupor, alternating with convulsions; irregular respirations, abundant sweats, coldness of the skin, and death. Such cases may terminate within twenty-four hours or last two or three days. During some epidemics patients have been seized with convulsions, complete collapse, or coma, and have died in a few hours, before there has been time for any of the characteristic features of scarlet fever to be developed. Except from the prevailing epidemic, such cases could not be distinguished from small-pox, cholera, or other intense blood poisoning.

In the following case convulsions occurred within thirteen hours of the attack, frequently recurring on the second, fourth, and seventh days. On the tenth day the patient died.

Scarlatina Anginosa, with Convulsions frequently returning.—Pneumonia.

William Howley, æt. 2 years, was well on May 31st until 11 p.m., when he became hot and uncomfortable. During the night he vomited frequently, and his bowels were much relaxed.

June 1st. At noon he had a fit which lasted ten minutes; the face became livid, his limbs twitched and were stiff; his eyes were fixed. His skin was hot, mottled about the neck, and he was thirsty. The next morning at 3 a.m., a succession of fits occurred. In the course of the day a red rash was seen on chest, neck and abdomen. He was delirious towards evening.

3rd. Delirium. Some pain in swallowing.

4th. Admitted into the hospital. Rash well out, but of a dusky tint. Skin very hot. Tongue furred and red. Uvula and tonsils swollen and red. Pulse 130, weak. Had two fits during the day; his left hand and left eye twitched.

6th. Discharge from nose. Sordes on lips and teeth. Tongue dry and brown. Tonsils and uvula much swollen, coated with muco-purulent secretion. Glands of neck considerably swollen. Rash pale. Pulse 160, weak. No delirium.

7th. Has had two more fits. Bowels relaxed. Pulse 152. Respiration 52.

8th. Pulse 160. Respiration 40, slight action of nares. When attempting to swallow fluids they return through the nose.

9th. Languid and drowsy. Glands of neck more swollen. A good deal of difficulty in swallowing. Fluids return through the nose.

10th, 11 *a.m.* Has been unconscious for four hours. Lying on back. Pulse 180, very weak. A hard swelling under left angle of jaw. He remained insensible, and died at 3 p.m. After death, much lividity of dependent parts.

Brain.—Superficial vessels of convexity generally injected, but more in posterior half. No excess of fluid in the ventricles. No softening, no lymph.

Tonsils and walls of pharynx generally ulcerated. Lymphatic glands below both parotid regions much enlarged, moist, red, and not very brittle.

Lungs.—Lower lobes of both, the seat of lobular pneumonia. Bronchial tubes congested, containing a good deal of yellowish-grey muco-purulent fluid. Weight of two lungs 8¾ oz.

Spleen, weight 2 oz., rather soft.
Kidneys, heart, stomach and intestines healthy.

Scarlatina without eruption has been described, and is not very rare. One often has children brought suffering from albuminuria and dropsy, succeeding a feverish attack with sore throat and swollen glands in the neck. On inquiry, the mother states that there has been scarlatina in the house or in other members of the family, but that this child has had no rash on the skin, though it was carefully looked for, because it suffered for several days from very much the same symptoms as those children who had undoubted scarlatina. In these cases, it is true, we have not had the advantage of a careful medical scrutiny, and in some cases the parent's report is not to be relied on ; but in others the mother is obviously an observant, trustworthy person.

The following case may be given as one in which there was little or no rash, in a severe case of the disease.

Ellen Mackay, æt. 11½ years, a girl who had enjoyed good health. Towards the end of November several children died of scarlatina in the house in which she lived.

Dec. 1*st*. In the morning she complained of sore throat, took no breakfast, and soon afterwards shivered and ached all over. She did not go to bed, but seemed ill all day, and during the night was restless and hot.

2*nd*. She got up, but very quickly returned to bed. She was more feverish and very ill during this and the following day.

4*th*. Some running from her nose was observed, and during the following night she was delirious, and constantly wanting to get out of bed. Her bowels only acted once in three days and she did not vomit. No eruption was seen on any part of her skin.

5*th*. She was admitted into the hospital.

6*th*, 10 *a.m*. She was found with flushed face, hot skin, dry lips. The nares slightly excoriated, and a thick muco-purulent discharge was exuding from the nose. Tongue inclined to dryness, coated with a very thick brownish-white fur, concealing all the papillæ. Fauces red, inflamed, and much swollen. Both tonsils swollen and ulcerated. Much muco-pus obstructing the fauces. Glands at the angles of the jaws swollen and tender. Pulse 116, of moderate volume, but weak. Temperature in axilla 102·8°. No rash. Bowels have acted five times since admission, from a powder containing 1 gr. of calomel and 8 of rhubarb. She passed a restless night, wandering at times. Urine is of a clear pale yellow colour, free from albumen, specific gravity 1025. At noon two pails of water (80° Fahr.) were dashed over her in an empty bath ; after which she was immediately wrapped in blankets. At 3 p.m. she says she has been more comfortable since the cold affusion. Her temperature now 102·6°. Pulse 118, full and soft. Skin not so pungent, has a slight degree of moisture. She has taken since admission 3 gr. of carbonate of ammonia every three hours, and is now to take 3 oz. of port wine in the day.

7th. Passed a quiet night and is decidedly better this morning. Tongue is moist, cleaning at the tip. Fauces less swollen. Pulse 106. Temperature 99·8°. Urine cloudy from lithates, free from albumen.

8th. Appetite returning. Throat scarcely pains her at all. Temperature 97·8°. No rash. Tongue clean at tip and edges ; fur on dorsum thinner. Fauces less red. Uvula œdematous. Ulcers of tonsils cleaner. Pulse 100.

9th. Much better. Pulse 88. Temperature 97·2°. Tongue clean, except in centre ; papillæ rather prominent. From this time she made steady progress ; no albumen appeared in the urine, or but the faintest trace on the 12th day. The ulcers of tonsils were healed on the 13th day ; the throat was relaxed for another week. The pulse was 72 and irregular on the 14th day. She was kept under observation till the 27th, but no desquamation was detected.

Remarks.—In this case we had a case of scarlatina anginosa, with all the symptoms well marked except those on the skin. The cold affusion was grateful to the patient, and the ammonia treatment seemed to answer well.

The *diagnosis* of scarlet fever requires a few observations. When the rash is typical, its characters are so distinct as to leave no room for doubt. When the rash is scanty or absent the sudden accession, the amount of fever, the sore throat, and the characteristic tongue, will generally enable us to make a diagnosis, especially if the patient has been exposed to infection. Cases of malignant sore throat are not always easily distinguished from scarlatina anginosa with but little or no eruption.

Diphtheria may resemble scarlatina, but the accession is usually less sudden, the temperature is not nearly so high ; the pulse is not so frequent ; the throat is at an early period lined with false membrane ; the redness is very intense, but usually involves one tonsil more than the other. In scarlatina both tonsils are often covered with an opaque white layer of no great thickness, which may extend to the soft palate and tongue. This never becomes much thicker, and is separated in shreds, leaving a red sensitive surface. In diphtheria the deposit becomes of considerable thickness, and separates in layers, leaving a superficial ulceration in some cases, but often producing no loss of substance; and the sensibility is diminished. For further diagnostic marks between diphtheria and scarlatina, see the chapter on diphtheria.

The rash of scarlatina may be mistaken for measles, typhus, roseola, urticaria, lichen, or miliaria. *Measles*, when diffused and uniform, may resemble scarlatina; the tint is more purple, and there is more elevation of the reddened spots. When scarlet fever rash is darker than usual and the puncta large, it may closely resemble measles rash. Measles usually begins behind the ears or near the forehead, and is at first distinctly papular, the papules not being very small or acuminated. They tend to arrange themselves in segments of circles.

Roseola in some of its forms closely resembles scarlatina, the puncta are larger, more irregular, and less disposed to be confluent; their colour is of a rosy tint rather than scarlet. Cases arise in which the diagnosis is almost impossible from the rash alone.

Urticaria ought not to be ever mistaken for scarlet fever, when its anatomical characters are remembered; the wheals are never studded with puncta, are distinctly raised under the finger, appear and disappear suddenly with more or less tingling sensations.

Miliaria is vesicular, but the vesicles may be very small, and the redness around them bright; still ordinary care will prevent mistake. It must be remembered that miliary vesicles sometimes accompany the rash of scarlet fever.

Herpetic angina, often mistaken for diphtheria, can scarcely be distinguished from scarlatinal angina *per se*. The redness is not so diffused, and the tongue does not usually present scarlatinal characters; fever does not run so high.

Tonsillitis usually affects one side more than the other, and is more or less limited to the tonsils; fever often runs high, but keeps pace with the local inflammation and does not precede it.

There is a disease which is said to be more frequent on the Continent than in Great Britain, but which has been seen here, and described by numerous observers. It partakes of the characters both of measles and scarlatina. It has been called *Rötheln* and *Rubeola* in Germany; in England *Rubeola notha*, *Epidemic Roseola, Rosalia* (Richardson), *Bastard Measles*, and *Bastard Scarlatina*. The eruption appears on the second or third day, and at first resembles that of measles, becoming subsequently more like that of scarlatina. There are both coryza and angina in some cases, with slight subsequent desquamation. In other cases there is no coryza and but little angina; there is often a little swelling of the lymphatic glands of neck.

In some cases the rash looks more like scarlatina at first, and like measles on the later days.

There has been much discussion as to the true nature of these epidemics; some regard them as a hybrid or combination of the two diseases, whilst others consider them as constituting a totally distinct disease. The latter view appears to me to be the more tenable one, for the following reasons:—

Epidemics of this kind occur without the prevalence of measles or scarlatina at the same time. A previous attack of either or both of the true diseases does not appear to be any protection against this anomalous disorder. None of the characteristic sequelæ of measles and scarlatina occur after this complaint. I think the term epidemic roseola is the best one in the present state of our knowledge.

Sequelæ.—One of the most constant of the sequelæ, one which may be almost considered a part of the disease, is albuminuria already referred to. In connexion with this are nephritis and dropsy. From the observations of Dr. Ringer it appears that in the second and third weeks of the disease there is a decided decrease in the amount of urea excreted by the kidneys; during the first week there is no increase, sometimes a decrease. It thus appears that there is retained in the system either urea or some products capable of being converted into urea or derived from it. The function of the kidneys appears to be impaired more or less throughout the disease.

During the third and fourth weeks of the disease the urine is very liable to undergo the following changes. It becomes less copious, is sometimes almost suppressed, urates are deposited, urea is diminished, and the chlorides still more so; a considerable amount of albumen is present.* At this period there are no blood discs, but casts of renal tubes, some transparent and fibrinous, and some granular. Soon the amount of urine is increased to even above the normal amount; the specific gravity is low, the colour indicates the presence of blood, being red, brown, or smoky; under the microscope are found blood discs, renal epithelial cells, with granular and epithelial casts. Albumen is still present in considerable quantity, but relatively less to the amount of water. There may be a large quantity of blood with very little albumen. Subsequently smokiness disappears, but may return from time to time after exposure to cold or indiscretion in diet. By degrees albumen disappears, and the urine returns to its normal amount, being passed in large quantity for a short time. In other cases albumen never disappears, and the foundation is laid for renal degeneration.

In connection with the diminution of the urinary secretion there is very commonly present anasarca and serous dropsy to a greater or less degree. Dropsy is a symptom which seldom occurs to any serious extent, when the patients are kept in bed, or thoroughly warm, until the end of the third or fourth week, as they should be in all cases of scarlatina, however mild.

I do not remember to have ever seen a case of severe dropsy after scarlatina in a patient who was kept in bed long enough. Slight œdema of the face and over the sacrum may be frequently observed even with this precaution.

Dropsy without albuminuria has been observed in some epi-

---

* I once attended a case in which no urine was passed for 36 hours before death; and there was found in the bladder, post mortem, a small quantity of highly saccharine urine.

demics. I have never met with more than slight œdema without albuminuria. This may be due to anæmia, which is often very great, and is sometimes induced with great rapidity.

Dropsy of the pericardium and peritoneum is not often a marked symptom; occasionally, however, hydropericardium may cause dyspnœa, lividity, and fatal syncope. Hydrothorax is more frequently serious, and may be accompanied with œdema of the lung, which of course aggravates dyspnœa and hastens the final issue. Œdema of the glottis sometimes occurs and proves fatal.

The appearance of albumen in the urine is usually preceded by an elevation of temperature to the extent of four or five degrees; there is sometimes, but much less frequently, a quick throbbing pulse, thirst, nausea or vomiting, headache and restlessness.

Dropsy usually appears after the middle of the second week; it may appear for the first time as late as five weeks from the attack.

Serous inflammations are sequelæ very much to be dreaded. Meningitis is rare; cases are given at pages 307 and 323. Three or four cases in all have come under my notice. Pericarditis is also rare. Pleurisy is more frequent and is often very insidious. Peritonitis is rare. I have seen it in one case, where there was a large collection of purulent effusion which was discharged by the umbilicus. Pleurisy I have seen begin on the 21st day, on the 26th, and on the 28th day from the commencement of the primary disease. Several cases have been brought to the hospital where empyema existed, following scarlatina, in which it was impossible to say exactly when the pleurisy began.

The following will illustrate this statement:—

W. R., aged 2½ years. He appears to have had an attack of scarlatina of moderate severity, with sore throat and swollen glands. He appeared to be recovering favourably, when about the 15th day he complained of pain in the stomach. This was ascribed to gastric disorder by two medical men who saw him. It was very probably the commencement of pleurisy, which is often attended in children with epigastric pain. Four days later his breathing was hurried, there was no cough or wheezing; two days afterwards he was admitted into the hospital, and it was found that he presented signs of effusion in the right pleura. There was great dulness on percussion at the lower part of the right side of the thorax posteriorly, which was less marked when the child was laid on his face. Respiratory sounds were not absent over the dull part, but they were of a distant blowing character. The next day there was fine crackling friction near the angle of the scapula, which it was not easy to distinguish from pneumonic crepitation. The child became rapidly worse; the left pleura became involved, the urine was very scanty and highly albuminous, and on the fourth day after admission he died. No less than 3½ oz. of pus were found in the right pleura, and 3½ oz. of sero-purulent fluid were found in the left pleura. There was neither shivering, chilliness, nor convulsions to indicate this intercurrent disease; nothing but a careful examination of the chest would reveal the nature of such a case. There was no pain in the chest and no cough.

Empyema following scarlatina is not invariably fatal. Recovery may take place after spontaneous opening or after paracentesis.

Pneumonia is another sequela not very infrequent, and is also a very serious one. It may set in at any time after the first week of the disease; it also is sometimes very insidious in its approach, not being always attended by any symptoms which direct attention to the lungs. There is usually, however, a dry cough, increased heat of skin, and hurried respiration at the outset, with dyspnœa and movement of the alæ nasi after a day or two. It may be confined to one lung or involve both. It is remarkable for running a very rapid course. It is either lobar or lobular; in the former case especially it is usually accompanied with a certain amount of pleurisy. It is sometimes accompanied with pericarditis.

Diphtheria is sometimes a sequela of scarlatina, occurring at any time after the beginning of the third week. I have seen it begin once on the 31st day and once on the 43rd day; in each of these cases it proved rapidly fatal by extension of the disease to the air passages. Diphtheria sometimes occurs as a complication, commencing about the end of five or six days, or even earlier. It is generally fatal. Cases of this kind are reported at page 311.

Convulsions occasionally occur in connection with the renal symptoms following scarlatina. They probably depend on the retention in the blood of the products of retrograde changes of azotised tissues, which the kidneys are unable to remove fast enough. They are not of very frequent occurrence. It is a less serious symptom in connection with renal disease in children than in adults. Of 13 cases mentioned by Barthez and Rilliet 10 recovered. Dr. West reports 4 cases, of which 3 recovered. They commonly usher in local inflammations, taking the place of rigors. One case, (S. B., æt. 10 years) came under my notice in 1861, which proved fatal from the occurrence of meningitis, pericarditis and pleuro-pneumonia. On the 30th day of the disease violent convulsions came on without any premonitory symptoms. There was albumen in her urine at the time, but this was diminishing in amount from day to day, and she seemed to be gaining strength. The convulsions were accompanied with lividity and much elevation of temperature, and lasted, with short intermissions, for several hours. Soon after this the symptoms of pleuro-pneumonia appeared and proved fatal in nine days. There was enlargement of the kidneys, and the epithelial cells exhibited under the microscope a quantity of granules in their interior; there was some opacity of the arachnoid, both on the upper surface and at the base of the brain, and the sulci of the convolutions were filled with clear semi-coagulated fluid. There was pleuro-pneumonia of the right lung with œdema and congestion of the left. There was also lymph in the pericardium. Brunner's glands in the duodenum were considerably elevated and injected. The mucous membrane of the jejunum and duodenum was redder than usual. Peyer's patches of upper part of ileum were of dark colour, as if from pigmentary deposit in them. The mucous membrane of the large intestine at its upper part was roughened and studded with minute ulcers. The stomach looked healthy, except that the prominent portions of rugæ were of a darkish tint.

Bronchitis, leading to lobular pneumonia, sometimes occurs and occasionally proves fatal. Coryza occurring after the seventh day indicates mischief going on in the nasal mucous membrane and the adjoining sinuses, which may extend up the eustachian tube into the tympanum, leading to suppuration, rupture of the membrana tympani, discharge of the ossicula, and permanent deafness. This result may ensue without rupture of the membrane; and, if the patient be under five or six years old, dumbness is usually consequent on the deafness.

Otorrhœa may be limited to the external meatus. In strumous children especially it often becomes chronic, leading to thickening of the walls of the meatus and of the membrana tympani.

Suppuration in the subcutaneous tissues near the joints, or at the back of the pharynx, sometimes follows scarlatina. There may be one abscess or many.

Rheumatism is another sequela of scarlatina occurring after the middle of the second week. It sometimes presents all the symptoms of ordinary subacute rheumatism. In other cases, what appears at first to be rheumatism turns out to be a suppurative inflammation in and around the joints, a "pyogenic fever." The relation between these cases and true rheumatism is a subject which requires further elucidation.

Cancrum oris following scarlatina is rare. I have seen very extensive destruction of the bones of the nose, the palate, and the teeth, in which recovery followed. Death usually occurs.

Scarlet fever is less prone to develop tuberculosis than is either measles or hooping cough. I have met with a disproportionate number of cases of tubercle of the brain following scarlet fever; whether this has been a mere coincidence, or whether there is any connection between the two forms of disease, is a point for future observation.

Scarlet fever may occur with other acute specific diseases. Small-pox and scarlet fever may co-exist; in some cases scarlet fever has first shown itself, in others small-pox has been first, and in a third set of cases the manifestation of the two diseases has been simultaneous. (Dr. Murchison, "Medico-Chir. Review," July, 1859.) Vaccinia and scarlet fever have been observed together. Some of the cases of "rubeola notha" have been probably combinations of scarlet fever and measles. Scarlet fever in the course of typhoid fever has been noticed in a number of cases. In a boy under my care, the rash of scarlet fever appeared on the 14th day of typhoid fever; on the previous day he had been sick, and the tongue became coated with a thick white fur. He did not complain of sore throat; there was, however, a little injection of the fauces, and some swelling for a couple of days. The intercurrent disease did not in any way appear to aggravate the primary disease; the temperature, which had previously been about 103°, continued about the same for some days after the accession of scarlet fever; the pulse remained at about the same rate of frequency, and was improved in force for several days. One rose-coloured typhoid spot appeared about the seventh day of the scarlet fever. A very few spots had been previously seen. In some cases there have occurred simultaneously copious typhoid and scarlatinal rashes.

In one case under my care, scarlatina set in on the 29th day of typhoid fever, and proved fatal in six days. There was pneumonia on both sides, and sloughing of the gums of the lower jaw. There were numerous ulcers in the lower part of the ileum. The case is referred to in the chapter on typhoid fever.

Sequelæ of Scarlatina. Albuminuria. Double broncho-pneumonia. Parotid bubo. Recovery, after his life was despaired of.

George Saunders, æt. 7 years, was attacked suddenly with a scarlatina rash and vomiting on the 6th May; the rash lasted four days; he soon recovered, and was allowed to go about as usual. He made no complaint of sore throat.

June 21*st*. Nearly seven weeks from the attack of scarlatina his face was observed to be puffy; he had sickness and diarrhœa, with hurried breathing. Towards evening he became delirious.

22*nd*. Went out of doors with naked feet. His dyspnœa was increased. Two small swellings were observed at the angles of the jaws.

24*th*. He was admitted to the hospital with great dyspnœa. No note was taken until the 26th. Very restless. Skin not very hot, moist. Tongue fissured, coated in patches with white fur. Is very thirsty. Was sick twice yesterday. Pulse 170, very soft, of moderate volume. Respiration 66. A constant hacking cough. No œdema. Soft elastic swelling about left angle of jaw, reaching below the ear and over masseter, seems to be rather in connective tissue about parotid than due to enlarged lymphatic glands.

Chest.—Sibilant and sonorous rhonchi audible both back and front.

SCARLATINA.

Vocal resonance rather strong at right apex behind. Tonsils are swollen, not very red. Urine turbid. Amount of albumen not large. In the last 24 hours has taken 1 pint of beef-tea, 6 pints of milk, 4 eggs, 12 oz. brandy (1 drachm every 15 minutes).

27th. Very restless last night. The swelling on left side of neck greater and harder than yesterday. No tenderness or swelling on right side. Pulse as before. Some lividity of lips. Urine turbid with pale urates; albumen increased. Food and stimulants continued.

29th. Sitting up, leaning forward in bed. Pulse 148, more distinct than it was.

Chest.—Percussion note over whole right back half tubular and half amphoric, resistance increased. Large and small bubbling over the whole of right side with occasional sibilus; at the angle of scapula during inspiration, crepitation almost pneumonic in character.

Left back percussion note good. The same bubbling as on right side except at apex, where the respiratory murmur is coarse and exaggerated.

Dyspnœa very marked; head moves with respiration, being thrown back on shoulders. Bowels loose. Parotid swelling decreased. The puffiness has disappeared, and there remains a hardness only in the position of the parotid. Urine of smoky tint; albumen in larger amount. Sediment consists of blood discs; granular bodies larger than white globules without nucleus; casts of tubes very granular, some about $\frac{1}{3000}$ inch wide, some about $\frac{1}{1500}$ inch. At 7 p.m. very drowsy, wandering. Continues stimulants and fluid food. To take every 3 hours, ammon. sesquicarb. gr. iv.; spir. ammon. aromat., m̃v.; tinct. lavand. co. m̃x.; syrupi, ʒj.; aquæ ad. ʒss.

30th. Passed a good night. Sweats at times profusely. Pulse 156. Tongue inclined to be brown and dry in centre. Urine much paler, not smoky; still very albuminous. Has now 6 oz. of brandy in 24 hours.

July 2nd. Much better. Sudamina on chest and abdomen. Coughs less and with more vigour. Had some fish yesterday. Pulse over 160, fuller but soft. Respiration 46. Physical signs in chest as before. Urine of higher colour, 3 pints in 24 hours. Specific gravity 1020. Anasarca less than a few days ago.

4th. Pulse 168. Respiration 48. Skin hot and dry. Heart beats one inch below and outside nipple. Right limit of dulness right margin of sternum. He looks better and is more cheerful.

8th. Improving. Colour of lips natural. Skin hot, moist. Can now lie down easily. The percussion note over back of lungs, which was extra resonant, higher pitched and harder than normal, is now becoming more natural; less rhonchus in chest. No parotid swelling.

10th. Had meat and potatoes yesterday. To omit the ammonia mixture and take liquor cinchonæ and nitric acid.

11th. Pulse only 108, firmer. Urine free from albumen. Sediment shows uric acid crystals and a few blood and pus cells.

24th. To go into the country.

The recovery in this case was, I believe, in great measure due to the unwearied perseverance and energy of the nurse in carrying out all the prescribed treatment, which consisted mainly in the frequent regular administration of nourishment and stimulants.

Scarlatina complicated with Diphtheria in two children and their parents. Recovery of the father only.

The first member of the family who fell ill was a very healthy boy of 7 years old. He came home with his sisters to a well-built new house in Bayswater on the 7th March. On the evening of 11th was seized with sickness and fever; the rash of scarlet fever appeared next morning. His sisters were at once sent away. The attack was tolerably severe, the throat was sore, and there was some glandular swelling on the 2nd day. On the 5th day the rash had nearly disappeared, he became rather low, and the right side of his throat was more swollen. Several times during the fever he had been delirious at night, and drowsy during the day. I was called to see him in consultation on the evening of the 16th. I found him dozing, picking the bed-clothes. Pulse 166, weak. A clear discharge from nostrils; right nostril excoriated. When roused he muttered some unintelligible words. The right side of his neck much swollen, two swollen glands to be felt. Tongue red, papillæ prominent. On the right tonsil and middle of hard palate was seen a white tenacious deposit, and much redness around it. Skin not very hot. Hydrochloric acid and honey was painted on the throat. He had a mixture of carbonate of ammonia (3 gr. every 2 hours), beef-tea, eggs and milk ad libitum, with 6 oz. of wine. The next day he seemed a little better; but the deposit had advanced forwards and backwards. Towards evening the hardness of neck had increased, and the glands could not be distinguished.

18*th*. Refused nourishment obstinately. Throat looks worse, cannot be well examined. Pulse from 160 to 200.

19*th*. Neck more swollen. He has evidently much difficulty in swallowing. Discharge from nares more tenacious. Tongue dry and brown. Throat and back of mouth covered with ash-grey slough, or false membrane.

20*th*. Efforts at swallowing intensely painful. Inspiration somewhat laryngeal, especially when struggling, and there is on inspiration some recession of the lower part of chest. A distinct return of scarlatina rash on trunk and legs. Towards night dyspnœa and restlessness great. Can only swallow ice. Has nutrient enemata.

21*st*. Seemed a little better. Less laryngeal breathing and less difficulty of swallowing. Towards evening he had convulsive movements of his arms, lasting for a minute; afterwards was constantly tossing arms about, but seldom moved legs, which seemed almost powerless. His urine was passed in moderate quantity, loaded with lithates, free from albumen.

22nd. Was about the same. Mouth very dry, lips black. Throat looks as if generally ulcerated, no thickness of deposit.

23rd. Was much worse. Breathing hurried, rather tracheal. Very cadaverous looking. Will now swallow anything.

He died at 1 p.m., being the 11th day of scarlet fever and 6th of the diphtheria.

His mother was seized with shivering and sore throat on the night of 17th March.

The next morning her pulse was very quick and weak. The tonsils, uvula, and pharynx of a dusky-red colour. Headache and aching of limbs. Much pain in swallowing. The rash was not seen till the morning of 20th. Her pulse was then 128, weak. Throat as before.

21st. Appeared a little better. Pulse still quick and weak. Rash a little more visible.

22nd. Much thick mucus about the throat. Tongue rather dry, papillæ prominent.

23rd (6th day). Has a very troublesome cough with a barking laryngeal sound. Inspiration shallow, after the cough distinctly laryngeal. False membrane seen on the throat low down. Somewhat confused in manner towards evening.

24th. Passed a restless night, frequently waking up frightened. Feet cold early this morning. Complexion dusky. Cough very croupal, induced by any effort at swallowing. Right lung imperfectly expanded ; dull on percussion towards base. Respiration mainly abdominal. She remained conscious, becoming gradually weaker until her death at 5 p.m.

Her husband became ill on the morning of 22nd. He had an attack of medium severity.

His throat at first was swollen and coated with patches of tenacious secretion, extending from right to left tonsil. The rash came out as usual.

26th (5th day). A distinct diphtheritic deposit was seen on the right tonsil, which was painted with hydrochloric acid and honey. He was very weak ; was supplied freely with beef-tea, milk, and wine. Quinine and chlorate of potash were now given. The tonsil slowly recovered its healthy appearance, and the patient made a steady but slow recovery ; requiring a free supply of nourishment, wine, and quinine.

One of his daughters, who had been sent away directly her brother became ill on the 12th March, was brought home on the 17th April with the rash of scarlatina out upon her. She was scarcely at all ill before the rash appeared.

April 18*th*. Her throat was said by her medical attendant to present a diphtheritic appearance.

19*th*. I saw her and found her very ill. Pulse 160. Conjunctivæ suffused. Glands swollen on left side of neck. Fauces generally swollen; both tonsils covered with a false membrane. Rash out copiously.

20*th*. There is now a white patch of diphtheritic deposit on the hard palate. When induced to swallow, which she now strongly objects to, sometimes draws in a large laryngeal inspiration. Conjunctivæ not quite so red.

21*st*. Was restless during the night, trying to get out of bed. The throat and palate looked rather better. Towards evening she became much worse, refused to swallow medicine and wine. To have brandy and egg per rectum. Clear discharge from nostrils.

22*nd*. No better. Pulse 130, very weak. General brawny swelling of neck. Conjunctivæ much injected of a dusky-red tint. At 7 p.m. much worse. Pulse hardly to be counted, over 150. Head thrown back from swelling of neck. Fluids taken into mouth return in great measure by nose. She is conscious. Pupils are very large and there is some strabismus. The next morning a croupy cough came on, and she died at 4 p.m., being the 7th day.

Remarks.—These cases are remarkable from the well-marked co-existence of scarlet fever and diphtheria in each. In the first case, diphtheria began on the 5th day of scarlet fever; in the second case, it probably began on the 5th day, and had extended to the larynx on the 6th day; in the third case also, although the throat was partially covered with exudation from the 2nd day, yet it was not of such a character that it could be pronounced diphtheritic until the 5th day; in the fourth case, the throat was decidedly diphtheritic from the 3rd day, if not on the second.

The mode in which the last patient contracted the disease was a subject of careful inquiry. She was sent away from home on the 12th March, with her sisters, and all communication between her and those who were known to be infected was cut off until three weeks after her brother's death. At this time, the nurse who waited on her brother having been to the sea side, came back and slept with the little girl. She had been strictly enjoined to bring nothing with her which she had worn when nursing the boy; it was found afterwards that she had brought with her some of his hair, and probably some articles of dress which she had used when nursing. The little girl slept with her on her return, and was taken ill within three days. Her two sisters escaped infection; they were with the nurse for a day or two, but were not quite so much with her as the sister who died.

Treatment.—In judging of the value of different plans of treatment, it must be always borne in mind that some cases are so virulent that no treatment will save the patient, that other cases are so mild that they will recover if left alone, or even if treated injudiciously; whilst a third set of cases, if well treated, will recover, and the speed and completeness of their recovery will be influenced by the mode of treatment. These last cases, if badly treated, will not recover at all, or recovery will only occur after more protracted illness and with greater damage to the constitution.

The *prevention* of scarlet fever is a matter of great importance. We know of nothing which will effect this, besides the avoidance of infection. Small doses of belladonna, which have been much vaunted, appear to be useless. Fully to settle the question, it might be worth while to put it again to the test in some large public establishments, in which the infection has been admitted. The only careful experiments made on a large scale, with which I am acquainted, are those of Dr. Balfour, and their evidence goes against the utility of the remedy. The disease is frequently spread by convalescents being allowed to associate with other children too soon. Amongst the poor the means of isolating the sick are not to be found. Appliances for purifying bedding and woollen materials which cannot be put into boiling water, stoves to insure a temperature between 200° to 250° Fahrenheit, ought to be supplied in all towns. The use of public cabs for the conveyance of scarlet fever patients from place to place no doubt spreads the disease. Until recently there was no legislation to discountenance this practice, and even now the law is so lax that practically it is almost a dead letter. Medical men, after visiting cases of scarlatina, should always carefully wash their hands, and should go some distance in the open air before seeing other patients. Medical men in attendance on parturient women should not attend patients with scarlet fever.

I believe that if this disease were treated by government with the same activity that was displayed to prevent the cattle plague, it might be almost extirpated.

Complete isolation of the sick, and careful disinfection of all articles used by them, and of places occupied by them, would effect this. It is a disease whose ravages ought to be arrested, if it be in any way possible, for it cuts down often the healthiest children in the community, and is in London, as we have seen, destructive to a larger number than is cholera itself.

When a patient has suffered from the disease, it is not safe to allow him to return amongst the healthy until desquamation is completed, or if there be not much desquamation, for a month from the beginning of his illness.

In mild cases of scarlatina the treatment consists of good nursing; drugs are not needed. The patient must be kept in bed, not too warmly clad; the air of the room should be kept pure by free ventilation, not above 60°. The diet should be fluid, milk and farinaceous; and if the patient fancy them, he may have animal broths or beef tea. The skin should be sponged with tepid water. The patient may drink freely of toast and water or barley water, or suck ice, if this is grateful. The body and bed linen should be frequently changed.

During convalescence warm baths are useful. The skin may be anointed with suet, or lard and suet mixed. A mild aperient may be given if the bowels are decidedly costive.

In severer cases the best medicine is carbonate of ammonia, in two, three, or four grain doses, according to the age of the child, every two or three hours. It may be given in milk and water. The throat should be freely steamed if the patient is old enough; or ice will be useful to the throat in other cases. In malignant cases, preceded by heat of skin and febrile excitement, the remedy introduced by Currie, and occasionally praised since, appears to be of use; I refer to cold affusion. The disease is so desperate, and the remedy strikes friends as so rash, that practitioners are naturally rather timid in resorting to it, especially seeing that a large proportion of the cases subjected to it will die on any plan of treatment.

I have used it in a few cases, and have reason to be very well satisfied with the effects. The patient has always been relieved for a time, and in a few cases permanent benefit has been received, and the recovery promoted. The patient is placed in an empty bath lined with a blanket, and then as quickly as possible two or three pails of water, at a temperature of 70° to 75° Fahr., are thrown over him. Currie used the water colder than this. He is then immediately wrapped in dry blankets, and placed back in bed. Reaction is usually established in ten minutes or quarter of an hour. The process should be repeated once or twice in 24 hours, according to the gravity of the symptoms. In cases of collapse and cold extremities, it would not be prudent to resort to the operation.

Another plan of treatment not quite so formidable I have also seen adopted with benefit; the process is known in hydropathic establishments as "packing." The patient is closely enveloped in sheets, wrung out of cold water, and then covered over with several blankets, also tucked well in at the sides and the bottom, or else with some waterproof sheeting. He is left in this for a couple of hours, and is then taken out and put to bed in the ordinary way. The process of packing is thus described by Hebra: "The mattress on which the patient is to be placed should be protected by waterproof sheeting; over this are placed transversely two long strong bandages or well-folded towels, upon which is spread out a thick hairy double blanket, which reaches beyond the top and bottom of the mattress; upon this blanket is stretched a sheet dipped in cold water, and well wrung out, and upon it is placed the patient quite naked. The patient is then so completely enveloped in the wet sheet, that it is applied over the head and forehead to the eyebrows; also over the ears and cheeks to the chin, whilst the sides of the sheet are bound over the trunk and extremities. In like manner, the blanket is then laid over the sheet, and only the mouth, nose, and eyes of the patient are left free. Then the cross-bandages are bound together, so that the sheet and the blanket are firmly pressed on to the body.

If the pulse be compressible, as well as frequent, and the patient remain low in spite of ammonia, wine should be given in small quantities often repeated. To the neck, spongio-piline, wrung out of hot water, or linseed meal poultices, should be applied, if the glands be swollen and tender, or the neck be puffy.

Coryza should be treated in very young patients by syringing with lime water, when the discharge is thin and acrid, or a solution of salt and water, a drachm to two ounces. When the secretion is tenacious, soap and water may be used. In patients over eight years of age, the plan of washing out the nostrils introduced by Dr. Thudichum is to be recommended ("Lancet," Nov. 1864). A simpler apparatus than that described by Dr. Thudichum will answer the purpose. A vessel filled with the required solution, at the proper temperature, is placed on a higher level than the patient's head; by means of a flexible caoutchouc tube, with a suitable nozzle acting as a syphon, the fluid is conveyed to the patient's nostril. The patient must breathe through the mouth, and not swallow during the operation; the fluid passes round over the posterior edge of the septum, and escapes from the other nostril. The force of the current is regulated by the height of the vessel above the head; at first it should be but little raised, and the elevation may be increased. The fluid may be sent through the two nostrils alternately. Whilst syringing, the patient should sit up with the head inclined a little forwards.

For the diarrhœa, which accompanies many malignant cases of scarlet fever, very little treatment is desirable; magnesia in small doses and mucilaginous drinks are the best remedies to give. There is a desquamative process with congestion going on in the stomach and intestines (as shown by Dr. Fenwick), and this is best treated by demulcents. If diarrhœa continue more than a few days, a little compound ipecacuanha powder may be given. Where an adynamic condition sets in without previous reaction, the case is very desperate. Warm salt or mustard baths may be used, or the skin well rubbed with flannels dipped in vinegar and salt. Warm wine and water may then be given, with ether and ammonia, if the extremities remain cold.

After the first four days, if the patient remain weak, quinine is of use. At this time, if the pulse is weak and there is much depression, wine may be given, if not given before. The effect of alcoholic stimulants on the urinary secretion should be carefully noted. If the urine is scanty it is important to give plenty of diluents, water in some form or other, and not much alcohol. When swelling comes on in the neck rapidly, a free allowance of stimulants is indicated with quinine internally, and poultices with hot fomentations locally. As soon as there is any evidence of pus, lancing should be adopted.

When the throat is ulcerated after the fifth day, one application of caustic is desirable; equal parts of hydrochloric acid and honey is the one which I prefer. This should not be repeated under three days; in the interval a concentrated solution of chlorate of potash may be used as a gargle, or with young children syringed into the mouth. If swallowed it does no harm.

If the tonsils and throat generally remain relaxed, a gargle or wash of tannin will be found useful.

For albuminuria and hæmaturia, with or without dropsy, the skin and the alimentary canal should be acted on. Whilst the urine is scanty, hot air baths alternating with warm water baths and preceded on each occasion by a dose of antimony. Wine should be given every night or every other night for a week or more according to the strength of the patient; plenty of fluids should be taken by the patient. To act on the bowels the compound jalap powder in doses of from fifteen grains to half a drachm should be given; or elaterium, if it can be obtained of good quality, is a good hydragogue purgative and less bulky. Infusion of digitalis may be used as a diuretic for a short time. Dry cupping to the loins may be used with benefit at the early stage. When the quantity of urine is large, and fever and dropsy are at an end, the most useful medicine is the tincture of the perchloride of iron. This sometimes does not agree, and hæmaturia remains to a considerable amount; gallic acid may then be resorted to. If no benefit is observable in a few days it should be omitted. Benzoate of ammonia in five-grain doses has been recommended, and has sometimes appeared to me to exert a good effect. When the albuminuria appears likely to become chronic, quinine is sometimes of service.

If convulsions occur during the progress of the albuminuria, elaterium or compound jalap powder with hot air baths should be given. Trousseau recommends blisters to the legs. I have tried venæsection, which was the usual practice until recently, but I have not seen any benefit such as to make me recommend this mode of treatment.

One of the most important points in treating scarlet fever is not to allow patients to leave their beds for three weeks in the mildest cases, and four weeks or longer in the more severe cases.

If albuminuria arise, the patient should be kept in bed for two months, if the urine be not free from albumen. When patients get up, they should wear flannel round the loins, and if the weather is cold, over the chest and limbs.

If pleurisy or pneumonia set in, the case is very unfavourable. Depletion is not allowable; a simple saline and plenty of light nourishment, beef-tea, milk and eggs, are the measures to be recommended. After the acute stage in pleurisy, iodide of potassium may be given if the patient is not too low and anæmic, and this should be followed by quinine.

In pyæmia large doses of quinine should be given, and they sometimes effect a cure.

For otorrhœa, if there is acute pain, leeches and hot fomentations; followed when chronic by warm water injections, and after a time weak astringent lotions of lead or zinc; counter-irritation behind the mastoid process may be kept up for some time.

Scarlatina simplex, without complication. Recovery.

Jemima Phelp, æt. 5, was quite well on the evening of 15th June. During the night she became feverish, was not delirious, and did not complain of a sore throat. On the following day a red rash was noticed on her neck and face; she was drowsy, hot, and thirsty, and had no appetite.

June 17th. She was admitted to the hospital. A slight flush of redness on forehead; a very distinct rash about upper part of chest; scarcely any on abdomen; more intense over pubes. Punctiform redness on each leg, and to a very slight extent on arms. Tongue moist, rather furred, not red. Fauces and tonsils very red and swollen, covered with muco-purulent secretion. Glands at right angle of jaw rather enlarged. Pulse 160. Temperature at 8.30 p.m., 100°.

(4th day.) Temperature, 9 a.m., 99·2°. Pulse 132. Appetite tolerably good, no thirst. Rash well marked, except on face. Tongue clean; fauces better; swelling almost subsided. 7 p.m,. Temperature 99·4°.

(5th day.) Temperature, 9 a.m., 98°. 7 p.m., 99·4°.

(6th day.) Appetite good. Pulse 136. Fauces still red and covered with muco-pus. Temperature, 9 a.m., 99°. 7 p.m., 99·2°.

(7th day.) Temperature 98·8°.

(10th day.) Sitting up in bed playing. Pulse 124 rather weak. Slight desquamation on neck, none elsewhere.

(13th day.) Pulse 104, rather irregular. Desquamation on chest. Seems quite well. The urine was examined daily until the 21st day and found free from albumen.

Scarlatina maligna proving fatal on the 8th day.

Mary Clarke, æt. 5. On the morning of 9th July she was chilly, and in the evening she complained of headache and was hot.

July 10th. She first complained of sore throat, there was more fever. The following day a rash was noticed. She was delirious during the preceding night.

(4th day.) She was oppressed, drowsy and heavy. Whilst under examination vomited grass-green fluid. Her lips were dry; some sordes on teeth and gums. Tongue dry, with a thick fur posteriorly. Eyes injected. Pulse 148 weak. Rash strongly out over the whole body; some ecchymoses at the top of chest. Fauces and back of pharynx red and much swollen.

(5th day.) Passed a very restless night. Vomited green fluid five or

CASES OF SCARLATINA. 321

six times in the 24 hours. Is very drowsy and oppressed. Pulse 152, weak. Tongue dry, free from fur, papillæ prominent. Glands at angles of jaw not enlarged. Rash out abundantly. Miliary vesicles on the neck, chest, abdomen, and thighs.

(6th day.) Was very delirious during the night, passed her urine under her. She seems worse, drowsy and moaning constantly. Alæ nasi dry and excoriated. Pulse 148, not so weak. Rash still abundant, especially on chest, where it is rather livid. Miliaria more marked. Fauces red and swollen, covered with muco-pus.

(7th day.) Again delirious during the night, but could be roused to consciousness by being spoken to. Still drowsy; frequently rolling about. Glands at angles of jaws enlarged and hard. Pulse 152, weak. Passes urine under her.

(8th day.) Looks more typhoid. Alæ nasi much excoriated; a thin mucous discharge. Some livid rash on face and on the trunk. Pulse 164, very weak.

She died at 3 o'clock. Her temperature examined twice daily, ranged from 103° to 104·6°. The urine was on all occasions free from albumen.

Post-mortem examination was not allowed.

Remarks.—In this case the malignant character of the case did not manifest itself very decidedly until the fourth day. She was treated with carbonate of ammonia and stimulants, with strong beef-tea; but I think if the ammonia had been given more frequently there would have been a better chance of her recovery.

Scarlatina. Pericardial friction towards the end of the 2nd week. Otitis 26th day. Recovery.

Ellen Kipping, æt. 11. She was attacked at noon on the 30th September with vomiting and diarrhœa; she complained of sore throat and lay down. Rash appeared next day over the whole surface. It became subsequently very intense, and did not disappear until the 7th day. Pulse 132. Tongue was furred; thickly coated on the 3rd day, becoming strawberry-like on the 4th until the 9th. Fauces very red, not swollen; after the 5th day they began to improve. Conjunctivæ were injected from the 3rd till the 7th day. Glands at the angles of the jaws were slightly swollen, those in the groins more so. There was delirium during the 3rd and 4th nights. Improvement commenced from the 5th day. Her appetite was very bad till the 7th, when it began to improve. Her pulse was never above 120 after the 3rd day; it fell to 76, and became irregular on the 11th day. Her temperature was taken four times daily until the 9th day. It rose to 103·4° on the 2nd, and

Y

to 104° on the 3rd day. The highest temperature on the 4th day was 103·6°; on the 5th day it did not rise above 101·6°; on the 6th and 7th days not above 101·1°. On the 8th day it ranged from 99·4° to 100·4°. On the 11th day from 97·4° to 99·2°. On the 12th, 13th, and 14th days it ranged from 98° to 99·4°; and on the 15th day it rose to 100·2°. From the 10th day she appeared quite convalescent; her pulse never rose above 90; it was at times irregular, but she felt well and her appetite was good. Peculiar signs, however, were observed about the heart. On the 11th day a loud systolic murmur was audible at the 3rd left and 2nd right costal cartilages, also at the apex. On the 15th day a murmur of a rubbing character was audible over the 4th left cartilage. On the following day a distinct friction sound was audible at the 4th left cartilage; this continued for three days and disappeared. By the 20th day all abnormal cardiac sounds had disappeared. She was allowed to get up. Her urine was examined throughout; it contained traces of albumen on the 5th and 6th days, and on the 15th. On the 26th day she was attacked with severe earache which lasted three days, and again returned at the end of a fortnight. After this she remained well.

Remarks.—This case is a good specimen of a tolerably sharp attack of scarlatina simplex. The occurrence of endocardial murmurs and pericardial friction during convalescence was notable. The absence of febrile disturbance and of subjective symptoms was remarkable. The elevation of temperature on the 15th day coincided with the friction sound which was heard. Whether the murmurs which had been previously heard were due to endocarditis is not very certain. Murmurs of this kind have been observed in several cases under my care; some of them have disappeared, whilst others have continued as long as the patient was under observation. I believe that they are due to the deposit of lymph on the valves, which sometimes leads to permanent roughness, and is sometimes removed by the current of the blood.

Anomalous Scarlatina. Followed by Albuminuria, Anasarca and Meningitis. Death about 18th day.

Eliza Rowe, æt. 19 months, was brought to the Children's Hospital on the 27th December. Her mother stated that a younger sister had died of scarlet fever on the 21st December after a week's illness. At the time of this child's illness the father and mother had sore throat. On the day her sister died, Eliza had a cold in her head, with running from the nose. For a day or two before this she had complained of pain on the left side of her head, and the glands of her neck were swollen. Her skin was hot, she was thirsty, and had no appetite. About the 21st December were noticed some red spots on her skin, nearly as large as peas, with clear fluid in them; there were in all about 20 of these vesicles. No rash of scarlatiniform character was noticed throughout.

Dec. 25th. Her feet and hands commenced swelling, and she was drowsy. Her urine was dark in colour, staining linen and making it stiff when dry.

27th. Her hands and feet swollen. Bowels relaxed. A purgative, warm bath, and diaphoretic mixture ordered.

30th. Œdema has now extended to face. Fingers and toes livid. She is constantly drowsy; at night wakes up suddenly with a shrill scream. Pulse 160, weak. Tongue pale, rather swollen. Yesterday and this morning vomited. She coughs a good deal, especially at night. Urine scanty, turbid, of a dark amber colour; becomes quite solid on the addition of nitric acid, and boiling.

Jan. 3rd. Œdema continues. Urine scanty; seems to have earache. She cries out when touched.

4th. In the night screamed very much, especially when moved. Put her hands frequently to her head as if she had pain. Slept at short intervals.

Jan. 5th. Dozing and frequently waking up screaming. When awake she could talk. Skin hot and moist; face flushed. During the night there were convulsive movements of arms and legs, and twitching of face and eyeballs. Urine passed into bed. Emplastrum sinapis nuchæ et cruribus applicetur.

6th, ½ *past* 3 *p.m.* Head inclined to left shoulder, eyes drawn up and to the left. Pupils of medium size. Sordes on teeth. When lifted out of bed keeps eyes and head fixed in the same way, until she is roused by an effort; she then looks straight down and cries. Pulse 144. Respiration 32, regular. Skin warm. Slight œdema of hands and legs. Nails of fingers livid. A leech was applied to left ear.

7th. After the leech came off she was quite sensible and looked about. Dozed through the night and did not scream. About 5 a.m. she became much convulsed for nearly 3 hours. Her body was hot and hands perspiring. Bowels acted once in night; motion dark in colour, not specially offensive. Has not passed any water since last night. At 3 p.m., after several hours convulsions, she died.

Autopsy 24 hours after death. Slight eczematous eruption behind left ear. No discharge from meatus. On removing calvaria a little clear fluid escaped from beneath the dura mater, which was wounded. On the left hemisphere, over its posterior half, the arachnoid presents a yellowish-milky aspect. The membranes at this part are much thickened and extra-vascular. There is no increased vascularity of the membranes elsewhere. The cerebral substance appears quite healthy at every part. No trace of tubercle. Some clear fluid in lateral ventricles, about 1½ oz. in the two. Lungs healthy. Some moderately firm adhesions in left pleura. Heart healthy. Liver large. Central portion of lobules injected, peripheries pale. Kidneys large. Cortex pale and rather wide. Under the microscope there was a considerable amount of granular degeneration of the epithelial cells.

Remarks.—The rapid development of renal disease in this case, after an anomalous acute attack, characterised by pain in the head, glandular swelling in the neck, and feverish symptoms, leaves little doubt that this was a case of scarlet fever, especially when it is remembered that she had been exposed to infection.

Meningitis is one of the rare sequelæ of this disease. It was in this case peculiarly circumscribed in extent.

TYPHOID FEVER.

THIS disease, whose most striking anatomical character is a peculiar ulceration in the agminated glands of the small intestines (" Peyer's patches "), is not unfrequent amongst children. Many cases of "infantile remittent,"* and of gastric fever, belong to this category. It assumes as a general rule a less severe form amongst children than amongst adults; a large number of cases occur in which the fever subsides at the end of from 12 to 14 days; but there is great danger in children, as in adults, of a return of fever, or of serious abdominal mischief, at any period before the end of the fourth week. It is of great importance that a correct diagnosis should be made in the mild cases, so that no indiscretion be allowed; the patient should be kept in bed, and solid food forbidden until the end of from 22 to 28 days.

The following case may be taken as a specimen of a mild form of the disease without complication :—

Oliver Turpin, æt. 3⅖ years. A moderately healthy boy, but rather weak in his chest. He seemed well on 9th September and went to school. The next day complained of frontal headache and could not go to school. For several days previously his bowels had been open too freely from medicine which his mother gave him. From the 10th September until 17th, when he was admitted to the hospital, he was feverish and his bowels were relaxed.

Sept. 18*th.* Boy does not look very ill; he has a dusky flush over both

---

* The term "infantile remittent" is one which it is well to discard, because it has been used to include a number of different diseases; amongst others typhoid fever, intermittent febrile conditions due to malaria, some forms of pneumonia, some cases of tubercular disease, and febrile disturbance from dentition or gastro-enteric catarrh.

cheeks. Skin dry. Pulse 124. Respiration 28. Tongue red at tip and edges, with a white fur in the middle, rather dry. No appetite. Bowels not open for 24 hours. Abdomen rather full and resonant. Spleen not to be felt. Two slightly elevated rose-coloured spots, which disappear on pressure, are to be seen on trunk. Temperature, at 9 a.m., 100·6°; 6 p.m. 100·6°. Sonorous râles audible over both lungs.

19th (10th day). Much the same. Temperature 101·5°. Has passed one loose stool of greyish-yellow colour.

20th (11th day). Did not sleep well, but lay quietly. Skin acted freely this morning. No sudamina. Herpes labialis. One large loose stool greyish in colour. Temperature, 9 a.m., 99·5°. 6 p.m. 101°.

(13th day.) Improving. Sweats a good deal. Four loose watery stools in 24 hours. Tongue cleaner. Temperature 100·2° at 9 a.m. 102° at 6 p.m.

(14th.) Much better. Pulse 112. Respiration 36. Five loose stools, two yellow, the rest greyish.

(15th.) Tongue clean, moist. Has vomited once. One fresh spot to-day. Temperature 100° to 101·4°.

26th (17th day). Much better. Temperature 97·6° to 99°.

(18th day.) Temperature 96·2° at 9 a.m. 98·4° at 6 p.m. From this time his recovery was rapid, and without a drawback. He was sent into the country on the 18th October, the 39th day from the commencement of his disease.

In the next case, although the symptoms were not very severe during the first three weeks, the case assumed suddenly a graver aspect on the 23rd day, and terminated fatally on the 33rd day

Benjamin Bevan, æt. 7 years. He lived in a very badly-drained house in St. Luke's. He was in his usual health until 1st July, when he seemed poorly, complained of headache over his eyebrows, and his bowels acted three or four times. From that time he remained feverish and talked in his sleep at night. He was brought to the hospital on the 8th July.

(8th day.) He has a white tongue, red at the tip and edges. Complains of headache. Abdomen too full and resonant, rather tender in right iliac fossa. Bowels very relaxed. Pulse 118, weak. No rash on skin. Sibilant râles in chest.

(9th day.) Sordes on teeth. Still has headache; no delirium. Tongue furred, red at tip, rather dry. Pulse 120.

(12th day.) Two rose-coloured spots on abdomen. Bowels loose. Spleen now just felt below false ribs, is tender on pressure. He sleeps well.

(15th day.) Seems better.

(21st day.) Pulse 114. Still some headache; has not been delirious.

TWO CASES. 327

Temperature has varied from 101° to 103°. Two or three loose stools daily. Tongue cleaner.
(23rd day.) Not so well to-day. Now feverish, very thirsty, and restless. Temperature 104°. Two loose stools in night.
(24th day.) Pulse 140. Temperature 104°.
(25th day.) Patient cries out when disturbed. Face pale. Tongue dry and glazed ; lips dry. Pulse 144. Respiration 42. Temperature 105°. No chest signs to account for the greater elevation of temperature. Sonorous rhonchi generally heard over lungs. Bowels loose.
(27th day.) Slept badly. Pulse 144. Temperature 104·25°.
(28th day.) Tongue dry, with transverse cracks. Sordes on teeth and lips. Temperature 104°. Tongue trembles when protruded, and arms shake when held out. Bowels open six times in 16 hours.
(29th day.) This morning a small quantity of blood was passed with stools. Pulse 148, very weak.
(31st day.) Pulse 156, very weak. Respiration 44. Temperature 102°. Tongue very dry.
Aug. 1*st* (32nd day). Pulse 160. Face pale. Pupils very large. Lies in a drowsy listless condition. Abdomen tender. Temperature 102°.
2*nd.* Pulse not to be felt. Temperature 98°. Face very pale. Respiration sighing. Skin cold, clammy.
Died at noon.
Treatment.—July 1*st.* Carbonate of ammonia and citrate of potash. Beef-tea and milk.
10*th.* Wine, 2 oz.
21*st.* Carbonate of ammonia discontinued.
23*rd.* Wine omitted.
27*th.* Acid sulphur. dil. ♏v.
Tinct. aconiti, ♏ii.
Æther. chlorici, ♏x.
Mucilaginis, ʒj.
Aquæ ad ʒss. 4tis die.
One drachm of brandy every hour.
28*th.* Enema opii with 5 minims of tincture of opium. One drachm of brandy every half hour.
29*th.* Tincturæ opii, ♏vij. in enema.
Plumbi acetatis, gr. j.
Acid. acetici dil. ♏ij.
Tinct. opii, ♏ij ; Mucilaginis, ʒij. 4tis horis.
Aug. 1*st.* Spirit. ammon. arom. ♏v.
Mucilaginis aquæ āā, ʒj. 4tis horis.

Remarks.—I believe that a mistake was made in omitting the wine on the 23rd day, when the fever became higher. The citrate of potash and ammonia mixture I do not now give, but prefer mineral acids and chloric ether; chlorate of potash is also useful.

Autopsy, 50 hours after death.
Heart weighed 3¾ oz. Much fluid blood escaped on removing heart. Blood not firmly coagulated in right ventricle.
Lungs.—Hæmorrhagic spots on pleura. The bronchial tubes contain a good deal of frothy serosity, and the lungs are congested.
Liver very pale. Near right supra-renal capsule hæmorrhagic effusion into cellular tissue about 1 inch square.
Kidneys rather pale, firm.
Spleen weighed 3⅜ oz., of uniform dark colour, firm.
Intestines.—On removing them small intus-susceptions of an inch in length are seen, readily reduced without any adhesions.
Mesenteric glands large; from size of bean to a large almond, light pink to dark purple in colour, rather soft.
Pancreas.—Interlobular structure extra vascular.
Stomach.—Mucous membrane generally pale. A circular patch 3 inches in diameter in the large curve is red with punctiform and capillary injection.
Small Intestines.—Upper part healthy. On reaching ileum small ulcers are here and there seen on Peyer's patches. On proceeding downwards they are larger and more numerous; near the ileo-cœcal valve they are very abundant. The patches here are elevated, pulpy, eroded in centre, with redness around them. There is one patch with slightly overlapping edges; the base of the ulcer pale, with a few streaks of injection.
Large Intestines.—A number of small ulcers, in size varying from pins'-heads to beans; margins red, rather sharply cut; base pale and sloughing in places. At a little distance from ileo-cœcal valve a portion of the intestine presents a puckered appearance; some portions of it elevated and of a dark slate colour, in transverse ridges. Other nodules, of an opaque white colour, on section exude puriform matter, they vary in size from a pin's-head to a large pea; there is no vascularity around them.

The occurrence of ulceration in the solitary glands of the large intestines, as met with in this case, is not at all rare. The appearance, as of small pustules, was an example of what was called by Cruveilhier, the "forme pustuleuse."

I do not propose to give anything like a complete account of typhoid fever, such as is given in a systematic treatise on medicine, but merely refer to the points in which this disease differs in children from what is met with in adults. As already mentioned, mild cases of this disease are probably more frequent in children than in older persons. In such cases, the *remittent* character of the fever is especially observed; during several hours of the day there is a more or less complete remission of all febrile symptoms, especially of the temperature, and to some extent of the quickened pulse.

At the onset *vomiting* is not a more frequent symptom than in adults. Dr. Murchison states that it is met with in 36 per cent. of all the cases at all ages. My own experience would lead me to say that it occurred in about the same proportion in children. It may take place on the first day, or any day during the first week. I have seen it occur daily for the first fortnight, and it sometimes occurs for the first time in the second week. When combined, as it occasionally is, with constipation at the onset, it may lead to an incorrect diagnosis from the resemblance of the symptoms to those of cerebral disease, especially tubercular meningitis. It sometimes comes on at the commencement of convalescence without any other bad symptom, the tongue being clean, the abdomen not distended and free from tenderness. In such cases, Trousseau recommends that a little solid food should be given. Vomiting after the end of the second week, accompanied with abdominal pain, tympanites, and great depression, is probably indicative of *perforation*. This is of rare occurrence in children, only having been met with once out of 232 cases.

The stools present the same characters as in the adult—a tendency to fluid consistence, and a pale ochre or drab colour; there is usually a supernatant fluid, and a flaky deposit consisting of epithelium, disintegrated sloughs, and undigested food. They tend rapidly to become alkaline. In some cases the bowels are confined throughout. Hæmorrhage from the bowels occurs in a certain proportion of cases ; in 4 out of 30 cases in which the motions have been specially noted. It is generally a grave symptom ; but I have seen it in a comparatively mild case that went on well.

Epistaxis is not a very frequent symptom. I have seen it in the first week and in the third. Headache is usually present; it is often so slight in degree that it is overlooked by parents, the children not calling special attention to it. It sometimes continues as late as the third week; and occasionally it precedes the attack for several weeks. As observed by Dr. Jenner, it is not present after the occurrence of delirium. This symptom is not often met with till the second week, and is usually worse at night; it is not of a violent kind, though it may be attended with restlessness and a constant desire to get out of bed; occasionally the patient cries out violently during a great part of the night, and sinks into more or less stupor in the day. The quieter forms of delirium, also, alternate with drowsiness and stupor, from which, however, the patient can usually be roused without difficulty. Drowsiness is occasionally observed as one of the initial symptoms. Deafness is not infrequent in the course of the second week; it may be purely nervous; or it may be due to inflammatory congestion of the eustachian tube, or the external meatus, or, it is said, to softening of the muscles of the internal ear.

Earache, with or without a discharge, is occasionally met with towards the end of the third week, for 10 or 12 days. Amongst the sequelæ, bronchitis and broncho-pneumonia are not infrequent. Pleurisy is rare. Tuberculisation is not specially frequent. Cancrum oris or noma pudendi is rare. Laryngitis is also rare, and albuminuria likewise.

The *eruption* is said to be absent in a larger proportion of cases in children than in adults, though, if the surface of the trunk is carefully examined, one or more spots will, in a large majority of cases, be found; and in some cases, usually severe ones, a large number of spots will be found; but there is nothing like a constant ratio observed between the severity of the case and the amount of eruption. In 30 cases, where the eruption was looked for, it was only quite absent in seven. The first spot was observed as early as the 6th day, and in one case as late as the 25th. In five cases, the first eruption appeared on the 12th day; and in 20 cases before the end of the second week. Successive eruptions of spots are noticed till the end of the third or fourth week; I have more than once seen them as late as the 40th day. In a large proportion of the 23 cases in which an eruption was seen, the total number of spots did not exceed half-a-dozen. In one patient, at the end of 30 days, the feverish symptoms subsided, and were absent for five days; at the end of which a relapse occurred, lasting six days, during which an eruption appeared of red papules, distinctly elevated, rather hard, and tender, having the characters of *erythema papulatum*.

Sudamina are often seen towards the end of the second week.

In children, as in adults, there is often great debility left for many weeks by an attack of typhoid fever, even when the symptoms have not been of the gravest kind. I have several times seen cerebral symptoms of a hydrocephaloid character following typhoid. When patients come under treatment with such symptoms, and without a satisfactory history of the previous illness, such as would enable one to make a diagnosis, the case may be mistaken for tubercular meningitis. These cases require to be treated with nourishment in a concentrated digestible form, stimulants and opiates in small doses. The diagnosis of hydrocephaloid from hydrocephalus is treated of in the lectures on tubercular meningitis (page 170).

Occasionally patients die, some weeks after the fever has subsided, from exhaustion consequent on ulceration of the intestines.

In the following case death occurred at the end of two months' illness, mainly from exhaustion caused by ulceration of the lower part of the rectum, and an ischio-rectal abscess :—

Thomas Richardson, æt. 10 years, was admitted into the hospital on the 12th November. It was stated that he had had typhoid fever 6 weeks before, a severe attack accompanied with troublesome diarrhœa and the passage of blood in the stools. He appeared convalescent on 3rd November, was allowed to eat freely of all kinds of food. On 8th November he became worse, diarrhœa returned with its previous intensity. On admission he was found to be excessively emaciated ; his face was pale, with a slight livid tinge on his cheeks. His temper was irritable. His tongue had a thick aphthous-looking fur on the dorsum, with the edges pale. His skin dry and harsh. His abdomen was very tender, walls considerably retracted ; he keeps his legs drawn up and his hands on his belly. The transverse colon could be seen as a flattened band moving up and down behind the wasted integuments. His bowels acted 15 times in the night ; motions of a greyish colour. Pulse 140, weak. Respiration from 14 to 30 in the minute. He was ordered starch and opium enemata, and a pill every 4 hours containing $\frac{1}{8}$ of a grain of acetate of lead, and $\frac{1}{4}$ gr. of extract of opium. Beef-tea very strong, 2 eggs, milk, and 3 oz. of brandy.

Nov. 14*th.* The dose of acetate of lead was increased to $\frac{1}{4}$ a gr. and the pill given every 2nd hour.

16*th.* A black slough appeared in the left ischio-rectal fossa, which soon separated. An enema consisting of 4 gr. of nitrate of silver and 4 oz. of water was administered.

19*th.* He was not any better. A pill containing $\frac{1}{4}$ gr. of powdered opium was given every 4 hours.

23*rd.* He has been better in respect to the abdominal pain and diarrhœa during the last 24 hours. His bowels act still 5 or 6 times in the day and night. His general appearance has not much altered. Mouth full of sticky mucus, and tongue covered with aphthous patches. Pulse 132, not quite so weak. Respiration 12 only. There is a deep ulcer with sharply cut edges measuring 2 inches by 1 in the ischio-rectal fossa. He became gradually weaker, and died on the 26th November.

Autopsy—Weight only 32 lbs.

Blood very watery.

Liver pale, weighing 38$\frac{1}{2}$ oz.

Spleen 4$\frac{1}{2}$ oz. enlarged, not particularly pulpy.

Intestines.—In the lower two yards of small intestine were numerous ulcers and cicatrices, some in the position of Peyer's patches, others not. The ulcers were of two descriptions :—

(1.) With sharply cut edges, slightly undermined, muscular coat of intestine exposed. Floor thin, but peritoneal coat normal.

(2.) With edges slanting and gradually lost in the floor of the ulcer. The cicatrices were perfectly smooth on the surface. In the large intestine there were about a dozen similar ulcers (open and cicatrized), chiefly in the ascending colon. The lower end of the rectum was almost wholly destroyed, opening into a large sloughy cavity in the left ischio-rectal fossa. This cavity communicated with the exterior by two openings, viz., the anus, and the opening left by the separation of the slough.

Bedsores after typhoid fever are not so frequently met with in children as in adults ; they ought never to occur, and are probably always due to want of care in the management of the patient.

SEX AND AGE OF PATIENTS.

Diphtheria is another sequela of typhoid fever, which is of very unfavourable omen.

Sex.—It has been said to be more common in boys than girls. Dr. Murchison gives, in 339 cases, 238 boys and 101 girls. In the Children's Hospital, of 255 in-patients, 133 only were boys, and 122 girls.

Friedleben gives, 46 boys to 52 girls. Probably both sexes are equally liable, if our statistics were comprehensive enough. It must be remembered that there are more boys living than girls.

Age.—Barthez and Rilliet say, it is rare between 4 and 8 years, rarer still from birth to 2 years. The cases in this hospital occurred at the following ages :—Under 2 years, 1 ; from 2 to 3 years, 4 ; from 3 to 4 years, 26 ; from 4 to 5 years, 21 ; from 5 to 6 years, 36 ; from 6 to 7 years, 24 ; from 7 to 8 years, 28 ; from 8 to 9 years, 35 ; from 9 to 10 years, 25 ; over 10 years, 34 ; from 11 to 12 years, 11 cases.

These figures do not correspond with Barthez and Rilliet's statement in reference to the period from 4 to 8 years of age, during which 109 cases occurred out of 244 patients between the ages of 2 years and 12.

The number of deaths amongst 255 cases, was 26, or about 10 per cent.; at all ages the mortality is nearly 20 per cent., according to Dr. Murchison. Amongst children the mortality would probably be less than 10 per cent., if all mild cases were included, an undue proportion of these not coming into the hospital.

The following case of typhoid fever was remarkable from its mildness till about the 17th day, and from the occurrence of scarlatina on the 29th day, which proved fatal in six days, with double pneumonia.

Susan Jessop, æt. 5 years, became sick and languid on the 14th September. She has been living in a badly-drained house, her brother is dying of typhoid now.

Sept. 15th. Was feverish ; her bowels acted once ; complained of headache, pains in her limbs, and has lost appetite.

16*th*, at 9 *p.m.* Temperature of axilla 102·4°.

17*th*, 10 *a.m.* 100·6°. Pulse 132. 6 *p.m.*, 100·8°.

18*th*, 4 *p.m.* Her appearance is not typhoid ; cheeks pale, eyes clear and expressive. Lips dryish. Tongue moist ; two lateral streaks of white fur, with redness at tip, edges, and centre. Pulse 132, small and weak. Respiration 32. No cough. Skin not dry. No eruption. Abdomen normal. No headache or delirium. Bowels open once in 48 hours after castor oil. Stool of semi-solid yellowish fœcal matter.

But for the history, it would probably at this period not have been regarded as a case of typhoid fever, so ill-defined were the symptoms.

Sept. 19*th*, 9 *a.m.* Temperature 100·1°. Nitro-muriatic acid mixture, beef-tea and milk. Slept well. Bowels open once, motion loose whitish. Looks a little more oppressed, disposed to cry. Pulse 132, weak. Respiration 18. Skin looks everywhere indistinctly mottled, almost as if a *typhus* rash were coming out. No spots. Tongue less furred, redder. Some sibilant rhonchi at apices of lungs. 6 p.m. Temperature 101·5°. Pulse 120.

20*th*, 9 *a.m.* (7th day). Temperature only 99·3°. Pulse 120, weak. Skin still a little mottled. 7 p.m. Temperature 100·5°.

21*st*, 11 *a.m.* Temperature 98·8°. Pulse 128. Looks almost well, but pale ; no typhoid spot. Tongue natural. 5 p.m. Temperature 100·1°.

22*nd*, 10 *a.m.* (9th day). Temperature 99°. Pulse 132. Bowels not open. Is more cheerful. At the evening visit, her temperature had risen to 103° ; nothing could be detected to explain this rise. Tongue a little dryer.

23*rd*, 11 *a.m.* Temperature 99°. Tongue moist. No tympanites. No eruption. To take a drachm of castor oil. 5 p.m. Temperature 101·5°.

24*th*. Temperature 101°. She looks pale but lively, sits up and plays. Bowels open 3 times after oil. Motions said to be pale. Spleen not to be felt. No fulness of abdomen. Pulse 128. Was allowed fish.

25*th*, 5 *p.m.* Much as yesterday. Bowels not open. Pulse 120, quiet. Temperature 97°.

26*th*, 9 *a.m.* Child is cheerful. Tongue, too, smooth, moist. Temperature 97·6°. Pulse 112. 6 p.m. 100·4°.

27*th*, 11 *a.m.* Temperature 99·2°. Tongue moister. Pulse 100. Still more lively.

28*th*. Temperature 99·6°. Pulse 98. Seems pretty well. Tongue still rather red and dry in the centre.

29*th* (16th day). Temperature 100·6°. Pulse 96. She seemed so well, and her appetite was so good, that she was imprudently allowed to have meat for dinner.

30*th*. Temperature 101·2°. Bowels open. Motions quite natural.

Oct. 1*st.* Not so well. Does not care to sit up. Temperature 102·6°. Pulse 110. Tongue moist.

2nd. Temperature 101·6°. Pulse 108. Seems better. Tongue red. Bowels not open. To take 2 drachms castor oil.

3rd. Tongue dry. She seems heavy and listless. Bowels open twice after oil. Motions said to be natural. Temperature 104·6°. Pulse 118.

4th, 9 a.m. Temperature 103·2°. Pulse 128. Tongue dry and red; has lost appetite. Bowels not open. 6 p.m. Temperature 104°. To leave off meat.

5th, 9 a.m. (22nd day). Temperature 104·8°. Pulse 136. Was delirious in the night. Tongue dry and red; lips brown. Bowels not open. Some moist râles at bases of lungs. To have a mustard plaster on back. Ammon. carb. gr. ij.; liquor cinchonæ, ℥x.; syrupi, ʒj.; aquæ, ʒij.; misce sextis horis sumend; olei. ricini, ʒij. statim.

6th. Temperature 104·2°. Pulse 140. Tongue dry and brown. She is restless. Bowels open once.

7th, 9 a.m. Temperature 103·2°. Pulse 148. Bowels open twice; stools loose of a light colour. 6 p.m. Temperature, 104°.

8th. Was restless and delirious during the night. Bowels open once; stool watery, light yellow. 9 a.m. Temperature 103°. Pulse 140. 6 p.m. Temperature 104°.

9th, 9 a.m. (26th day). Tongue dry, brown in centre. Pulse 160, weak. Temperature 102·8°. Was more delirious last night. Several typhoid spots have appeared for the first time, although carefully looked for daily. 6 p.m. Temperature 103·2°. To take 4 oz. of wine. After the first 2 oz. the pulse was of the same frequency but less compressible.

10th, 9 a.m. Has had a quieter night. Temperature 103·4°. Pulse 154, not so weak. Bowels open twice; stools loose yellow ochre coloured. Some more spots.

11th. A better night. Pulse 140, not so weak. Tongue dry but less brown. Abdomen seems generally a little tender on pressure.

13th (30th day). Has passed 2 quieter nights. Lies in a listless condition, half-dozing. She has a short cough with a tendency to sickness. Left side of face on which she lies is red and a little swollen. There is the appearance of a bruise on this cheek, said to be from having struck it against the bed a few days since. Pulse 152 distinct, rather sharp, of small volume. Alæ nasi move in inspiration. Respiration 48. Thighs and abdomen covered with branny desquamation, and scattered over the trunk is a fine punctated brown mottling, not disappearing entirely on pressure. On the back it is much redder. It reminds one of scarlatina rash on the 3rd or 4th day. Lips and tongue covered with sticky dark-coloured secretions; some fulness of glands of neck with tenderness. Bowels have acted twice in 24 hours. Motions not loose, of a darker colour. Dry rhonchi over bases of lungs, with weak respiration; no dulness on percussion.

14th. Tongue moister, lips also. Temperature 103°. Pulse 140. Some tenderness at angles of lower jaw. This morning the rash of scarlatina was well out on chest, abdomen, back and thighs. An inclination to sickness continues, which is aggravated by the child forcing her fingers into her mouth. Bowels acted twice. Motions solid.

15*th*. Tongue and lips moister, less tenderness at angle of jaw ; throat on inspection seen to be generally red, not swollen or ulcerated. Will not take her food well.

18*th*. Gums on left side of mouth deeply ulcerated and sloughy.

The next day she died. The notes of the last 3 days of life are incomplete.

On *post-mortem* examination, the chief lesions found were lobar pneumonia of lower and middle lobe of right and lower lobe of left lung. Recent adhesions in right pleura.

Liver fatty.

Spleen large, weighing 2 oz. 6 drachms.

Kidneys pale, opaque ; not very notably changed.

Gums of lower jaw on both sides sloughed, the bone exposed. Cheeks not ulcerated.

Stomach pale. Glands of duodenum large. The lower end of the ileum in a length of two feet contains about 25 ulcers with free overhanging edges, exposing the muscular coat ; they were obviously seated in the agminated and solitary glands. They did not seem to be extending in depth. Mesenteric glands large, free from typhoid deposit. The muscular tissue of adductors of thigh and recti-abdominis was carefully examined microscopically ; but was not found to exhibit any of the changes described by Zenker as occurring after typhoid fever.*

Remarks.—A great mistake was made in the case in allowing the patient to take meat so early as the 16th day. On the 12th day the temperature fell below 98°, and the fever seemed to be at an end ; there had been no diarrhœa, no typhoid spots, and the child's appetite was good, and she did not look ill. Still we might have been sure that Peyer's patches were ulcerated, and that the fever had not run its course. The late appearance of eruption, for the first time after the third week, is remarkable.

* Zenker describes two changes, which he says are almost constantly found in some of the muscles of patients who die from typhoid fever.
(1.) A granular degeneration in the contractile substance of the muscular fibre, either albuminous or fatty.
(2.) A transformation of the contractile substance of the primitive fibre into an entirely homogeneous, colourless material, which is wax-like and glistening. There is an entire disappearance of striæ, and destruction of nuclei, whilst the sarcolemma remains. The muscles most constantly affected are those above named. Rupture of muscular fibre and hæmorrhage into its substance after typhoid fever, had been described by Rokitansky as of frequent occurrence, and Virchow observed that these lesions usually occurred in muscles undergoing degeneration.

Typhoid Fever lasting 30 *days; an interval of* 5 *days followed by a relapse of* 5 *days.*

William Nye, æt. 7, had not been in good health for several months. During the last month he has had pain in abdomen, has lost his appetite and become thinner.

Sept. 11*th.* He became much worse. His bowels were relaxed and he had frontal headache.

18*th.* Was admitted to the hospital with tympanitic tender abdomen, hot skin, and weak pulse. Tongue red. Wine 3 oz.

21*st.* Much depression. Looks heavy, and is very languid. Pulse 132. Respiration 48. Tongue too red, moist. Abdomen tub-shaped, tender near umbilicus. Bowels very loose. Is frequently sick. Six rose-coloured spots seen on abdomen and chest. Increase wine to 4 oz.

24*th.* Many spots. Abdomen still swollen, extra-resonant. Bowels relaxed, motions of typhoid character. Pulse 144.

27*th.* Weak and pale. Deafness first noted. Pulse 152. Two fresh spots.

28*th.* Abdomen less swollen. Pulse 144.

30*th.* Profuse sweating. Pulse 144, large and soft. Spleen not enlarged.

Oct. 2*nd.* Pulse 136. No sweating. Brandy 4 oz.

6*th.* Bowels not open since the 4th; motions then solid. Tongue smooth, dry, and glazed. Has been more feverish the last 2 days.

7*th.* One fresh spot on back. Bowels have acted. Motions solid. Pulse 140 weak.

10*th.* Pulse 124. Bowels confined. Temperature in the morning 97·2°, evening 99°. He seems convalescent; has no fever. Appetite good. Tongue clean.

17*th.* Fever has returned since the evening of 15th. Fresh spots appeared to-day. To-day is low and pale, has pain in forehead and over his right eye. Right pupil a little larger than left; and there is a slight squint. Bowels confined. To take cod liver oil, and citrate of iron and quinine.

18*th.* Abdomen swollen. Frontal headache and pain in eye continue. Numerous small elevated spots of the size of peas, which are tender on pressure, like erythema papulatum. Urine contains albumen to-day and yesterday.

19*th.* Seems low and depressed. Pulse 144. Headache and pain in eye persist. Slight squint. Pulse 144 not very weak. To take 2 oz. brandy.

23*rd.* Better, more lively. Pulse 124. Bowels confined. There is slight dulness at apex of right lung, with harsh respiration.

27*th.* Since last note has continued to make steady improvement.

This case was peculiar in several respects. The long antecedent illness, the distinct remission at the end of a month, and return of symptoms in about five days; the occurrence of an erythematous papular eruption during the relapse; the pain in one eye, with squint, during the same period, were not usual phenomena. I am inclined to believe that the boy was tubercular, and that is confirmed by the examination of his chest on the 23rd October. The peculiar cerebral symptoms must be regarded as mere neuroses dependent on debility.

His sister Charlotte, æt. 9 years, was brought to the hospital on the 22nd October, just as William was recovering. She became ill on the 14th October, feeling alternately hot and chilly. An aperient was given her, which acted freely.

On admission she did not look very ill; was flushed. Pulse 132. She had no headache. The next day, the 11th of her illness, one rose-coloured spot was found. She seemed better; her pulse was 112. Was treated with nitro-muriatic acid mixture.

(13th day.) Scarcely looks ill. Pulse 136. Six more spots noted.

(17th day.) Looks more ill. Complexion pale, muddy. Six additional spots. Pulse 132, not very weak. Tongue coated with light white fur. Spleen felt an inch below false ribs. Bowels act 3 or 4 times a-day. Starch injections given occasionally.

(19th day.) Pulse 120, full and soft. Four more spots. Opiate enema. 1 oz. brandy.

(21st day.) Pulse 144; face flushed.

(24th day.) Temperature higher than it has been. In axilla 103·6°. Pulse 112.

(27th day.) Pale and turbid complexion. Seems low. Pulse 152, very compressible. First sound of heart scarcely audible. Tongue inclined to be dry. Bowels open; stools very loose, of a brownish colour. To take 3 oz. of brandy, egg, and beef-tea, with carbonate of ammonia and chloric ether. At night ¼ a gr. of powdered opium.

(29th day.) Fever still higher. Temperature 104·2° and 105·2°. Abdomen rather tympanitic. Fresh spots appear daily. Brandy 4 oz. Pulv. ipecac. co., gr. iss, bis in die; and gr. iij. omni nocte.

(31st day.) Pulse 168, very weak. Delirious; sitting up in bed and talking nonsense. Abdomen less swollen.

(34th day.) Much in the same condition. Pulse 144, very weak. Eruption still continues. Stools copious, resembling yolk of egg. Temperature 103·5°. Now to resume nitro-muriatic acid mixture with liquor cinchonæ.

(37th day.) Cheeks livid. Pulse 152, very weak. Takes fluid nourishment well and 6 oz. of brandy. Abdomen full and rather tender. Temperature in the morning 102·8° ; in the evening 104°.

The next day, being the 38th day of illness, her pulse was still 152, weak, and abdomen as on the previous day. There was a considerable remission of fever in the morning, the temperature falling to 99·6°, rising in the evening to 104·6°.

On the 2 following days it ranged from 99° to 102·6° ; on the 5 following days from 98·6° to 102° ; only reaching 100·2° in the evening of the 44th day. After that day it never rose above 99° ; and on the 47th day it did not reach 98° even in the afternoon.

On the 46th day she looked better, but still very weak ; her pulse was 132, and very soft.

On the 49th day pulse was 120, dicrotous. Bowels open once daily ; stools of loose consistence.

On the 52nd day much improved. Pulse 120, soft ; every other beat is weaker than the alternate one.

This is a good illustration of a protracted case of typhoid fever without complication. Until the end of the first fortnight, during which she was treated with beef-tea and milk, with nitro-muriatic acid mixture, she did not seem very ill; subsequently great weakness, with abdominal symptoms, came on and lasted until the end of seven weeks ; the fever remitted to a great extent after the 38th day ; there was no pulmonary or bronchial complication.

The stimulant was not commenced until the 19th day ; it ought, perhaps, to have been given two or three days earlier, and a little more freely ; but I was at the time making observations to determine whether it is advisable to withhold stimulants in the typhoid fever of children ; but I am quite satisfied that in the majority of cases after the end of 12 or 14 days, especially if the pulse becomes more frequent and is compressible, it is not wise to dispense with wine or brandy. Very small quantities, often repeated both night and day, should be administered; the amount to be regulated by carefully watching the pulse and the general condition after and before the stimulant. As a general rule I do not think any benefit is gained by giving stimulants earlier than about the 12th day.

Treatment.—In the vast majority of cases of typhoid fever

in children, drugs are scarcely needed. The points to be attended to are : to keep the patient in bed in a well-ventilated room, to attend scrupulously to the removal of all discharges, to bathing the skin once or twice daily with tepid water, and to a frequent changing of the bed and body linen. Diet is also of supreme importance ; it must be light fluid or pultaceous, and nutritious. Milk and beef-tea should form the chief articles of nourishment; they may be combined with arrow-root, semolina, vermicelli, or ground rice. Eggs may also be given, beaten up in milk, or combined with wine or brandy. If there is great thirst, cold water, in small quantities at a time, or cold tea, may be given. No solid food should be given until the end of from 21 to 28 days, and must be forbidden for a longer period, if diarrhœa continue. I have already spoken on the use of stimulants. They are indicated in the majority of cases from the end of the 10th or 12th day.

Special symptoms may require special treatment. *Diarrhœa*, when severe, should, if possible, be checked. At first chalk and catechu may be given, and the occasional employment of a starch and opium enema is useful. (For a child six years old two ounces of mucilage of starch, with five minims of tincture of opium.)

If diarrhœa is excessive, acetate of lead in doses of from $\frac{1}{4}$ to $\frac{1}{2}$ of a grain, or sulphate of copper, in doses of from $\frac{1}{8}$ to $\frac{1}{6}$ of a grain should be used.

Abdominal pain and tenderness may be treated by hot bran poultices, or turpentine stupes. I have never found it necessary to resort to leeches.

It is well to support the abdominal walls with a bandage, if there be a tendency to great distension. An assafœtida enema may be given in case of great tympanites.

Cerebral Symptoms.—In the earlier stages cold lotions to the head may be used to relieve pain and excitement. In the later periods delirium will be usually best treated by the use of wine and opium. I have never met with symptoms requiring depletion or blisters.

The occurrence of *bronchitis* may call for a little ipecacuanha, with a mustard plaster under the shoulder blades ; if the patient is very low, and mucus accumulates in the bronchi, carbonate of ammonia and senega should be given. Turpentine is recommended in these cases, with chloric ether and yolk of egg or mucilage. Instead of diarrhœa, there may be constipation, which will call for an occasional dose of castor oil.

During convalescence the diet must be carefully regulated, and the return to ordinary food must be very gradual. Cod-liver oil may be used with advantage, if the patient is much emaciated, and diarrhœa has ceased.

In the uncomplicated cases, a mixture containing a few drops of mineral acid and chloric ether, is what I prefer. Chlorate of potash is sometimes given instead of the acid. I have lately been giving moderately large doses of quinine with the acid in a number of cases, and it has appeared to be useful in reducing the intensity of the fever. It is recommended by Barthez and Rilliet.

Typhoid Fever complicated with Broncho-Pneumonia.

Adam Anderson, æt. 5 years, had been subject to a cough from his early infancy. During the 2 weeks preceding his admission on the 13th November, the cough had been worse than usual. He did not, however, appear ill until the 6th November. His mother and another of her children had recently suffered from typhoid fever.

Nov. 14*th.* He was pale with an anxious expression. His lips were dry ; and tongue rather dry. Pulse 190, small and tolerably distinct. Respiration 50. Coughed a good deal ; sputa swallowed. Chest of moderately good shape, with some recession at the lower part. Percussion note generally rather full-toned. Over both lungs, fine bubbling rhonchi were audible both with inspiration and expiration, except at the apices, where the respiratory sounds are coarse. Temperature 102·4° and 103°. To have a mustard plaster to his chest and take ammoniæ muriatis, gr. iij. ; tinct scillæ, ₥vj. ; misturæ ipecacuanhæ, ʒij. every 4 hours. The bowels have not acted for 5 days ; a castor oil injection is to be administered. His diet, good beef-tea and milk.

17*th* (12th day). Looks much the same. Pulse 152 weak. Respiration 52. Nares dilate with inspiration. An occasional loose cough. Bowels not open for 2 days. His appetite is tolerably good. Tongue clean, moist. Abdomen a little swollen, not tense. Spleen not palpable. Two typhoid spots seen on chest. Physical signs in thorax as before.

To take 3 gr. of carbonate of ammonia in decoction of senega every 4 hours. The temperature of the axilla has ranged since admission from 101° to 103·8°.

18th. Weaker. Is pale with slight lividity. A frequent loose cough. Some dulness on percussion under right clavicle. In other respects physical signs about as before. Abdomen swollen and hard. Liver reaches nearly to nipple level. Spleen felt 1¼ inch below ribs. A rose-coloured spot on left thigh. Bowels open 4 times in 24 hours. Stools, mortar-like, passed in bed. Takes food pretty well. To have 3 oz. of wine.

20th (15th day). Looks very wretched, pale and low. Pulse 160, very weak. Tongue moist and clean. Additional spots. Lymphatic glands enlarged. ℞. Liq. cinchonæ, ♏x.; mist. acidi nitro-muriatici, ʒij. Ter die. Temperature, a.m., 101·2°; p.m., 104·2°.

(16th day.) Much the same. Pulse 148. Respiration 48. Is deaf. Is inclined to be sick after drinking anything.

23rd (18th day). Temperature 101·0° and 102·6°.

25th (20th day). Pale and exhausted. Pulse 140, small and weak. Skin began to be moist yesterday; chest covered with sudamina. Over front of chest, there is full-toned resonance everywhere. On the right side, a little bubbling with inspiration. On the left, respiration full. With expiration sonorous rhonchi on both sides. Over back, percussion note everywhere extra-resonant tympanitic; resistance increased. Fine bubbling sounds over lower half, both inspiratory and expiratory. On the right side, general bubbling is not very abundant, except about the middle of the back, where it is small and sharp. Spleen is receding, now felt only ½ an inch below ribs. Abdomen flatter. For the first time to-day stool consisted of a solid mass with much fluid. Has taken pulv. ipecacuanhæ co. gr. iij. for 2 nights past. Is to have 5 oz. of wine.

28th. From the 24th the stools have been solid; once daily. Brandy 4 oz. Fish. Temperature fell yesterday to 98·7° in the morning, and rose to 100·4° in the evening; this morning it was again 98·4°.

29th (24th day). Is pale, with a light violet colour in cheeks, easily increased by exertion. Pulse 140, weak. Percussion note notably dull over left back from base to the angle of scapula. Over this part moist crackling sounds with inspiration; and with expiration a creaking sound, conveying strongly the notion of friction and causing a faint fremitus to the hand, over the whole lungs are more or less fine moist crepitant sounds, with inspiration; and sonoro-sibilant rhonchi with expiration. Nares dilate with inspiration.

Dec. 6th (30th day). Much better; looks comparatively well. Pulse 120. Coughs a good deal at times. There is still slight dulness at left base. Moist crackling sounds still heard over chest.

12th. Cough still troublesome. Otherwise pretty well. To take in a mixture morphiæ acet. gr. $\frac{1}{35}$ and vin. ipecac. ♏iij. occasionally.

29th. Gaining flesh and very cheerful. Pulse 108. Respiration 30.

Cough nearly gone. Percussion note extra-resonant over whole of both lungs. Respiratory sounds normal, but over right back and at left base, there is still to be heard the creaking sound before described, chiefly with inspiration, but partly with expiration.

Jan. 31*st.* Discharged quite well.

Remarks.—This patient, the subject of chronic cough, had some emphysema of his lungs, and suffered a good deal from capillary bronchitis, with slight lobular pneumonia, which protracted the case and added much to its gravity. The extreme rapidity of the pulse (190), on the 14th November is noteworthy. Great frequency of the pulse in a child is not so serious an element in prognosis as in an adult.

Typhus fever is usually a very mild disease in children. It is a disease which we see comparatively seldom in our hospital. It is seldom met with, except from direct infection in families, where the elder branches have first suffered. It requires to be treated on the same principles as in adults. One case of cancrum oris, which proved fatal in the hospital, appeared to have been due to a previous attack of typhus.

The following case may be taken as an illustration of the course generally run by typhus in children :—

Harriet W., æt. 7 years, one of five children, living in Shoreditch. Her mother has just recovered from typhus, for which she was in St. Thomas's Hospital 8 weeks. Her father has been short of work lately, so that the family have suffered from the want of necessaries. He only earns 20*s.* a week when fully employed.

Harriet was a stout girl till 4 months ago, when she became gradually thin and peevish. For 2 months has suffered from nausea and vomiting. For 4 or 5 months has complained occasionally of headache, which within the last day or two has become very severe.

Aug. 6*th.* Was quite light-headed, and took to her bed. During the previous week, she had several attacks of chilliness. On the previous day (August 5th) she was running about the street, but was worse than usual.

7*th.* Became very hot ; her headache was very severe over the eyes and in the forehead. Her nose bled. She was admitted into the hospital under my care on the 9th of August. There was a central flush over the malar bones ; all her skin was injected, dusky looking. There were innumerable petechiæ over her surface (chiefly from fleas) ; on the inner side of her thighs there was a faint mottling such as is seen in typhus. Lips were dry and scaly. Tongue red at the tip, was covered by a thinnish white fur. Skin very hot. Abdomen not swollen or retracted. Spleen was felt, deeply under false ribs. There was distinct gurgling in right iliac fossa. The bowels had been rather confined lately. There was notable tenderness in left hypochondrium. She had no headache now. Pulse 148, regular and weak. Wine 3 oz., beef-tea and milk.

10*th*. Slept well without delirium. This morning looks dusky but less flushed; a flush is easily caused by the least excitement. The skin presents more of the faint mottling referred to yesterday. It is especially visible about the thighs. Tongue as before. Bowels acted three times yesterday. Motions small, loose, and dark in colour. Spleen reaches half an inch below the costal margin. The lymphatic glands are everywhere slightly swollen. Pulse 102, weak and small. Respiration 32. Temperature 102·6°.

11*th*. Looks much the same. Tongue as before. Bowels open once in the night. Stool dark-coloured, very small. The rash same as yesterday. Pulse 140, small and weak. The region of spleen still tender. Has a slight cough; no physical sign of disease in chest. Temperature 101·5.°

12*th*. Still drowsy; complexion very dusky. Rash very copious; no petechiæ. Tongue is becoming dry and brownish.

13*th*. More lively. Tongue moist; clean in the anterior third, with a white fur behind. Rash paler, except in a few places, where slight ecchymosis is present. Bowels open twice; small black stools. Spleen as before. Pulse 140, fuller.

From this time she rapidly improved, and was discharged on the 31st of August.

Remarks.—This case presented a symptom which is considered to be almost pathognomonic of typhoid, namely, gurgling in the right iliac fossa. It is often absent in typhoid, and I have met with it in one other case of typhus, so that it is not of much diagnostic value. According to Dr. Murchison it is a rare symptom in typhus, having been found only once in 43 cases. In children diarrhœa in typhus is not infrequent; from my small experience I should say that it occurs in a larger proportion of children than of adults. At all ages, diarrhœa has been found as a symptom in about one-tenth of the cases. The abdomen is, however, very seldom enlarged or extra-resonant. The patient takes to bed at an earlier period of illness in typhus than in typhoid; the course of the disease is usually run within a fortnight, and it may be completed within eight or nine days.

The mortality of typhus fever is much lower in children than in adults; according to Dr. Murchison, however, under five years of age the mortality at the Fever Hospital amounted to 17·6 per cent.; between 5 and 10 years of age, it was 7·6 per cent., and between 10 and 15 years of age, only 4·9 per cent. At all ages the mortality is nearly 21 per cent. It is probable, however, that the cases of typhus in young children admitted into the Fever Hospital are above the average in severity. According to Dr. Steele at Glasgow, the mortality of typhus under 10 years of age, is only 4·4 per cent., between 10 and 15 years 5·1 per cent.

The *treatment* of typhus fever in children does not differ from that which is indicated for adults. Alcoholic stimulants are usually required at an earlier period after the commencement of typhus than of typhoid fever.

SKIN DISEASES.

I PROPOSE to devote this chapter to some of the cases of skin affection which have come under my treatment in this hospital. The most frequent skin diseases met with in children are eczema, impetigo, strophulus, lichen, prurigo, urticaria, syphilitic rash, xeroderma, ecthyma, psoriasis, scabies, and ringworm. The less common diseases are purpura, molluscum, pemphigus, alopecia, lupus, ichthyosis, and favus. I do not propose to enter into descriptions of these diseases, having done this in my "Handbook of Skin Diseases." I will only refer to the notes of a few of the rarer diseases, and give some therapeutical observations on some of the more common affections.

Pemphigus is a disease occasionally met with in children. It is a disease running a very chronic course, and liable to frequent relapses. It is attended with much debility in some cases, and with a good deal of constitutional irritation. It is not a fatal disease in children; it generally yields to the persevering administration of arsenic.

The following was a severe case of the kind.

G. F. S. W., aged 5 years, an orphan, whose mother died of phthisis and syphilis. The person who has had charge of him says he had snuffles and a rash when an infant. Has had plenty to eat and drink. He has had the usual children's diseases and small-pox; from all of them he recovered easily. He has usually had tolerably good health until 2 months ago, when some blebs with clear fluid in them appeared on his fingers; these burst and crusted; the crusts separated and other blebs followed. These successive eruptions continued until his admission; the blebs becoming larger as time went on. On admission the following notes were taken. A light-haired boy with fair skin. Skin generally harsh, cool. Over the front of his trunk are a number of patches having a rounded outline. The central portions are slightly yellowish; the margins are elevated and about ⅛ of an inch wide, of a pink or brownish colour. The patches vary in diameter from half an inch to 2 or 3 inches. The colour disappears on pressure, they are not distinctly elevated, are not tender, and present no desquamation. (Erythema circinatum.) On the left elbow is a large livid discolouration, and in the centre of it a thickish crust, the remains of a bleb. The middle and fourth fingers of left hand are the seats of blebs as large as horse-beans. On the right arm are 2 blebs with reddish serum, several crusts, and some stains of former blebs. On the thighs are several stains of blebs and several rings of erythema similar to those described on the trunk. Immediately below the right knee is an enormous bleb measuring 6 inches in diameter, half filled with a turbid fluid in which flakes are floating; this bulla is surrounded by a red margin. The lymphatic glands of the groin on this side are tender. There are blebs or traces of them on the feet and left knee. Tongue furred behind. Pulse 132, weak. To take pulv. ipecac. co. gr. v. at night and vini ferri, ℨij. Ter die. Zinc ointment to be applied to the broken surface.

Dec. 3rd. Ten days after admission. Fresh bullæ appear from time to time without any previous redness. He is now to take 3 minims of liquor arsenicalis in each dose of steel wine.

20th. Blebs have appeared from time to time on mouth, wrist, fingers, feet, thigh, external ear and penis. On the crown of head, amongst the hair, are crusts like those of impetigo, but they have originated in blebs whose contents became rapidly turbid and did not distend their walls.

28th. To take liq. pot. arsen. ℟iv.; aquæ ℨss. Ter die.

Jan. 4th. Blebs still make their appearance, some with a bright red halo, some without any areola; some have clear light-coloured fluid in them, others have reddish fluid, in some the contents are flaky, and in some they rapidly become purulent. The clear fluid is slightly alkaline, the purulent fluid is acid. By an oversight the patient's diet has not been so liberal as it should have been. Pulse 120, weak. He is now to have meat every day, an egg, and 3 oz. wine. ℞. Liq. arsenici chloridi, ℟vj.; mist. quinæ, ℨiij. Ter die.

13th. General health improving. Increase the dose of liquor arsenici chlor. to 7 minims. The skin of trunk generally dry, covered in many parts with fine papular elevations, almost translucent, like obstructed sebaceous follicles. To have a soap-and-water bath every other day. Very few blebs appearing now.

15th. Omit medicines.

21st. In consideration of his mother being reported to have suffered from syphilis, he was ordered pot. iodidi, gr. ij.; mist. quinæ, ℨij. Ter die.

28th. Fresh blebs appear almost daily. To take now liq. arsenici chlor. ℟ix.; aquæ, ℨij. Ter die. .

Feb. 11th. No new blebs. He seems stronger and much better.

14th. Omit arsenical mixture, his gums are a little ulcerated. To take cod liver oil and steel wine, and a lotion of chlorate of potash. Yesterday another small bleb appeared. He seems stronger. Appetite not very good. Pulse always too quick and weak.

April 3rd. His gums are healing. Has had no fresh eruption lately. Some of the old surfaces are not quite healed.

After this he continued free from pemphigus for two years, when he was brought to the hospital in consequence of epileptic fits, of which he had had several. In other respects appeared well.

The urine passed by this boy in 24 hours was tested on January 16th, 23rd, 24th, and 26th. The mean daily quantity of urine passed was 456 cubic centimetres (a little more than 16 ounces), the mean amount of urea was 17·4 grammes (or nearly 270 grains), of chloride of sodium 3·89 grammes (about 60 grains), uric acid, ·514 grammes (between 7 and 8 grains). The boy's weight was 20·4 kilogrammes (45 pounds). It would appear that the water of urine was deficient, that the urea was about normal, the chlorides and uric acid in rather large amounts.

There was no uric acid found in the fluid contents of the blebs; this ingredient has been met with in some cases.

This patient would seem to have been syphilitic in infancy; but the pemphigus cannot be ascribed to this circumstance. This form of pemphigus does not appear to be at all connected with syphilis.

Several other cases of pemphigus have come under treatment, which have resisted all remedies used except arsenic, under the influence of which they have got well.

The following case is a somewhat remarkable one. The disease has repeatedly recurred in the warmer months of the year. It first began when he was three years old, one month after vaccination. Treatment has not been fairly tried.

William B., aged 6 years, is said to have had good health till 3 years old. His mother died when he was 6 months old. He has several brothers and sisters, who have no affection of the skin. He was vaccinated at the age of 3 years and the pustules healed well. About a month after this, the eruption from which he now suffers appeared; and has returned every year through the summer months. Each attack is preceded for a day or two by feverishness and loss of appetite. The eruption comes out on the face, neck, and hands in the form of blebs which contain yellowish fluid and are clustered together. They lead to scabs of the size of large split peas, the skin beneath being excoriated. He was treated with liquor potassæ arsenitis for a short time; but his friends became impatient, and he was handed over to an old woman, who gave him ground-ivy and elder-flower tea and some drops said to be like bark. He got better towards the cold weather, but the disease returned in the following summer as usual. I do not believe that the vaccination was in any way the cause of this eruption; but the coincidence in point of time is worthy of record.

In the following case the curative effect of arsenic was well-marked.

Emma Wilson, æt. 5, a tolerably healthy-looking child. She is the 4th of 7 children, of whom 5 have died. Last January an eruption of blebs appeared on her nose and spread over her face. Six weeks later a similar eruption appeared on her legs. From that time till the end of May successive crops of blebs came out every week or 10 days, either on her face, near the pubes, or upper part of thighs. She was admitted on 29th May. She was weak and poorly. She was ordered quinine and mineral acids; with oxide of zinc and starch powder to absorb the fluid from blebs.

June 3rd. No improvement. Was ordered 2 minims of Fowler's solution 3 times a day.

6th. Very feverish. A saline mixture of citrate of potash and colchicum wine given.

11th. Fever is gone. Eruptions recur from time to time. To take quinine and liquor arsenicalis 2 minims, which was increased on the 18th to 3 minims, and on the 8th July was combined with vinum ferri.

July 14th. Not much improved. Took decoction of ground-ivy with spiritus ætheris co. till 4th August, when the disease was as active as ever.

Aug. 13th. To take liquor potassæ arsenitis, ♏iv. Ter die. She rapidly improved, and went out well on 2nd September.

Oct. 5th. Having omitted the medicine the disease returned. She again recovered on taking the medicine.

Several times the experiment was repeated, and each time the disease returned after the medicine had been omitted a week or ten days. By continuing the medicine a month after all rash had ceased the cure was at length permanently established. Fluid from one of the blebs was found to contain uric acid.

The smaller dose of arsenical solution did not effect a cure, but the dose of four minims three times a day was in every case successful.

Purpura Hæmorrhagica occasionally attacks children, and sometimes resists all treatment. In fatal cases the skin, mucous membrane, and sometimes serous membranes, are found to have been the seats of hæmorrhage.

It may come on without any previous illness, or as a sequel of measles, scarlatina, rheumatism, or other acute specific disease. A very severe case of it occurred in the course of chronic abscess of the femur. In several cases oil of turpentine has been used with good results, as in the following case.

Louisa Larrey, æt. 11 years, the daughter of a journeyman butcher, who is often out of work. She has lived chiefly on bread and butter or dripping.

Nov. 20*th.* Her mother noticed some "red lumps" on her legs.

30*th.* She was admitted to the hospital. She was pale, cachectic-looking, nose red and shining, mucous membranes pale. Pulse 88, weak. Tongue clean. On her face, arms, and legs were numerous hæmorrhagic spots ; also on her tongue and inside her lips. Below the knees there were large elevated ecchymotic patches, looking like bruises. She had also many papules and vesicles from scabies on her hands. Her bowels were acted on by magn. sulph. ʒj. ; acid. sulph. dil. ♏v. ; ferri sulph. gr. j. ; aquæ carui, ad ʒj. twice a day for 2 days.

Dec. 3. To-day an attack of epistaxis, which was checked by plugging the anterior nares with lint soaked in a solution of perchloride of iron. The urine has become bloody, and there is some bloody serum in stools. No fresh purpuric spots.

5*th.* She took two drachms of oil of turpentine in half an ounce of castor oil.

6*th.* Bowels acted twice ; a semi-solid grey stool with bloody serum. Last night before taking the turpentine, she vomited fluid, in which were a few streaks of blood. A good many fresh purpuric spots, especially on legs. The scabies is much better ; has been treated by frictions with benzole. Urine still bloody. To take at night the same dose of turpentine with 3 drachms of castor oil.

7*th.* Nose bleeding again. Blood very watery in appearance. Urine as yesterday. Pulse 120, regular. Diet : eggs, milk, broth, and wine an ounce. At noon 2½ drachms of turpentine in mucilage.

8th. To take a drachm of turpentine in mucilage and syrup twice a day, and once a day an enema, consisting of oil of turpentine and castor oil, half an ounce of each, and half a pint of decoction of barley. To-day her urine is quite normal, free from albumen. The stools free from blood. All the spots are fading ; no fresh ones. From this time she rapidly improved. Her purpura came to an end, and she rapidly gained a comparatively healthy appearance.

In the next case there was an appearance of erythema papulatum which became hæmorrhagic, and the cuticle was raised in some spots giving rise to blebs.

Turpentine was also of service in this case.

Emma Isaacs, aged 10 years and 9 months, a pale delicate child ; she has had meat every day and a fair quantity of vegetables. Spots began to appear on her legs 5 weeks ago ; she has been languid, and her legs have ached.

When admitted on 13th December, there were numerous light purple spots, looking as if they had been darker, but fading ; they did not disappear on pressure. There were also numerous blebs filled with bloody serum ; also spots on soles of feet, on arms, and between the fingers. She had epistaxis yesterday. She was ordered ol. terebinth, ol. ricini aa ʒiij. ; mucilaginis ʒij. at once ; and to take twice a day a drachm of oil of turpentine in mucilage.

Dec. 22. To take vini ferri, ʒij. Ter die.

24th. Is taking turpentine 2 drachms with half an ounce of castor oil every morning.

29th. Dose reduced to 1 drachm.

Jan. 2. The child frequently complains of pain in stomach. There are now scattered over her arms, legs, and face, spots and patches from the size of pins' heads to one-third of an inch in diameter, and in colour from pink to purple ; some disappearing on pressure, others not at all affected by it. They are irregular in outline, some distinctly elevated; others not so. The seat of the former blebs is ulcerated. Pulse small and feeble. Is now to take quinine.

16th. Till to-day has been steadily improving. But this morning there are a few bright purpuric spots about each internal malleolus. She was again put on turpentine treatment.

26th. No fresh purpura. Pulse 68, irregular. The ulcers are not healed ; to be dressed with· solution of nitrate of silver. To take 5 minims of Fowler's solution three times a day, and a turpentine enema daily.

Feb. 1. Add 1 grain of ammonio-citrate of iron to each dose of the medicine.

19th. Is wonderfully improved. Looks quite healthy. Has taken Olei morrhuæ ʒj. Ter die.

In the next case a boy, 12 years old, was in a very weak condition from an abscess near the hip joint, which had existed four years. On 30th October very profuse epistaxis came on, and lasted five hours, leaving him very faint and sick.

Two days later his neck and legs were covered with spots of purpura. In the evening he was unable to pass his water until late at night; after many ineffectual attempts, his urine was passed involuntarily and contained much blood.

Nov. 1*st.* On admission very pale and thin. His neck, trunk, and limbs, studded with petechial spots, varying in colour from bright red to purple, and in size from that of a flea-bite to ¼ inch in diameter. Some of the larger ones are slightly raised from the surface. Some clots passed per urethram. To take gallic acid 4 grains, and diluted sulphuric acid, 8 minims every 3 hours. There is a large abscess pointing in the upper and anterior part of left thigh; and outside of this is the scar of a former opening, around which the cuticle is raised into vesicles with bloody fluid.

5*th.* Plumbi acetatis, gr. ⅓ every 3 hours.

6*th.* Yesterday had spitting of blood and vomiting of black clots. The urine also was loaded with blood and clots. Much depressed. Ice to be constantly swallowed.

9*th.* Boy feels and looks much better. Pulse 140, fuller than it was. Now to take tinct. ferri sesquichloridi ♏x. Ter die.

13*th.* Still improving. No hæmorrhage since the 11th. From this time he went on steadily improving.

Dec. 23*rd.* Very much better. Has a slight cough. Pulse 104, when standing up. Though he gains colour, is still very thin. Very little discharge from thigh.

Jan. 11*th.* Is much improved. Pulse 88, occasionally intermitting. Colour much improved. Acetate of lead followed by the sesquichloride of iron were the remedies in this case.

Remarks.—It is not at all unlikely that the hæmorrhage in this patient was due to albuminoid degeneration of the small blood-vessels, brought on by prolonged suppuration. It has been observed in cases of secondary syphilis in which albuminoid disease existed.

Favus is a disease which is now becoming very rare in this country. In ten years I have met with five or six cases only at this hospital. It is a disease bred by poverty and filth, and flourishes most in strumous subjects. It is important not to confound impetigo with favus. The error has been often made, and is encouraged by the name which Willan and Bateman gave to impetigo capitis, namely *porrigo favosa*. Favus was called by them *porrigo lupinosa*. Impetigo capitis is pustular, leading to thick, greenish-brown or dark-coloured crusts, in which pediculi congregate; it does not cause destruction of the hair, and is readily cured. Favus, on the other hand, is a parasitic disease in which sulphur-yellow crusts appear; it attacks the hair in its follicle, leading to permanent baldness, if left to itself. Both diseases are contagious, both are favoured by dirty habits; the lymphatic glands become enlarged and may suppurate in both diseases, but more in impetigo than in favus. In cases of doubt the microscope will decide; the crust of favus is found to consist of epidermis with a vegetable growth consisting of oval cells larger than blood globules, attached to each other at their narrow ends; the appearance is very much like that presented by the yeast-plant; there are also undulating filaments with dichotomous ramifications, and fine granular matter.

Favus especially affects the hairy scalp, but it is not limited to this part; it sometimes involves the trunk and limbs, as in the following case.

Emily Thatcher, æt. 9 years, has had skin disease since the age of 6½ years. She is pale, with a dusky complexion and grey eyes. The first spot appeared on her head near the vertex; soon afterwards two spots showed themselves on the trunk. The spot on the head gradually spread; on the trunk additional spots appeared. On admission her general health seemed tolerably good. Nearly the whole scalp was covered with a thick yellow crust, matting the hair together and swarming with lice. The crust on the head is irregularly diffused and of different thickness in different parts. Where thickest it is of the consistence of dry mortar. The margins of the hairy scalp are almost free from disease. It has a peculiar mousey odour. The lymphatic glands at the back of neck are hard but not much enlarged. On the left eyebrow are crusts almost like impetigo. On the trunk are numbers of brown stains, about the size of a large pin's head, round, slightly elevated, and covered with an epidermic scale. On the left shoulder and near the right axilla, are several well marked *favi* about the size of peas, umbilicated, sulphur-yellow, with some concentric markings in the concavity, and surrounded by a narrow red halo. On the arms are numerous thick crusts, some nearly an inch in diameter, others not larger than a pin's head. The smaller ones are surrounded by a red ring, the larger by a white fringe of cuticle; the small ones are cupulated with concentric markings, the larger ones are irregular and rough on the surface. In most cases the crusts can be detached without making the part bleed, leaving a concave reddened surface. Crusts of the same character are met with on the legs, but larger and more irregular. On the nates are a number of red papules of a pruriginous character. All the crusts were speedily removed by the constant application of a solution of sulphurous acid. On the trunk and limbs most of the parts remained free from crusts for some time, leaving a livid purple tint. In a few cases the crusts again appeared, but after a second or third application they continued quite clear. On the scalp all traces of disease were easily removed, but some parts were permanently bald, and the scalp was glazed in appearance and too adherent to the subjacent tissues.

After a week or ten days the favous growth begins to appear around separate hairs, in the form of a round yellow spot, at first convex, but soon becoming concave on the surface, and rapidly growing till it coalesces with adjoining growths.

Recourse is now had to epilation, which is a very tedious process, but the only certain way of eradicating favus on the scalp. Every day a portion of the scalp, about an inch square, is completely cleared of hairs, which are drawn out by the roots by means of a suitable forceps. Most of the hairs are loosened by the disease, which very much diminishes the painfulness of the process. After each sitting a little weak ammonio-chloride of mercury ointment is rubbed in. This process takes eight weeks to complete, and requires to be repeated again in some parts of the head once or twice. The short hairs are often rendered brittle by the disease, which makes epilation more difficult. Occasionally, as in this case, the cure is retarded by suppuration at different parts, which makes it necessary to poultice and to change the ointment, using zinc instead of mercury. At the same time the patient should have a liberal diet and take tonics—quinine or iron and cod-liver oil.

This patient was nearly twelve months under treatment, when she was sent out apparently quite cured. Six months later, having been neglected at home, she returned with a relapse of the disease, which yielded to treatment in a month.

Full details of other cases are given in my "Handbook of Skin Diseases," pp. 268—276.

The disease is contagious but not liable to spread much, if ordinary care is taken. Only one child contracted the disease from this patient during the twelve months she remained in the hospital. A small circular branny patch appeared on the forehead, on which favous crusts soon formed. It was readily cured by sulphurous acid lotion.

Another disease of the hairy scalp, much more common than favus, is *common ringworm* or *tinea tonsurans*. This, unlike favus, is not limited to the poor and dirty section of the community, but it is frequently met with in boarding-schools and private families of the middle and upper classes. Like favus, it is especially troublesome when attacking the hairy scalp; but it also occurs on other parts under the name of *herpes circinatus* and *ringworm* of the body or limbs. It very rarely attacks the heads of adults. It is undoubtedly contagious, and may be contracted from cats, dogs, and other domestic animals. Like favus, it is a parasitic disease, being characterised by a microscopic fungus, which infiltrates the cuticle, the hair, and the root-sheath of the hair. The most striking feature is that it renders the hairs brittle, so that they are broken off usually at the distance of about one-eighth of an inch from the scalp; the fragments of hairs thus broken are rendered opaque and darker in tint, and as they emerge from the follicle are surrounded by a white fringe of cuticle. Before these changes are visible in the hair, the scalp is slightly reddened over a more or less circular patch, and there is branny desquamation around individual hairs. After the disease has existed some time the hairs break off as they emerge from the follicle, quite close to the scalp, and there is a prominence of the follicle causing a kind of goose-skin appearance; the scalp, in the parts affected, has a bluish slatey colour. When examined by the microscope it is found that the hairs are infiltrated with small round globules, that the normal striation of the hair is destroyed, and the hair rendered opaque. The fracture is fibrous, like that of a bundle of sticks of unequal length. These round globules, which are spores of a fungus, are smaller than those of the fungus in favus; they vary in diameter from that of a blood globule to less than half that size. A number of the same bodies, together with short fragments of mycelium will be found infiltrating the cuticle, and may be made visible by the addition of diluted liquor potassæ or ammoniæ.

There is one form of disease closely allied to tinea tonsurans in which the hairs become loosened in the follicles instead of becoming brittle; the scalp is raised into tubercular swellings, which are red and painful, and on them the hairs are loosened and come out, leaving red bald patches. The disease is much like sycosis of the beard. It is parasitic; the spores are small, and are found infiltrating the cuticle of the root-sheath and the cuticle round the hair. The hair itself is opaque and more granular than in the normal condition.

Tinea tonsurans, when treated in its early stages, may be speedily cured; but if allowed to remain for some time the cure is more tedious. On the body and limbs an ointment, containing half a drachm of the red oxide or the ammonio-chloride of mercury in the ounce of lard, applied twice daily for five or six days, will suffice. The same plan will seldom succeed on the hairy scalp. An ointment containing sulphur and the ammonio-chloride of mercury, a scruple of the former and half a drachm of the latter to an ounce of the menstruum has been recommended, and will in recent cases often effect a cure. A blistering solution may be applied, but this is painful, and often requires to be repeated several times. The result often is that subcutaneous abscesses arise, and there is more or less extensive obliteration of hair follicles, causing permanent baldness in the affected part.

The most satisfactory plan of treatment which I have seen is the use of a mixture of oil of tar and iodine, introduced by Dr. Coster, of Hanwell Schools. A colourless solution distilled from coal-tar of specific gravity ·853, is gradually and carefully mixed with iodine in the proportions of four to one; some heat is generated in the mixture, and a dark coloured solution of thick treacley consistence is obtained. The more iodine that can be dissolved the better. The hair round the affected part, for a distance of a quarter of an inch, should be cut quite short, and the solution rubbed in firmly with a piece of sponge on the end of a piece of wood or whalebone. It is allowed to dry on the part, and left until the cuticle and the black crust separate at the end of a week or ten days. In recent cases *one* application is often sufficient ; in cases of long standing, the application may require to be repeated two or three times. It is a remarkable property of this preparation that it does not blister or cause pain, though containing so much iodine.

The subjects of this disease should be allowed a wholesome nourishing diet, and if they are strumous cod-liver oil and the iodide of iron will be of service.

Another affection of the hairy scalp which is common to children and adults is alopecia areata, sometimes called *tinea decalvans*. In this disease the hairs drop out by the roots over a limited area or several circumscribed patches ; this is occasionally preceded by some itching of the scalp and slight branny desquamation ; but usually there is neither itching nor roughness of the affected part at any period. The bald patches are pale, and have less sensibility than the healthy scalp, and are less acted on by rubefacients or blistering solutions. The bulbs of the hair near the bald parts are found to taper to a point, instead of being rounded and club-shaped ; and usually at early periods of the disease around the bald spots a few club-shaped hairs are seen about $\frac{1}{8}$th of an inch in length, with their free extremities enlarged ; but unlike the stubby hairs left in tinea tonsurans, there is no fringe of cuticle around them, and the hair itself is not more opaque than usual, or altered in structure, except in the smallness of its bulb, and in a fibrous fracture of the free extremity.

This disease is believed by many modern dermatologists to be parasitic ; but the parasite is so seldom found that it is doubtful whether it is an essential condition of the disease. It is, if ever contagious, much more rarely so than the two previously described diseases. I have once seen an outbreak of it in a large parochial school; the disease attacked 40 or 50 of the elder girls within a very short period, whilst all the boys and younger girls educated in the same establishment, and under exactly similar management, but not associating freely with the elder girls, escaped. It seemed impossible to account for this outbreak, except on the theory of contagion. Mr. Hutchinson has also recorded some cases of apparent contagion.

The treatment which I have found most useful in this disease is to apply, at long intervals, acetum cantharidis to the bald patches, and to paint them every other day with tincture of iodine; to wash the head twice a week with soap and cold water, and apply a wash (consisting of rum 1 pint, tincture of cantharides 1 ounce, aromatic spirits of ammonia half an ounce, and water 10 ounces) to the parts of the head which are not bald twice a week. The cure has been, I believe, sometimes promoted by the internal administration of arsenic. In some cases iron is indicated, and will be useful.

Molluscum contagiosum is not often brought under our notice. It is more frequently met with in parochial schools. It is characterised by round elevations, varying in size from a pin's head to a large pea or hazel-nut, with a depression on the summit of each. The elevations have a semi-translucent appearance; the skin over them is white or pinkish. They gradually increase in size, and after a time they ulcerate on the surface.

They generally occur on the face and neck. This form of molluscum consists in an ulceration and hypertrophy of the sebaceous glands. Their contagiousness has been said by M. Hardy to be due to a parasite, consisting of branched tubes and spores of a cryptogamic plant. This parasite has not, however, been found by observers in this country. I have several times looked for it, but in vain.

The treatment consists in squeezing out the contents of the tumours; it is seldom necessary to touch the interior with lunar caustic; sometimes it is necessary to incise the skin over the tumours before squeezing them out.

No internal treatment seems to be of any service. The disease may be propagated from a child to its mother, or nurse's breast, or from one child to another occupying the same bed; but the contagion does not extend to any distance.

The disease is quite distinct from molluscum fibrosum, which consists mainly of hypertrophy of connective tissue, is not seated in the sebaceous glands, and is not contagious. This latter affection more frequently occurs in adults.

Before leaving the consideration of parasitic skin diseases, I may mention that *pityriasis versicolor* is almost unknown amongst children. I have seen one case of it in a boy six years old, the subject of phthisis.

Scabies, on the other hand, is probably more common in children than in adults. It is unnecessary to dwell here on this disease, although its diagnosis leads to innumerable mistakes. I may just mention that on the feet and hands of infants under 12 months of age, are often to be met with the best specimens of long and perfect cuniculi of the acarus scabiei; because the little patient does not direct its scratching to the seat of itching, consequently the acarus is left to burrow in the cuticle, and to deposit her ova undisturbed. The irritation of the itch insect, however, often causes suppuration in children, and the abscesses formed often encroach upon the channel made by the insect.

There is one disease frequently met with in children, from the age of six months to four or five years, which is constantly confounded with scabies, the diagnosis of which sometimes requires considerable care. It is an affection which has received several different names—*lichen urticatus, prurigo*, and *strophulus pruriginosus*, which is the name that I prefer. It is characterised by the appearance of erythematous patches and moderate-sized papules, usually isolated; they are at first red, and their colour is gradually lost in the surrounding skin. Sometimes the papules merge into pomphi or wheals with white centres. They are the seat of intense itching, severe in the day but worse at night; this leads to violent scratching; many of the papules bleed, and get a black scab on their summits. The redness soon subsides, and the papule presents all the characters of a papule in prurigo. Interspersed with the papules are pustules with hard base, which may lead to troublesome ulceration, and leave brownish-red stains. The parts chiefly affected are the back, the thighs, and the arms. The wrists, the palms of the hands, and the soles of the feet, are usually comparatively free, as well as the front of the abdomen, and the inner aspect of the thighs; in this respect contrasting with scabies. It differs from this disease also in its course; being very chronic, not always getting gradually worse, but at times without treatment almost disappearing and then suddenly breaking out again.

The sudden appearance of the erythematous spots and small pomphi, the absence of the characteristic acarine cuniculus, the freedom from contagion, the age of the child, the localisation of the eruption, and the absence of vesicles, will usually enable one to make a correct diagnosis.

The treatment of this disease consists in attention to diet and the state of the bowels, and the administration of tonics, especially the syrup of the iodide of iron. Locally, a lotion containing carbolic acid and glycerine, a drachm of the former, half an ounce of the latter, to half a pint of water, will be sometimes useful. Glycerine and tar in the proportion of 2 to 1, or unguentum picis liquidæ, will sometimes allay irritation better than anything else.

Eczema in its various forms is the most common skin disease in children. It begins frequently at the age of three or four months, about the time at which vaccination is performed in this country. It is from this circumstance often ascribed, by the mother and others, to vaccination; but on inquiry it will often be found that the eczema began before the operation was performed, or that other members of the family have suffered from eczema before they were vaccinated. In other cases, however, it would seem that the occurrence of eczema has been excited by vaccinia; but any other little febrile disturbance or error in diet would probably have acted in the same way. It is certain that the eczema is not due to any specific action of the vaccine virus.

Eczema *capitis*, commonly known as "scalled head," is the most frequent aspect of the disease; when it extends to the face it has been called *crusta lactea*. It often spreads to the trunk and limbs. The anterior aspects of the joints are in some children the only parts affected, and these often remain in a state of chronic irritation, which is from time to time liable to exacerbation.

Eczema of the nates from acrid discharges, and eczema of the groins (beginning in erythema, which gradually runs into a vesicular condition) are also very usual forms of disease.

The treatment of eczema, as of most other skin affections should be both constitutional and local. Whether there is a distinct diathetic condition, of which eczema is one of the main anatomical characters is, I believe, uncertain. This skin affection occurs in nearly all the members of some families, and not in any of the members of others; some children with eczema are fat and apparently well-nourished, but more frequently the subjects of it are thin and show signs of malnutrition. Some eczematous children are scrofulous, but many are not so; some of them are tuberculous, but very many are not so. It is sometimes brought on by errors in diet, or by mental emotion, or irregularities in the mother or nurse when the child is still nourished at the breast. A deficiency of animal food in older children will encourage it; but a liberal meat diet will not always cure or prevent it. Attention to the bowels and hepatic secretion is necessary; occasional doses of rhubarb and soda, and more rarely of rhubarb and mercury, are required. Steel wine and cod-liver oil are useful in most cases. In chronic cases these should be combined with arsenic in small doses. For young children under two years, milk should form the staple diet, with the addition of some farinaceous food; and if ossification and dentition be retarded, animal gravies and broths should be supplied in addition.

After this age a plain diet of bread, milk, farinaceous food, fresh meat, and vegetables, should be given. Saccharine matter in a state of change, as in malt liquor and preserves, salted meats, cheese, and spices, should be avoided. Locally, in the inflammatory stage, bread and water poultices, followed by simple cerate with a little camphor, or the benzoated zinc ointment, may be used. It is always important to remove the crusts before any ointments are applied. This may be done by poulticing with bread and warm olive or almond oil, through the night, and if the crusts do not come away in the morning, the part should be treated with fresh oil and the crusts removed half an hour afterwards by the nail or a blunt instrument. If the crusts re-form they must be again removed from time to time.

If the discharge is very profuse, a powder of oxide of zinc, camphor, and starch, may be dusted over the part in the following proportions, camphoræ, ʒi ; spirit. vini rect. q. s. zinci oxidi, pulv. amyli, āā ʒvj. Camphor tends to allay the burning sensations.

In the chronic stage, where there is itching, but not much heat, the skin being thickened and more or less papular, strong alkaline lotions are useful, the strength being made to depend on the amount of exudation into and thickening of the affected skin. Liquor potassæ undiluted, or with one, two, three, or four parts of water, should be painted over the part with a firm brush twice a day ; if the smarting caused by this process is great, it may be allayed by cold water. When there is great thickening of skin, and the part affected is not extensive, stronger solutions of potassa fusa, as 30 or 40 grains to the ounce of water, may be used with advantage.

Another plan which is very useful in the chronic stages, when the discharge is not profuse, and the infiltration of skin is not so great, is to use tarry preparations, either the unguentum picis liquidæ or tar plasma, made of tar, glycerine, and starch, recommended by Brady. A combination, recommended by Hebra, of tar, methylated spirit, and soft soap, in equal proportions, is also very useful.

Where expense is of less consequence, pure rectified spirit should be used, the pyroxylic oil of juniper substituted for common tar, and oil of lavender added to make it more agreeable.

In some subjects tar causes too much inflammation, and cannot be used; it is desirable for this reason to apply it in the first instance over a small surface only, lest its action should be too irritating; if it does not cause much heat and swelling, it may then be applied freely to the whole of the affected parts.

In the later stages, mercurial ointments, especially the nitric oxide, 15 grains to the half ounce of olive oil and half ounce of lard, or the dilute nitrate of mercury ointment, are of service.

It is important in using ointments not to apply more than enough to give a glistening appearance to the skin, and from time to time they should be washed off to prevent the fats becoming rancid. This may be done by means of soap and water, to which I have no objection in eczema in its chronic stages. A medicated soap, containing petroleum or huile de cade, may be used, if it is preferred.

Occasionally, but very rarely, children are met with in whom eczema resists all treatment, and the skin disease never leaves them until they fall victims to tuberculosis or some other fatal malady.

In eczema impetiginodes of the scalp, the only thing necessary to effect a cure in many cases, is to remove the crusts, destroy pediculi with an ointment containing mercury, and to keep the part clean.

The disease described by Bateman as porrigo favosa, and now often called porrigo, is allied to impetigo, and requires chiefly local treatment, with tonics internally.

Psoriasis is a disease which often commences in childhood; once I saw it set in soon after birth, and it was thought to be syphilitic; but this was found out to be an incorrect supposition. It got well, and has not returned; the child is now 6 years old. In five cases I have seen it under 5 years, in five between 5 and 10 years, in ten between 10 and 15 years of age. It is often ushered in by gastric disturbance, but remains after this has subsided. It requires the same treatment in children as in adults, and is liable to relapse at one age as well as at the other. I have seen scarlet fever in a boy aged five years, the subject of psoriasis; the acute disease did not materially affect the course of the chronic one. In another child very severe inflamed psoriasis came on as a sequel of scarlet fever. I have also seen it make its first appearance immediately after an attack of small-pox.

It has been stated that psoriasis is especially prone to occur in families of a rheumatic or gouty constitution; this observation has not been corroborated by my inquiries on this head amongst children.

If there is gastric disorder this should be treated by dieting, an occasional dose of rhubarb and soda, with a mixture containing citrate of potash and tincture of orange three times a day. Afterwards arsenical preparations in moderate doses are to be given, and tar ointment applied locally. If this is objected to, citrine ointment, with a little carbolic acid, may be used instead.

It is useless to apply ointments until the scales are removed, which may be done by keeping the part in wet rags covered with oil silk or gutta percha.

Alkaline baths will be found useful where the cuticle on the affected parts is much thickened. If the skin be irritable, baths containing gelatine will be serviceable.

There is a form of skin disease presenting some resemblance to psoriasis which is occasionally seen in children. It has been called "lupus psoriasis" by Mr. Hutchinson. It has very much of the naked-eye appearance of psoriasis; it differs from it in the thinness of the layers of cuticle, and in the fact that there is a tendency to the production of a scar from interstitial absorption of the cutis vera. The colour of the patches is rather that which is characteristic of lupus. It is described by M. Hardy under the name of "scrofulide tuberculeuse disseminée."

The following case afforded a good illustration of the

disease. The patient was also suffering from caries of the spine.

Susannah C., æt. 7½ years. Was admitted under my care on the 28th November, 1864. Her mother has had 7 children, of whom 5 are living and free from skin disease. There is no history of phthisis in the family. She had measles 2 years ago, and continued weak for 3 months afterwards. About 21 months since, the eruption made its appearance. It commenced on the face as a number of spots like nettle-rash; they were at first pale, but gradually became red. A few spots appeared on the chest nearly at the same time. One spot made its appearance on her arm so lately as last week. For 12 months past her mother has not observed much change in the character of the eruption. Six months ago she fell down and broke her collar-bone; since that time her back has been getting bent; there is now a sharp antero-posterior curve in the lower part of her dorsal region. On admission she was moderately well-nourished. She has not the aspect of tuberculosis, scrofula, or syphilis. Her complexion is tolerably good. The eruption is situated on the nose, cheeks, upper lip, chin, shoulders, upper part of chest, arms, fore-arms, nates, thighs (especially behind), and slightly on the legs. The distribution of it is very symmetrical. The eruption consists of flattened elevations of the skin from about one-sixth to a quarter of an inch wide, of a purple colour, covered with thin dry scales; a slight blush surrounds the elevation. A few of the spots are surmounted by a scab; a few present a depression, as if a scab had been removed, with elevated rounded edges. A few are slightly elevated and look very glistening, the cuticle very thin. There are a number of white scars on the chest, which the mother states were formerly red, and exactly like the existing eruption. She was ordered cod-liver oil, a liberal diet, and porter. She remained in the hospital a month, during which no change was observed. She has since then been lost sight of.

Remarks.—The case must, I think, be looked upon as essentially lupus: the colour, the course, and especially the scars left by the eruption, were quite characteristic. The rather sudden appearance of numerous patches, and its widely diffused character, seemed to bring it into alliance with psoriasis. The treatment which I should adopt in a similar case would be to give cod-liver oil, an arsenical mixture, and a liberal diet.

Single patches of a very similar character are not infrequently seen on the cheeks of children. They run a very chronic course, spread at the circumference, become flattened, glazed, and depressed in the centre. They are of the nature of "lupus erythematosus." In such cases the constitutional treatment is the same, and locally I recommend the application of a lotion consisting of carbolic acid 1 part and glycerine 2 parts; occasionally the margin may be touched with pure carbolic acid.

The *syphilitic* rashes of infancy are chiefly erythema, psoriasis, and mucous tubercle. These are usually combined with coryza and a general cachectic condition. The plan of treatment which I prefer is to give grain doses of grey powder twice or thrice daily, with the syrup of the iodide of iron. Chlorate of potash without mercury will sometimes do good, but is not so effectual as mercury. It must be borne in mind that congenital syphilis is contagious; the mucous membrane of the mouth may infect the nipple of a nurse, and mucous tubercle may convey the disease to other children. Cases of this kind have come under my notice.

EPILEPSY AND CONVULSIONS.

EPILEPSY not unfrequently comes under our care at this hospital, and presents very many of the same aspects amongst children as amongst adults. The younger the child and the more frequent the attacks, as a general rule, the more seriously is the intellectual power of the child impaired.

The disease is sometimes only one feature of general deficiency of cerebral development; such cases are occasionally brought to the hospital, but are not retained, because they are only fitted for an idiot asylum.

Other cases date back to the period of dentition; some are ascribed to fright, some to blows on the head, some appear to be hereditary, some are due to a syphilitic taint, others cannot be at all accounted for.

It is not my intention to enter at any length upon this subject, but simply to select a few cases from my note books to illustrate some of the features of the disease, as they are presented to us here.

The following cases are good illustrations of the benefit sometimes derived from the administration of the bromides in large doses:—

Ann Barnes, æt. 5 years, enjoyed good health until last October, when she began to have attacks of giddiness, lasting not more than a few seconds. The attacks have become of longer duration, lasting 3 or 4 minutes. She becomes quite unconscious, and if standing would fall, but that she has warning of an imminent attack. The warning appears to be a sort of vertigo. During the attacks she becomes rigid in her limbs, not dark in her face, nor foaming at the mouth. She now (June, 1864) has from 8 to 10 attacks in 24 hours. After the fits she is drowsy. Is now never a day without a fit. Has been under medical treatment without benefit.

June 4th. From noon yesterday till 10 p.m. had 7 fits; from 10 p.m. till 10 a.m. this morning about 12 fits. Just seen in a fit by Dr. Gee. She was sitting in bed, when suddenly her head fell back against the top rail of the bed. The left limbs were rigidly extended and not moved; right hand grasped the rail of bed; right leg moved about slowly. There was powerful opisthotonos of head, the face looking over the left shoulder. The features of the face were drawn to the left, the muscles of that side twitched a good deal; both eyelids twitched continually; the eyeballs were drawn up and to the left; the left pupil was decidedly larger than right. Face very little congested; no foaming; no cry or biting of tongue. The fit lasted from 30 to 45 seconds; a drowsy condition followed for the next minute, and she then sat up. Ten minutes later she looked vacant. Pulse 136, small and regular. No loss of power in limbs; no difference between the two sides. With the exception of an aperient powder, no drugs were given for 6 days, to see whether the change of regimen in the hospital would at all reduce the frequency of the fits. She continued to have from 10 to 20 fits in each 24 hours.

June 9th. She was ordered 5 gr. of bromide of potassium every second hour.

10th. Had 30 fits in 24 hours.
11th. Twenty-four fits.
13th. Twenty fits.
14th. Eighteen fits.
15th. Seventeen fits.
16th. Sixteen fits. The dose of bromide to be increased to 7 gr.
17th. Nineteen fits.
18th. Thirteen fits.
19th. Seven fits. Pulse frequent, small and regular. Right leg a little fuller and firmer than left.
20th. Two fits.
21st, 22nd, and *23rd.* One fit each day.
24th. Two fits.
25th. One fit.
26th to *29th.* No fits. On the evening of 28th observed to have a "cold."
29th. Sneezed much; was sleepy and ill-tempered. The next day a measles rash appeared. The attack was very mild. She remained in hospital till 19th July; during her stay *had no fit after the 25th June.*

In January, 1865, I heard from her mother that she had no fit till Christmas, when, after plum pudding and mince pies, she had one fit, but has had no recurrence of the attacks.

In the next case the improvement of the general condition and the arrest of the fits for some months were very remarkable. On admission she had fits almost every two hours, and was quite paralysed.

Small doses of the bromides do not appear to have the desired effect.

M. A. Margett, æt. 6, has had fits for 3 years and a half. When they first occurred they were not very frequent, but have gradually increased in frequency, until she now sometimes has as many as 14 in an hour; each fit lasts from a minute to 2 minutes and a half. She was admitted on 21st December, complaining of headache. She was not sick, her bowels were confined ; could not feed herself, and could not stand. Between 7 and 8 p.m., on the 21st, she had 12 fits.

Dec. 22nd. Tolerably well nourished, lively, with some colour in her cheeks. No deviation of face. Pupils equal, no squint. Pulse 120, small and weak. Tongue furred, protruded straight. Bowels regular. *She cannot raise herself in bed, or sit up when raised; her head falls backwards or forwards.* She can lay hold of an object with her right hand, making the kind of studied but deviating movements which are seen in chorea. The left upper limb she cannot raise from the bed ; there is no rigidity of either arm. The legs are very weak ; she cannot stand ; can hold her legs in any position in which they are placed, but cannot raise them from the bed. There is more power in the right leg than in the left. No rigidity. Sensibility perfect. She has had several fits (undoubtedly epileptic) since admission ; they begin with rigidity followed by convulsions, and last nearly 2 minutes. To take one calomel and jalap powder, and to take 10 gr. of bromide of ammonium in water every 2 hours night and day.

23rd. In 12 hours had 29 fits.

24th. In 12 hours had 16 fits.

25th. In 12 hours had 5 fits.

Jan. 2nd. Has had no fits since 9 p.m. on December 25th. She is to take the bromide now every 6 hours, and take a drachm of cod-liver oil 3 times a day.

6th. Omit all medicine.

10th. Nitro-muriatic acid mixture every 6 hours.

16th. During the past fortnight there has been a great and rapid improvement; has had no fits. Her present state is as follows :—Looks well, is rather dull, but at times laughs heartily. Bowels slightly costive. Pulse 108, regular. She sits up in bed all day long. She can stand pretty firmly, if she has hold of any object. She can move her legs freely, though the movements are rather disorderly. She can grasp firmly and equally with both hands. To take one drachm of cod-liver oil 3 times a day.

25th. For 2 or 3 days has been dressed. Walking power not good. Intellect brighter. A slight tendency to strabismus.

Feb. 8th. Except slight hesitation in her walk she is apparently quite well. Has had no more fits. She had no fits for several months, and I hoped that she was cured. The benefit derived from very large doses of the bromide without any obvious ill effects was very striking. In September she came again under my care, having now had fits for 6 weeks about once a day. She is not paralysed. The fits were again arrested by the bromide of potassium in 5 gr. doses 6 times daily. In March, 1866, I note that the fits are again returning once a day. The bromide now appears to have lost its efficacy. I am giving her cod-liver oil and oxide of zinc without any marked result.

There can be no question that in some cases of epilepsy the bromides (especially bromide of potassium) exercise a most remarkably beneficial effect. Unfortunately there are other cases in which they fail to be of any service; and in some cases they do good for a while, and then lose their effect, as is so often the case with other remedies in this disease.*

I have only seen one case of acne from its employment in children. In this case large doses had not been given.

In the following case there was incomplete hemiplegia with epilepsy. The fits were here also arrested whilst under treatment.

Ada Madam, a girl, aged 3½ years, came under my care in August, 1865. In the winter of 1863 she had a fit which was followed by paralysis of the right arm and leg. She then had no fit for 12 months; but since last Christmas the fits have become very frequent; as many as 30 occurred on the 17th August. From August 19th to 24th had about 15 fits. She was then ordered to take 5 gr. of the bromide of potassium every 2 hours. The fits ceased on the day on which she began this treatment, and did not return again until 4th October, when she was discharged. The hemiplegic symptoms continued as before. She limped on her right leg; the face was very slightly drawn to left. Her right arm was drawn up against the side, and flexed at the elbow, her hand flexed at the wrist, thumb drawn in on palm. These contractions are due to muscular shortening, for when the wrist is more flexed by an observer, the fingers become limp. There is much wasting of the muscles of the right arm but none of the leg.

---

* Since writing the above, the second volume of "Reynolds' System of Medicine" has been published, in which the editor speaks of the value of bromide of potassium in epilepsy. He recommends large doses, long continued.

Eclampsia Nutans or Salaam Convulsions.

In the following case a child with a distinct history of syphilitic taint exhibited a peculiar nodding movement of the head from the age of four months, with want of intellectual power. After these movements had existed for 12 months, they were accompanied with screaming, and sometimes with lividity and grating of his teeth.

Unfortunately the patient was lost sight of after he had been under observation only nine days. It is probable that he would have been much benefited by the use of iodide or bromide of potassium, but his cerebral development was no doubt much injured, and the fontanelles having closed early, its growth was to some extent arrested.

Henry Smith, aged 2 years, was admitted under my care on the 17th March, 1862. His father and mother had syphilis about 6 years ago. Since then the mother has had a seven-months' still-born child, and an eight-months' child, which survived its birth only a few minutes, and two abortions. When 4 months old Henry had abscesses in his neck and spots on his body ; and about the same time he was observed to move his head and body backwards and forwards in a strange manner. It was at first taken as a sign that he was noticing people ; but the movement increased so much in degree and frequency that this idea was abandoned. At the same time the skin eruption disappeared, the child's appetite was good, and there was no loss of flesh. His intellectual power did not increase, and he seemed not to care more for his mother than for a stranger. Since the age of 18 months, the automatic movements have become much more frequent, being noticed on an average 8 or 10 times in a day. They are almost always observed when he awakes from sleep, even if but a short doze. He has latterly screamed and become black in the face, and has grated his teeth when the nodding comes on.

March 18*th.* The following notes of his condition were taken :—A well-nourished child, with small head, tolerably full face, light eyes and hair. He has a vacant look and takes no notice of any one. His lips have a pale fissured look, but are not actually cracked. His limbs and chest are well formed without trace of rickets. He is decidedly small for his age. Tongue moist, slightly white. Pulse 120, of moderate volume and power. On waking out of sleep, the child sits up, looks more vacant than usual ; and then suddenly and at short intervals bows his body and head forwards so that his forehead touches the bed, and as suddenly raises it. At the same time his elbows are moved outwards, his fore-arms flexed on arms and his fingers contracted, with thumbs drawn across his palms; whilst the backs of his hands are rubbed against his nose and eyes. His bowels have been open once since admission. He passed a quiet night, but this morning has already had 3 or 4 attacks such as that described. He has not become livid or screamed in any of the attacks.

21st. During the past 3 days he has had each day 8 or 9 fits of bowing; they have been always most marked in character when he has been waking from sleep. Soon afterwards the parents took him away and he has not been heard of since.

A number of cases of this kind are on record. Their tendency is to pass into ordinary epilepsy.

In the next case, peculiar rotatory movements occurred paroxysmally; after a time the tendency to rotation was less marked, but the child fell down. She was cured apparently by extract of belladonna and sulphate of zinc, and a short sojourn in the hospital.

Marian Bradford, æt. 4 years and 6 months, was admitted on the 31st March, 1862. Her mother's family is consumptive; she was one of twins; the other twin died at the age of 3 months with diarrhœa. Within the last 2 years she has had measles, scarlatina, and hooping-cough. The last complaint she had 8 months ago; she was very ill at the time, and made a slow recovery. She was apparently getting on well until 6 weeks since, when her mother noticed one morning that the child suddenly turned rapidly round from right to left about 6 times, as if in play. On being spoken to, immediately afterwards, she seemed dull, and not quite able to understand. During the same day she had three similar attacks whilst in the midst of her play. For the next 10 days these fits increased in frequency to about 12 or 14 in the day. During the nights she has also had attacks, in which her hands are clenched, her arms fixed, but with her legs she kicks violently. Just before the fits she bites her clothes in an excited manner. During the last fortnight the rotatory movements have ceased, but in the attacks the patient falls down, and usually on her back. She came as an out-patient on the 18th March, and was ordered to take the eighth of a gr. of extract of belladonna 3 times daily, which was increased to the sixth on the 21st. She was also to take a powder of scammony and calomel every other night for ascarides, from which she slightly suffers.

March 25th. She had an injection of lime-water administered which brought away no worms. She was to take with the belladonna 1½ gr. of sulphate of zinc. There was no improvement until the 29th March when she had 1 attack in the day and 6 in the night. During the first 2 days after her admission to the hospital, on March 31st, she had no attack.

April 2nd. The following note was taken of her state :—The child is fat and healthy-looking, with well-formed head and chest. The ends of the long bones are rather large, but the bones are not bent. Tongue clean and moist. Pulse 120, of moderate strength and volume. Heart and lung signs healthy, except that in left supraspinous region the percussion note is duller than on the right, and both in- and ex-spiration are accompanied with sonorous creaking sounds. In the evening of this day she had one fit which lasted 3 or 4 minutes. Her legs moved violently in a convulsive manner, her arms were somewhat raised above her head, firmly and rigidly, with some jerking spasmodic movement. Immediately after the attack, the patient was quite sensible; she uttered no cry, nor was her face red or livid before or after the fit.

From this time she had no other fit whilst in the hospital, and I heard of her in February, 1868, as having quite recovered.

Convulsions of different kinds are, as is well known, much more frequent in infancy than in adult life.

Under 12 months of age about three-tenths of the deaths occur from diseases of the nervous system, and of these nearly three-quarters are ascribed to convulsions.

Under five years of age about one quarter of the deaths are from nervous disease, and of these more than half are referred to convulsions.

The causes of the greater frequency of convulsions in infancy are the active growth of the nervous system and the disproportionate development of the spinal cord to the brain, and the consequent want of controlling power in the latter; so that reflex movements of all kinds are very frequently excessive from slight causes.

Causes which produce rigors or delirium in an adult will often produce convulsions in an infant.

Whilst convulsions appear on the tables of mortality as a frequent cause of infantile mortality, yet they are comparatively seldom fatal *per se*. Convulsion is no doubt the prominent symptom in many fatal diseases, and often shortly precedes death; it is frequently, however, a symptom of much less serious disease. It must always be remembered that it is a symptom and not a disease.

When the disease of which it is the symptom is not known, we speak of the convulsions as essential; but we must not rest contented with this, but should endeavour to ascertain what organ of the body is primarily at fault, and the nature of the lesion from which it suffers.

The paroxysm itself will not give this information; convulsions of very different characters may be due to one disease, and very different internal diseases may give rise to convulsions of exactly the same form. All that the paroxysm teaches is, that some portion of the nervous system is weakened and its irritability increased. This weakening may depend on loss of blood, or on general mal-nutrition, which may be manifested by anæmia, by emaciation, by rickety malformation and imperfect growth of bone; or it may be inherited from an hysterical, highly nervous or epileptic mother, or a drunken father, or from tuberculous parents in whose children nerve tissue is often imperfectly developed and ill-nourished. A poison in the blood, such as opium or belladonna, the virus of small-pox, measles, or scarlatina, or the products of metamorphosis of tissue from arrested renal secretion may produce the same result.

The exciting cause of the attack may be undigested food, the pressure of growing teeth on branches of the fifth pair of nerves, or the presence of worms in the alimentary canal, or pain from almost any cause; in many cases no exciting cause can be traced.

Until recently nearly all the convulsions of infants were ascribed either to teething or to worms; it is very doubtful whether in a healthy child these causes can produce convulsions at all; in a predisposed subject they no doubt often excite them.

When called upon to treat a case of infantile convulsions, it is important, in order to ascertain the nature of the case, that we should learn whether there have been any premonitory symptoms; how the child has been fed lately, and what is the state of its nutrition; the character of its fæcal evacuations, whether worms have been present in them, whether they are unduly offensive, slimy, too dark, or too pale in colour, very hard, or too relaxed.

It will very often be found on careful inquiry that there have been previous signs of disorder, and very often slight indications of a convulsive tendency.

A very common symptom in infancy is what is often called by nurses and others, "inward convulsions." The muscles of the face are twitched as if smiling ; the eyes are often but half closed, and wink in a spasmodic manner. This condition may be aggravated ; the respiration appears somewhat obstructed, and the neighbourhood of the lips becomes livid. The child starts on the slightest noise, sometimes waking up with a shrill cry. These symptoms, during the first 18 months of life, often indicate slight gastro-intestinal disturbance, and lead to nothing more serious. A more aggravated state of things exists when the thumbs are flexed rigidly on the palms constantly or only during sleep ; when all the muscles of the face and the limbs become twitched; or when the child wakes up suddenly, its face becoming red or livid, its eyes retracted under the upper lids, and the face looks alarmed and anxious. Sometimes a shriek or cry succeeds.

These symptoms often lead on to a *fit of general convulsions*, in which consciousness is arrested; the trunk at first stiff, soon becomes rapidly jerked to and fro, the limbs are alternately rigid, and in rapid vibration ; they may be either flexed or extended; the muscles of the face also are involved ; the eyes are either fixed or more or less distorted, at other times in more or less rapid movement; the pupils are contracted or dilated ; the respiration hurried and irregular, and the skin bathed in perspiration. This condition may last from half a minute to eight or ten minutes ; or, at short intervals, the convulsions may return for several hours.

In other cases, the child may fall into a drowsy state, during which partial convulsions are observed.

The patient may now fall asleep, or remain in an apparently confused state for a short time, or rapidly return to full consciousness. In rarer cases the child may die in the fit.

The distinction between convulsions of children and epilepsy has been drawn much too abruptly. There is no real distinction between them. It has been shown by M. Herpin that a considerable proportion of cases of epilepsy (about one quarter of his cases) commenced during the first five years of life.

In some children a violent fit of convulsions occurs without any premonitory symptom, and such fits may recur frequently or at long intervals, or there may be no recurrence of the fit. This may be quite independent of ascertainable cerebral disease; when such disease is present the fit is often preceded by vomiting and constipation; the abdomen is often flat or retracted; the convulsions are more or less unsymmetrical, sometimes quite unnatural. If there is inflammation of the brain or meninges, the head is hot, the pupils are small, and there have been usually vomiting and intolerance of light and sound before the attack. If there is tubercle of the brain, there will very likely have been pain in the head and some sign of local paralysis.

Paralysis, and especially hemiplegia, may follow convulsions in children as in adults, even when there is no palpable disease in the brain. The so-called essential paralysis is but rarely ushered in by convulsions as mentioned in the chapter on paralysis.

Hemiplegia is more commonly observed when the convulsions affect one side of the body more than another; the hemiplegia may last only a few days, or it may be permanent. If rigidity sets in there is reason to fear that the paralysis will not pass off.

The following case may be taken as an illustration of unilateral convulsion and hemiplegia :—

Mary C. Shepherd, aged 3 years and 8 months, had an attack of scarlatina 4 months previously. Five weeks since, having well recovered from the scarlet fever, she became convulsed and continued so more or less for 8 hours. Only her right side was convulsed. On recovering, her right limbs were paralysed, her face drawn to the left, she could not speak, and her skin was jaundiced. After a fortnight she began to regain her speech, but cannot yet speak so well as before the fits. When admitted she was tolerably well nourished. Her teeth were carious. She had a discharge from right ear. The left angle of her mouth is slightly drawn up in laughing. She talks very indistinctly. Her right arm slightly resists flexion; when flexed she can extend it. Fore-arm pronated resists supination slightly; fingers flexed slightly resist extension. Has considerable power over muscles of right shoulder. Can move right lower limb freely in bed; drags it slightly when attempting to walk, which she cannot do without assistance. Her foot is drawn upwards and inwards. There is no wasting or obvious difference of temperature in the paralysed limbs.

Remarks.—Loss of speech, which was present in this case, is a sequela of convulsions more common in children than in adults. It is usually temporary, but it may be permanent. According to Dr. H. Jackson, "the association of defects of speech with symptoms pointing to disease of the left hemisphere, is not so striking as it is in adults, but it will be found that defects of speech are more likely to occur with convulsion and paralysis of the right than of the left side of the body." This case, so far as it goes, lends support to the latter part of the statement.

Mental defects often follow convulsions; they may be loss of memory, general indifference and apathy, or they may be chiefly emotional and moral, such as great spitefulness, excessively bad temper, want of affection, and loss of the sense of decency. The greater the frequency of the attacks, the more does the mind suffer; but in very young children a single attack may be followed by idiotcy, or an approach to it. In these cases we must regard the convulsions as a phenomenon of disease which previously existed in a latent form, and was very likely congenital.

Amaurosis does not often follow convulsions, unless there is disease of the brain. Deafness is an occasional, but a rare, sequel of convulsions. It is, perhaps, more frequent after convulsions occurring in the course of hooping cough than after other cases. Squinting is said to be more commonly a result of congenital defect in the eyeballs (hypermetropia), than of spasm or paralysis of the muscles, but is not rare after convulsions.

One of the most dangerous forms of convulsions in children

is spasm of the muscles of respiration, and it is especially apt to occur in rickety subjects. It is generally ascribed to teething, and is no doubt most frequent at the age when teeth are actively growing, and pushing through the gums, from six months to two years. This is a period of life at which the spinal nervous system is very active, and the pressure of the teeth is no doubt one cause of irritation; but it must not be supposed that it is the only, or even the most frequent cause. Spasm of the glottis generally comes on gradually, and usually in children whose digestive functions are out of order, and very frequently in children whose dentition is delayed and whose ossification is imperfect, from the presence of rickets. The first sign is the occurrence of a peculiar crowing sound, from narrowing of the glottis, which occurs after a slight delay in respiration, from spasm of the diaphragm and other respiratory muscles. It is more frequent on waking from sleep, sometimes when crying, or even when sucking. At first, the crowing is not loud, and occurs at long intervals; it subsequently becomes louder and more frequent. After a time graver symptoms appear; there is greater delay in respiration, the head is thrown back, the face and lips become livid, and there is some spasmodic twitching of the muscles of the face. After some seconds, respiration is recommenced with a crowing sound, and the child recovers its former condition. Another symptom which often accompanies these attacks is a rigid contraction of the hands and feet; it is sometimes confined to the thumbs or to the great toes, or is extended to all the fingers and toes. The upper limbs are usually attacked sooner than the lower; the feet are seldom contracted without the hands. This state is usually at first temporary, and is often independent of the attacks of crowing, though it may be aggravated by each attack. These carpo-pedal contractions are sometimes attended with swelling, tension, and shining of the hands and insteps. A case of this kind was under my care in the hospital a few weeks since. Only one attack of convulsions occurred in this case; the hands and feet remained flexed for a few days, and then regained their former condition, and the swelling subsided.

As the disease becomes more aggravated the crowing sound is less marked, the dyspnœa is greater, and general convulsions may ensue. The very slightest causes will bring on an attack, such as a sudden change of temperature, a draft of cold air, any mental excitement, or the act of swallowing; they are also very frequent during sleep. Death may occur during an attack.

The general state of health of a child with laryngismus stridulus is not often satisfactory; the discharges from the bowels are usually of an unhealthy character, dark coloured, and offensive, or pale and deficient in bile; they may be relaxed or costive. In a large majority of cases the child exhibits signs of rickets. Occasionally the mouth is hot, and the gums swollen and tender.

The treatment of spasm of the glottis and of other forms of convulsions is founded on the same principles. During the attacks it is of little use to do anything; the child should not be held, but left quietly on a mattress, care being taken that it has an ample supply of pure air, and that nothing tight is left on its neck, chest, or abdomen. A tepid bath, from 95° to 100°, which is the popular remedy, will do no harm, and it may do good. It has no doubt rather a general soothing effect on the peripheral nerves. The feet and legs may be put in hot water, or mustard plasters may be applied to them. When attacks occur frequently, and with great severity, the inhalation of chloroform may be used to ward off the attacks. It is difficult to obtain the benefit of this remedy, unless a trained person be constantly in attendance. It is too dangerous an agent to be entrusted to any one who is not medically educated. If used at all, it should be when an attack is impending. It has been recommended to introduce the finger into the throat and thereby induce efforts to vomit, which relax the respiratory muscles.

When the paroxysm is over, the diet must be especially inquired into. If the patient has recently eaten heartily, and especially if he has taken what was unsuitable, an emetic of ipecacuanha should be given. If the gums be swollen and tender, they should be lanced.

The usual diet of the child should be carefully ascertained, and strict rules laid down for the future, the nature of which will depend on the age of the child and the state of its digestive functions.

Farinaceous food should not be given until the child is eight or nine months old; breast milk should, if possible, be the chief diet; but if this be impracticable, cows' milk, sweetened with sugar of milk, and diluted with one-third of water or lime water, should be given; or asses' milk, if this can be obtained, and cows' milk do not suit. When older, some good juice of meat and beef-tea should be added. If the bowels be costive, castor oil should be given, or compound decoction of aloes or powdered aloes. If aperients by the mouth fail, an injection per anum may be used. If the bowels be relaxed and slimy, drachm doses of a mixture containing, in an ounce of caraway water, one drachm of castor oil, half a drachm of sugar and powdered gum, and four mimims of laudanum will be found very useful.

In obstinate constipation, very small doses of extract of belladonna, the $\frac{1}{24}$ or $\frac{1}{32}$ of a grain, will be found useful, and the abdomen may be rubbed with soap liniment and castor oil.

If worms be seen in the motions, an enema of lime water may be administered from time to time.

The main indication is to nourish the child; for this purpose the diet must be as nutritious and digestible as possible, and on no account must crude articles of diet, such as grocers' fruits, pickles, cheese, or salted meats, be given. Cod-liver oil will promote nutrition, and in a large number of cases is of service. Occasionally, however, it induces diarrhœa and cannot be continued.

Stimulants may sometimes be required, when the child is exhausted by the fits and cannot take nourishment from its feeble circulation and deficient digestive powers. The bromide of potassium should be given, if the attacks be frequent, in full doses, as mentioned before. Iron may be given as a tonic with advantage, either vinum ferri, vinum ferri citratis, or syrupus ferri phosphatis.

All hygienic measures must be enforced, baths, pure air in the nursery and bedrooms, and frequent walks, or drives in the open air.

A change of air to the sea side, or other dry bracing locality, is sometimes of great service.

Exposure of the head to the direct rays of the sun must be carefully avoided.

Paroxysmal attacks of an abnormal character are not infrequently observed in children; they are more nearly allied to hysteria than to epilepsy; they are oftener seen in girls than in boys, but they have no obvious connection with the uterine or ovarian functions.

A delicate girl, aged 8 years, was brought to the hospital with the following history:—

She is said to have had "fits" almost every day for 12 months. She used to have 2 or 3 in a day. Now has from 20 to 30. The fits are said sometimes to last an hour. During the so-called fits she thumps her forehead for about a minute. She looks conscious but says she is not so; afterwards has pain in her forehead. She was taken into the hospital and kept a fortnight, during which she had only 2 or 3 fits. There was nothing peculiar about the child's manner to raise the suspicion of malingering. There was no spinal tenderness or neuralgia at any time except after the fits. The only medicine given was one dose of calomel and jalap.

APPENDIX.

CASE REFERRED TO ON PAGE 15.

Broncho-Pneumonia running an acute course.—Recovery completed in 24 days.

Emily Thacker, æt. 6 years, was admitted into the Hospital for Sick Children on the 28th April. Her mother's father died of phthisis. Emily had suffered from hooping-cough and measles; her convalescence from the latter disease was very tedious. She was, however, in fair health, and went for a long walk on the evening of 21st April; she was put to bed very tired. The next day she was feverish, refused her food; and in the evening she began to cough and was more feverish.

April 23rd. A powder was given which caused sickness; but the child was no better, complained of pain at the epigastrium, had no appetite, and was thirsty; her skin was hot and dry. She continued in much the same state until her admission on the 28th April. She was then somewhat flushed, had a patch of herpes on her lower lip commencing to dry up. Had a frequent short cough. Pulse 140, of moderate volume, rather sharp but feeble. Respirations 60, with some action of the alæ nasi. Skin rather hot. Tongue coated on dorsum with yellowish brown fur.

Thorax. Anterior aspect: percussion sounds normal on both sides. From an inch below clavicle to base on right side is audible rather sharp medium-sized crepitation with sibilus, chiefly at the end of the inspiratory murmur, and to a less degree in expiration. On the left side from the upper border of 3rd rib downwards medium-sized crepitation and some crackling are heard, especially at the end of inspiration. Posterior aspect: on percussion there is dulness on the left side from the angle of scapula to base, and over this portion of the lung there is heard at the upper part fine sharp crepitation at the end of inspiration, and towards the base both inspiration and expiration are of a blowing character almost tubular. The vocal resonance is also much increased in the same region. Over the back of right lung there is audible abundant fine bubbling of a sharp character, both with in- and ex-spiration.

The urine was found to be free from albumen, and to contain chloride abundantly. A grain of calomel and one grain of Dover's Powder were given night and morning ; and a mustard plaster applied below the left scapula until permanent redness was produced.

30th. A little better. Skin warm and soft. Cough still very frequent but looser in character. Pulse 132, very weak. Respiration 60. Physical signs in chest much the same, except that the dulness at left base was rather less, and the respiration less tubular. To discontinue the powders.

    R Ammoniæ sesquicarbonatis gr. ij.
    Vini Ipecacuanhæ . . ɱiv.
    Pulveris acaciæ . . . gr. v.
    Pulveris sacchari . gr. v.
    Aquæ ad . . . . ·ʒiij. Ter die.
Wine 3 ounces in the 24 hours.

May 1st. Has had a restless night. Pulse 132. Respiration 36, with less action of alæ nasi. There are still moist sounds heard over right lung behind ; the signs of consolidation on the left side are much less marked. She has not yet recovered appetite.

5th. Still coughs a good deal. Tongue clean and moist. Appetite improving. Pulse 124, rather jerking. Respiration 32. Percussion signs are almost normal over the whole thorax. There is still heard at the right base sonorous rhonchus, with some sibilant sound. At the left base inspiration is soft, but of a higher pitch than normal, and expiration is still too loud and a little prolonged.

16th. Is now quite convalescent.

Remarks.—In this case we had distinct consolidation of the left lung accompanied by capillary bronchitis of the right side. Resolution occurred at the end of less than a fortnight, and the patient made a good recovery.

FORMULÆ FOR MEDICINES IN CHILDREN'S DISEASES.

Aperient Mixtures.

1. ℞ Potassæ Sulphatis, gr. 40 ;
   Syrupi Rhei, ʒss ;
   Aquæ Carui, ad ʒij. M.
   Dose, a tablespoonful for a child six years old.

2. ℞ Magnesiæ Sulphatis, ʒij ;
   Syrupi Sennæ, ʒss ;
   Aquæ Anethi, ad ʒij. M.
   Dose, two teaspoonfuls to a tablespoonful.

Saline Aperient.

3. ℞ Magnesiæ Sulphatis, ʒij ;
   Potassæ Sulphatis, ʒss ;
   Potassæ Nitratis, gr. xxiv ;
   Syrupi Limonum, ʒij ;
   Aquæ, ad ʒij. M.
   Dose, two teaspoonfuls to a tablespoonful.

4. ℞ Decocti Aloes Co., ʒij ;
   Extracti Glycyrrhizæ, ʒss. M.
   Dose, two teaspoonfuls to a tablespoonful.

For Tape-worm.

5. ℞ Olei filicis maris, ʒj ;
   Pulveris Tragacanthæ Co., ʒj ;
   Aquæ Cinnamomi, ad ʒss. M.
   To be taken with an equal quantity of warm milk.

APERIENT POWDERS.

Very gentle Aperient.

1. ℞ Pulveris Rhei ;
Sodæ Carbonatis Exsiccatæ, āā gr. iij ad vj. M.

Or, ℞ Pulveris Rhei Compositi (Ph. B.), gr. viij.

2. ℞ Pulveris Rhei, gr. ij ;
Hydrargyri c. Creta, gr. iv. M.

3. ℞ Hydrargyri Subchloridi, gr. j ;
Jalapæ Resinæ, gr. ij ;
Pulveris Zingiberis, gr. j. M.

4. ℞ Hydrargyri Subchloridi, gr. j ;
Pulveris Scammonii Co., gr. iv. M.

For a child two years old :

5. ℞ Pulv. Aloes ;
Hydr. c. Cretæ, āā gr. ij. M.

ASTRINGENTS.

1. ℞ Tincturæ Catechu, ℳ 40 ;
Misturæ Cretæ, ad ℨij. M.
Dose, two or three teaspoonfuls.

2. ℞ Extracti Hæmatoxyli, ʒij ;
Tincturæ Catechu, ʒiij.;
Syrupi, ʒij ;
Aquæ Cinnamomi, ad ℨiij. M.
Dose, for child two years old, two teaspoonfuls.

3. ℞ Plumbi Acetatis, gr. viij ;
Acidi Acetici Diluti, ℳ xij ;
Tincturæ Opii, ℳ viij ;
Mucilaginis Tragacanthæ, ʒij ;
Aquæ, ad ℨij. M.
Dose, two teaspoonfuls for a child two years old.

4. ℞ Acidi Gallici, gr. xij ;
Tinct. Cinnamomi Co., ℳ 80 ;
Tinct. Opii, ℳ viij ;
Aquæ Carui, ad ℨij. M.
Dose, two teaspoonfuls for a child two years old, with chronic diarrhœa and irritable stomach.

COUGH MIXTURES.

5. ℞ Pulv. Cretæ Aromatici (Ph. B.) gr. v ad xv.

6. ℞ Pulv. Cretæ Arom. c. Opio (Ph. B.), gr. v ad xv.

7. ℞ Bismuthi Carbonatis, gr. xx ;
Spir. Chloroformi, ℥ xxx ;
Mucilaginis ;
Syrupi, āā ʒj.   M.
Dose, two teaspoonfuls.

COUGH MIXTURES.

1. ℞ Pulv. Ipecacuanhæ, gr. viij ;
Pulv. Acaciæ ;
Pulv. Sacchar. Alb., āā gr. xij ;
Aquæ, ad ʒij.   M.
Dose, one or two teaspoonfuls.

2. ℞ Potassæ Citratis, gr. 40 ;
Vini Ipecacuanhæ, ℥ xxiv ;
Syrupi Tolutani, ʒij ;
Decocti Hordei, ad ʒij.   M.
Dose, one or two teaspoonfuls.

3. ℞ Vini Ipecacuanhæ, ℥ xxx ;
Tinct. Camph. Co., ℥ xxv ;
Mucilaginis Acaciæ, ʒss ;
Aquæ, ad ʒij.   M.
Dose, one or two teaspoonfuls.

4. ℞ Sodæ Bicarbon. gr. xvj ;
Spir. Æther. Nitrosi, ʒj ;
Tincturæ Opii, ℥ viij ;
Vini Ipecacuanhæ, ℥ xxxii ;
Syrupi, ʒij ;
Aquæ Anethi, ad ʒij.   M.
Dose, two teaspoonfuls for a child two years old.

5. ℞ Pulv. Ipecacuanhæ, gr. iv ;
Pulv. Acaciæ, gr. x ;
Oxymel Scillæ, ℥ 80 ;
Tincturæ Hyoscyami, ʒj ;
Mist. Amygdalæ, ad ʒij ;   M.
Dose, two teaspoonfuls.

Stimulant Expectorant.

6. ℞ Ammon. Sesquicarb., gr. viij ad xij ;
   Tincturæ Scillæ, ℥ xx ;
   Syrupi, ʒij ;
   Decocti Senegæ, ad ʒij.   M.
   Dose, two teaspoonfuls for a child three years old.

DIURETIC MIXTURES.

1. ℞ Potassii Iodidi, gr. viij ;
   Potassæ Nitratis, gr. xxxij ;
   Extracti Taraxaci, gr. xl ;
   Infusi Digitalis, ʒj ;
   Syrupi, ʒij ;
   Aquæ, ad ʒiv.   M.
   Dose, a tablespoonful for child six years old.

2. ℞ Potassæ Bitartr. gr. 60 ;
   Potassæ Nitratis, gr. 40 ;
   Spir. Juniperi Co. ʒij ;
   Syrupi, ʒss ;
   Decocti Scoparii, ad ʒiv.   M.
   Dose, one tablespoonful.

OLEAGINOUS MIXTURE.

℞ Olei Ricini, ʒij ;
Pulv. Acaciæ, ʒj ;
Tincturæ Opii, ℥ viij ;
Syrupi, ʒij ;
Aquæ Carui, ad ʒij.   M.
Dose, two teaspoonfuls for child six years old. Useful in dysenteric diarrhœa.

NITRO-MURIATIC ACID MIXTURE.

℞ Acidi Nitro-hydrochlorici dil., ℥ xx.
Spiritûs Chloroformi, ʒj ;
Infusi Aurantii, ad ʒj.   M.
Dose, two to four teaspoonfuls.

SALINES AND TONICS.

SALINE MIXTURES.

1. ℞ Potassæ Citratis, gr. 40;
    Syrupi Aurantii, ʒij;
    Aquæ, ad ʒij.   M.
    Dose, two teaspoonfuls.

2. ℞ Potassæ Chloratis, gr. 20;
    Potassæ Citratis, gr. 30;
    Syrupi Limonum, ʒij;
    Aquæ, ad ʒij.   M.
    Dose, two teaspoonfuls.

TONICS.

1. ℞ Liquor. Cinchonæ, ℳ 60;
    Syrupi Aurantii, ʒij;
    Aquæ, ad ʒij.   M.
    Dose, two teaspoonfuls.

2. ℞ Ferri et Quiniæ Citratis, gr. 20;
    Syrupi Limonum, ʒij;
    Aquæ, ad ʒij.   M.
    Dose, two teaspoonfuls.

Chalybeate Tonics.

Vinum Ferri, or Vinum Ferri Citratis.
Or, Syrupus Ferri Phosphatis of the British Pharmacopœia.

℞ Tincturæ Ferri Perchloridi, ℳ xxv.
    Aquæ, ad ʒij.   M.
    Dose, two teaspoonfuls.

Tonic and Alterative.

℞ Sodæ Bicarb. Exsicc., gr. xxiv;
    Extracti Taraxaci, gr. 30;
    Syrupi Aurantii, ʒij;
    Infusi Calumbæ, ad ʒij.   M.
    Dose, two teaspoonfuls.

℞ Acidi Nitro-hydrochlor. dil. ℳ xxiv;
    Syrupi Aurantii, ʒij;
    Infusi Calumbæ, ad ʒij.   M.
    Dose, two teaspoonfuls.

The lozenges introduced into the British Pharmacopœia are convenient for children; especially the trochiscus ipecacuanhæ, containing a quarter of a grain of ipecacuanha; the trochiscus ferri redacti, containing one grain of iron; and the trochiscus bismuthi, which contains two grains of subnitrate of bismuth. The trochiscus morphiæ et ipecacuanhæ, containing the thirty-sixth of a grain of morphia and the twelfth of a grain of ipecacuanha, if used at all, must be very carefully regulated; and is too strong for young children.

INDEX.

A.

Abscess in skull from caries of sphenoid bone, 201
Accoucheurs ought not to attend cases of scarlatina, 315
Adams, Mr. William, on paralysis, 251
Ægophony, in pleurisy, 50
Age of mothers of rickety children, 82
Ages of rickety children, 83
Albuminoid disease in rickets, 95
Albuminuria in scarlatina, 297
  ,, after scarlatina, treatment, 318
Albuminuria, diphtheritic, 142
  ,, ,, how differing from scarlatinal, 142
Albuminuria in croup, 128
Alkaline lotions in eczema, 362
Alopecia areata, *see* Tinea decalvans, 357
Amaurosis after convulsions, 377
  ,, in cerebral tubercle, 180
  ,, ,, tumours, 184
  ,, after diphtheria, 136
Ammonia, carbonate of, in scarlatina, 315, 321
Amyloid disease of organs from empyema, 68, 72
Angina of fauces, diagnosis from diphtheria, 143
Antimony in chorea, 236
Aperient mixtures, 385
  ,, powders, 386
  ,, saline, 385
Aphasia in chorea, 242
Arachnoid hæmorrhage, 200
Arsenic in chorea, 236, 237, 238
  ,, in pemphigus, 346, 349
Ascites, chapter on, 277; cases of, 279—281; diagnosis of, 283; from cirrhosis, 284; from cardiac or renal disease, 277; from inflammation of hepatic veins, 277; from lymphatic obstruction, 281; from tubercular peritonitis, 277; prognosis of, 284; treatment of, 284; simulated by hydronephrosis, 283.
Astringent mixtures, 386
Atelektasis, 9
Atrophic paralysis, *see* Paralysis.
Atrophy progressive, muscular, 253
  ,, ,, ,, cases of, 255
Atrophy, after paralysis, 251

B.

Bastian, Dr., on degenerations of spinal chord, 261
Belladonna, to prevent scarlatina, 314
Blisters to be avoided, 7
Blood-letting discarded in pneumonia, 31
Bones in rickets, 84
  ,, growth of, 84, 85
  ,, chemical composition of, 84
Bowditch's syringe for thoracentesis, 76
Botrel, on chorea and rheumatism, 225
Brain, hypertrophy of, 96, 98
Brain, tubercle in, 179; anatomy of, 179; ages of patients, 180; case of, with recovery, 182; cases, 204—207. Diagnosis of, 190; from dropsy of ventricles, from epilepsy, 193; from hypertrophy of brain, 191; from intra-arachnoid hæmorrhage, 191; from inflammation of lining of ventricles, 192; from malformation, 191; from paralysis infantile, 193;

from rickety enlargement, 191; from tumours, 190. Duration of, 181 & 185; frequency of, 179; prognosis, 181; retrograde changes, 180 & 184; symptoms of, 180; absence of symptoms, 180.
Bright's disease, after empyema, 72
  ,,   ,, case of, 200
Bromide of potassium for epilepsy, 368—370
Bronchitis, increased fatality of, 9
  ,, in rickets, 96
  ,, after scarlatina, 307
Bronchial respiration in pleurisy, 49
Broncho-pneumonia, 9
  ,, cases of, 19, 383
  ,, in typhoid fever, 341
Buchanan, Dr., of Glasgow, cases of tracheotomy, 147

C.

Cancer of omentum, 112
  ,, of kidney, 121
Cancrum oris after scarlatina, 308
Canulæ used in tracheotomy, 149
  ,, removal of, 150
Carbolic acid in tapping chest for empyema, 77
Carbolic acid as disinfectant, 215
Cardiac murmurs of dynamic origin, 226
Cardiac murmurs of organic origin in chorea, 226
Cardiac valves affected in diphtheria, 143
Caries of bones, causes of, 203
  ,, sphenoidal bone, 201
Case of arachnoid hæmorrhage, 200
Cases of ascites, 277, 279, 280, 282
Cases of broncho-pneumonia, 19, 383
Case of cancer of kidney, 120
Cases of cerebral paralysis, 266, 267
Case of cerebro-spinal meningitis, 170
Case of caries of sphenoidal bone, 201
Case of congenital diaphragmatic hernia, 53
Cases of circumscribed empyema, 69, 70

Cases of diphtheria, 123, 126, 133, 138
Cases of chorea, 237, 238, 239, 240—244
Case of facial paralysis, 268
Cases of hydrocephalus non-tubercular, 155, 159, 192, 194, 195, 198
Case of hydrocephalus tubercular, 158; or tubercular meningitis, 161, 172—178
Cases of infantile paralysis, 246, 254, 270—276
Case of muscular atrophy, 255
  ,, neuralgic headache, 199
Cases of pleurisy, 55, 57, 58, 60, 63, 64
Case of pleuro-pneumonia, 33
Cases of pneumonia, 34, 35, 37, 38, 39, 40, 42
Cases of pemphigus, 346, 348, 349
  ,, purpura hæmorrhagica, 350, 351, 352
Cases of pyæmia, 213, 216, 217
  ,, pyogenic fever, 210, 211
  ,, rickets, 88; statistics of, 102
Cases of tubercle in brain, 182, 185, 204—207
Cases of tubercle in cerebellum, 188
  ,, tubercle of bronchial glands, 106; tubercular disease arrested, 108; of mesenteric glands, 109; tubercular peritonitis, 118, 119
Cases of scarlatina, 299, 300, 305, 309, 311, 312, 313, 320, 321, 322
Cases of typhoid fever, 325, 326, 331, 333, 337, 341
Cases of typhus fever, 343
Caustic to throat in diphtheria, 145
  ,, ,, scarlatina, 318
Cerebral congestion distinguished from tubercular meningitis, 169
Cerebral symptoms due to disease of other organs, 160
Cerebral tubercle, *see* Brain, tubercle in
Cerebellum, tubercle in, 188
  ,, ,, cases of 188, 190
Chalybeate tonics, 389
Chest, *see* Thorax
Chest, deformity of, in rickets, 92
Chloroform, use of, in tracheotomy, 148
Chorea, chapter on, 223; ages, 224; causes, 229; fear, season, and climate, 230; sex, 224; hereditary

INDEX.

transmission, 225. Diagnosis, from chorea major, 233; muscular tic, 233; nervous rhythmical affections, 233; duration of, 233; connexion with rheumatism, 225, 237, 239; in an epileptic subject, 238; causes of death in fatal cases, 242, 243; pathology, 228. Symptoms of, 230; premonitory, 230; accession, 231; impairment of sight, 232, paralysis, 232; effect of sleep, 232; effect of intercurrent acute diseases, 232. Treatment of, 234; arsenic, gymnastics, narcotics, purgatives, steel, shower baths, strychnia, 236
Clinical examination of children, 3
Coagulation of blood in cerebral sinuses, 218
Cod-liver oil in rickets, 101
Cold affusion in scarlatina, 316
Cold, application of, 32
Collapse of lungs in rickets, 94
Coma in tubercular meningitis, 165
Coma vigil ,, ,, 177
Congenital cases of rickets, 83
,, diaphragmatic hernia, 53
,, deficiency of brain, 373
Constipation, treatment by belladonna, 380
Convulsions, 373
,, great fatality of, 373
,, causes of frequency in children, 373
Convulsions, a symptom rather than disease, 373
Convulsions, sign of weakness of some part of nervous system, 374
Convulsions, variety of causes, 374
,, premonitory symptoms, 374
,, "inward," 375
,, general, 375
,, sequelæ of, 376
,, and epilepsy, often not separable, 376
Convulsions, treatment of, 380
,, in cerebral tubercle, 190
,, after scarlatina, treatment of, 319
Convulsions in tubercular meningitis, 165
Cord spinal, microscopic changes in, 261
Cornea, sloughing of, 203
Coryza in scarlatina, treatment, 317

Cough Mixtures, 387
Cough sometimes absent in pleurisy, 52
Croup epidemic, identical with diphtheria, 128
Croup, not distinguishable from laryngeal diphtheria, 127
Croup, case of, 127
,, not merely local, 128
,, relative mortality of, in boys and girls, 131
Croup, early accounts of, 130
Cruveilhier's atrophy, 253
,, see Atrophy, muscular
"Cynanche maligna," 297

D.

Deafness after convulsions, 377
,, in pneumonia, 40, 42
,, in typhoid fever, 330
Deformity of chest from pleurisy, 71
,, ,, rickets, 91-93
Delirium in pneumonia, 23, 37, 40
,, in tubercular meningitis, 165
Dentition as a cause of convulsions, 374
Dentition retarded in rickets, 91
Desquamation in scarlatina, 296
Diaphragm, irregular contraction of, 174
Diarrhœa in typhus, 344
,, in scarlatina, treatment of, 317
Diet in broncho-pneumonia, 18
,, lobar pneumonia, 32
,, rickets, 81, 100
Digitalis in pneumonia, 32
Diphtheria, cases of, 123, 126
,, insidious mode of attack, 123
,, without glandular swelling, 125
Diphtheria, its relation to croup, 125, 127
Diphtheria, contagion of, 134
Diphtheria, causes of death, 142
,, ,, asphyxia, 142
,, ,, virulence of general disease, 143
Diphtheria, sequelæ of, 143
,, diagnosis of, 143
,, ,, from catarrh, 143
,, ,, tonsillitis, 143
,, ,, scarlet fever, 144

Diphtheria, diagnosis from thrush, 144
,, ,, herpetic angina, 144
Diphtheria, treatment of, 144
,, ,, by caustic, 145
,, ,, hydrochloric acid, 145
Diphtheria, complicating scarlatina, 144
Diphtheria, following scarlatina, 306
,, fatal from asthenia, with renal symptoms, 138
Diphtheria, relative mortality of in boys and girls, 131
Diphtheria, proportion of cases with laryngeal symptoms, 131
Diphtheria laryngeal, rate of mortality, 131
Diphtheria laryngeal, in adult, 132
,, followed by cutaneous eruptions, 135
Diphtheria, nasal, 132
,, insidious, 133
,, on the skin, 134
,, ,, from contagion, 134
Diphtheria, inoculated, 134
,, nervous sequelæ of, 129
,, false membrane of, 130, 134
Diphtheria, false membrane, microscopical characters, 135
Diphtheria, false membrane, chemical analysis of, 135
Diphtheria, treatment of 145; calomel, 145; chlorate of potash, 146; carbolic acid, 147; Condy's fluid, 147; ice, 146; iodide of potassium, 146; ipecacuanha, 145; tartar emetic, 145; tracheotomy, 147; treatment after tracheotomy, 151; nervous sequelæ, 151; prognosis of, 152; recurrence in the same subject, 152
Diphtheria with scarlatina, 311—313
Diphtheritic paralysis, 129, 135
,, ,, more frequent in adults, 136
Diphtheritic paralysis, order of frequency in which different parts are affected, 136; of heart, 138; of pharynx, 137; of phrenic nerves, 141; of sensation, 137, 141; of special senses, 141; date of accession, 140; duration, 140
Diplopia in diphtheria, 141
,, in tubercular meningitis, 166
Displacement of heart in pleurisy, 48
Diuretic mixtures, 388
Drainage effect of bad, in scarlatina, 290
Dropsy after scarlatina, 304
Duchenne, Jun., on infantile paralysis, 249, 250
Dyspnœa in rickets, 97

E.

Ear, diseases of, 219
,, ,, not to be neglected, 221
,, ,, extension to brain, 221
,, ,, symptoms of, 222
Eczema, 360'
,, great frequency of, 360
,, capitis, 361
,, of nates, 361
,, treatment of, 361
,, constitutional characters, 361
Eczema chronic, treatment of, 362
,, impetiginodes, 363
Eclampsia nutans, 371
,, case of in syphilitic child, 371
Ecthyma, following diphtheria, 135
Empyema, cases of, 57, 58, 60
,, treatment of, 75
,, ,, by counter-opening, 76
Empyema after scarlatina, 306
,, from disease of bone and pyæmia, 68
Empyema, circumscribed, 68, 70
,, chronic, risk of tapping in, 62
Emphysema in rickets, 94, 97
Endocarditis in children, 226
,, not always rheumatic, 226
,, in scarlatina, 322
Enema, directions for administering, 111
Epilepsy and Convulsions, chapter on, 367
Epilepsy, effects on intellect, 367
,, from birth, 367

INDEX. 395

Epilepsy, hereditary, 367
,, syphilitic, 367
,, treatment by bromide of potassium, 367, 370
Epilepsy, treatment, cases of, 368, 369
Epilepsy with paralysis, 369, 376, 370
Epistaxis in typhoid fever, 329
Eruption, see Rash.
,, after diphtheria, 135

F.

Fall preceding meningitis, 159
Family constitution, influence in scarlatina, 290
Family history to be ascertained, 6
Faradisation, definition of, 249
Fatality of children's diseases, 1
Father health of, effect in causing rickets, 80, 82
Fatty degeneration of heart in diphtheria, 140
Favus, 352 ; cases of, 353 ; treatment of, 353 ; contagion of, 355
Fæcal fistula from tubercular disease, 113, 119
Fæcal accumulation distinguished from tubercular mesenteric glands, 111
Fistula, broncho-pleural, 55
Fistula in side leading to pleura, 57, 65
Flushing of face in tubercular meningitis, 163, 177
Formulæ, 385
Fractures of bone in rickets, 88
Friction sounds in pleurisy, 50

G.

Glands, lymphatic, in tubercle, 113
,, ,, in struma, 113
,, ,, albuminoid disease of, 113
Gölis on water-stroke, 169
Graves on scarlatina, 286
Growth arrested in rickets, 88
Gurgling in iliac fossa not peculiar to typhoid fever, 344
Gymnastics in chorea, 235
,, French system described, 235
Gymnastics set to music, 236

H.

Hæmaturia, intermittent, 122
,, in cancer of kidney, 122
Headache, neuralgic, 194
,, ,, case of, 199
,, ,, treated by quinine, 199
Headache in renal and arterial disease, 200
Heart, disease of in chorea, 227
Heart, fatty, in diphtheria, 140
,, white patches on, in rickets, 94
Heart, position of apex beat in rickets, 94
Hemiplegia after epilepsy, 370, 376
Hereditary predisposition to be ascertained if possible, 6
Herpetic angina, diagnosis from diphtheria, 144
Hilton, Mr., on absorption of contents of abscesses, 72
Hughes, statistics of chorea, 224, 227
Hydrocephalus, acute, 153; see Meningitis, tubercular
Hydrocephalus, case of non-tubercular, 155
Hydrocephalus, chronic, 185
,, of obscure nature with intense headaches, 194, 195
Hydrocephalus, congenital, 191
,, ,, case of, 192
Hydrocephalus, congenital, tapped, 192
Hydrocephalus, with wasting, case of, 199
Hydrocephalus, diagnosis from spasmodic contractions, 193
Hydrocephalus, diagnosis from neuralgic headaches, 194
Hydrocephalus, diagnosis from hydrocephaloid disease, 170
Hydrocephalus in rickets, 98
Hydronephrosis, congenital, 283
,, case of, tapped, 283
Hydropathic treatment in scarlatina, 316
Hydropericardium after scarlatina, 305
Hydrothorax, 305.

I.

Ice, for throat in scarlatina, 315
Inhalation of steam in scarlatina, 315
Inhalation of steam in pneumonia, 18
Iodide of potassium in chorea, 236
Ipecacuanha in broncho-pneumonia, 18
Iron in pneumonia, 33
Ischuria renalis in diphtheria, 139, 141

J.

Jackson, Dr. Hughlings, on loss of speech, 377
Jenner, Sir Wm., on albuminoid disease, 95
Jenner, Sir Wm., on the points of distinction between croup and diphtheria, 128
Jenner, Sir Wm., on the varieties of diphtheria, 132

K.

Kirkes on pathology of chorea, 227

L.

Laborde, M., on atrophic paralysis, 248, 250, 259
Laryngotomy in adult, 132
,, not advisable in children, 149
Laryngismus stridulus in rickets, 98
Lichen urticatus, 359
Liebig's food, how prepared, 100
Lister's plan of opening abscesses in empyema, 77
Lobular pneumonia, *see* pneumonia, 9
Loss of speech after convulsions, 377
,, ,, in chorea, 242
Lozenges, medicated, 390
Lupus-psoriasis, 364
,, case of, 365

M.

Maingault on diphtheritic paralysis, 146
Mastoid process, disease of, 216
,, ,, leading to abscess of brain, 216

Mechanical apparatus in rickets, 101
,, ,, in paralysis, 265
Medicines, choice of, 7
Meningeal granulations, nature of, 153
Meningitis after scarlatina, 305
,, tubercular, diagnosis from idiopathic, 169
Meningitis, cerebro-spinal, 169
,, ,, case of, 170
Meningitis, tubercular, 153
,, ,, anatomy of, 154
Meningitis, tubercular, without premonitory symptoms, 158
Meningitis simulating tubercular, 159
Meningitis, tubercular, without tubercle in lungs, 161
Meningitis, tubercular, in course of phthisis, 162
Meningitis, tubercular, with premonitory symptoms, 162
Meningitis, tubercular, symptoms of, 163
Meningitis, tubercular, pulse in, 163
Meningitis, tubercular, respiration in, 164
Meningitis, tubercular, vomiting in, 164
Meningitis, tubercular, retraction of abdomen, 165
Meningitis, tubercular, intellectual functions, 165
Meningitis, tubercular, prognosis of, 166
Meningitis, tubercular, recovery from, 166
Meningitis, tubercular, causes of, 166
Meningitis, tubercular, ages of patients, 167
Meningitis, tubercular, diagnosis, 167
Meningitis, tubercular, diagnosis from typhoid fever, 168
Meningitis, tubercular, diagnosis from cerebral congestion, 169
Meningitis, tubercular, diagnosis from pneumonia, 171
Meningitis, tubercular, treatment of, 171

Meningitis, tubercular, treatment of, prophylaxis, 172
Meningitis, tubercular, treatment of, ice, 172
Meningitis, tubercular, treatment of, mercurial aperients, 172
Meningitis, tubercular, counter-irritation, 172
Meningitis, tubercular, cases of, 172—178
Meningitis, tubercular, with capillary hæmorrhage, 175
Meningitis, tubercular, with diarrhœa, 177
Meningitis, tubercular, after scarlatina, 177
Meningitis, tubercular, after bite of a dog, 173
Mental defects after convulsions, 377
Mercurials, use of, 7
Mercurial ointment in eczema, 363
Mercury in diphtheria, 145
Meryon, Dr., on progressive muscular atrophy, 253
Metallic tinkling, 50, 57
Mollities ossium and rickets, 79
Molluscum contagiosum, 358
,,   ,, description of, 358
Molluscum contagiosum, treatment of, 358
Molluscum fibrosum, 359
Mother's health, effect in causing rickets, 80—82
Muscular fibres changed after typhoid fever, 336

N.

Narcotics in chorea, 236
Nitro-muriatic acid mixture, 388

O.

Ogle, Dr., on fatal cases of chorea, 227
Ophthalmoscopic examination in amaurosis, 184
Otorrhœa leading to disease of brain and lung, 217
Otorrhœa, varieties of, 219

Otorrhœa, after scarlatina, 308
,, treatment of, 319
Overcrowding, effect of in scarlatina, 290

P.

Paracentesis thoracis, 55
Paralysis, chapter on, 244
,, varieties of, 244
,, reflex, 244
,, "essential" or "infantile," 246
Paralysis, essential, two illustrative cases, 246
Paralysis, essential, symptoms of, 247
,, ,, premonitory, 247
,, ,, subjects of, healthy, 247
Paralysis, essential, as a sequela of acute disease, 247
Paralysis, essential, pain as a symptom, 248
Paralysis, essential, not affecting sphincters, 248
Paralysis, essential, not affecting sensibility, 248
Paralysis, infantile, not progressive, 248
Paralysis, infantile, electro-muscular contractility, 249
Paralysis, infantile, temporary, 249
,, ,, premonitory fever, 250
Paralysis, infantile, ages of patients, 250
Paralysis, infantile, parts affected in, 251
Paralysis, infantile, atrophy from, 251
Paralysis, infantile, deformity from, 257
Paralysis, infantile, causes of, 252
,, ,, diagnosis of, 252
Paralysis, atrophic infantile, a disease of spinal cord, 261, 262
Paralysis, atrophic infantile, anatomy of, 258
Paralysis, atrophic infantile, prognosis, 262
Paralysis, atrophic infantile, treatment of, 263
Paralysis, atrophic infantile, treatment of by baths, 263

Paralysis, atrophic infantile, treatment of by counter-irritation, 263
Paralysis, atrophic infantile, treatment of by strychnia, 263
Paralysis, atrophic infantile, treatment of by tenotomy, 264
Paralysis, atrophic infantile, treatment of by faradisation, 264
Paralysis, atrophic infantile, treatment of by galvanism, 265
Paralysis, atrophic infantile, treatment of by orthopœdic apparatus, 265
Paralysis of cerebral origin, cases of, 266—268
Paralysis, atrophic infantile, distinguished from progressive muscular atrophy, 253
Paralysis, atrophic infantile, distinguished from rickets, 257
Paralysis, atrophic infantile, distinguished from cerebral, 252
Paralysis, atrophic infantile, distinguished from myelitis, 252
Paralysis, atrophic infantile, distinguished from caries of spine, 252
Paralysis, atrophic infantile, distinguished from reflex paralysis, 252
Paralysis, atrophic infantile, tabular abstract of cases of, 270—276
Paralysis, diphtheritic, 129—135
,, facial, 268
,, ,, effect of Pulvermacher's chain, 268
Paralysis, facial, diagnosis from cerebral paralysis, 269
Paralysis, progressive, with enlargement of muscles, 253
Paralysis, progressive, with enlargement of muscles, case of, 254
Paralysis in tubercular meningitis, 166
Parotid bubo in scarlatina, 295, 309
Paroxysmal attacks anomalous, 381
Peculiar features presented by children's diseases, 2
Percussion of children, 5
Periostitis in chorea, 239
Pericarditis in scarlatina, 305—322
,, with pleurisy, 65
Peritonitis after scarlatina, 305
Phosphoric acid in urine of scarlatina, 296
Pleurisy; statistics, 45; symptoms, 46; urine, 47; pain in side,

46; pulse, 47; temperature, 46; headache, 47; delirium, 48; drowsiness, 48; physical signs, 48, 49. Diagnosis, 50—54; from pneumonia, 51; from collapsed lung, 51; from abdominal disease, 52; from cerebral disease, 52; from hydrothorax, 52; from pericarditis, 53; from rheumatic fever, 52; from diaphragmatic hernia, 53. Modes of termination, 70
Pleurisy, treatment of: by antiphlogistics, 72; by cold, 73; by mercury, 73; by salines, 73; by iodide of potassium, 73; iodide of iron, 73; by bandages of plaster, 73; by cod-liver oil, 73; by change of air, 73; by diet, 73; by paracentesis, 73
Pleurisy after scarlatina, 305
,, ,, treatment of, 319
Porrigo favosa, 353, 363
,, lupinosa, 353
Pneumonia, chapter on, 8—44; mortality, 8
Pneumonia, *lobar*, 20; ages of patients, 20; anatomy, 20; course of symptoms, 21; crisis, 22; cerebral pneumonia, 23, 40, 42; complications, 24; digestive functions, 27; pulse, 25; skin, 27; sputa, 26; temperature, 25; physical signs, 28; diagnosis, 29, 42, 44; prognosis, 30; treatment, 31
Pneumonia, *lobular*, disseminated and generalised, 10; anatomy of, 11; almost peculiar to childhood, 12; relation to tubercle, 13, 17; symptoms of, 13; physical signs of, 14; course of, 14; fatality, 15; diagnosis of, 16; prognosis of, 17; treatment of, 17
Pneumonia after scarlatina, 306
Prevention of scarlatina, 314
Prurigo, 359
Psoriasis, 363
,, treatment of, 364
Ptosis in tubercular meningitis, 166, 174
Ptosis in cerebral tubercle, 180
Pulmonary artery, pulsation of in pleurisy, 48

INDEX. 399

Pulmonary collapse, 9 ; anatomy of, 9
Purpura from albuminoid disease of small vessels, 352
Purpura with chronic abscess, 352
,, treatment by acetate of lead, 352
Purpura, treatment by sesquichloride of iron, 352
Purpura hæmorrhagica, 350
,, ,, cases of, 350, 351
Purpura hæmorrhagica, treatment by turpentine, 350
Pyæmia, article on, 208 ; and septicæmia, 208
Pyæmia, idiopathic, 209
,, chronic, 211
,, ,, cases of, 210, 211
,, with albuminuria, 212
,, after ozœna, 212
,, after otorrhœa, 212
,, ,, case of, 213
,, treatment of, 215
,, ,, by potash, 215
,, ,, sulphurous acid, 215
Pyæmia after scarlatina, 308
,, ,, treatment of, 319
Pyogenic fevers, 209
,, treated with quinine, 211

Q.

Quinine in pyæmia, 210, 211
,, in scarlatina, 318
,, in typhoid fever, 341
,, in periodical pain, 198

R.

Rash in pneumonia, 39
,, scarlatina, 292
,, absent in scarlatina, 300
,, typhoid, 330
Relapse of typhoid fever, 337
,, of scarlatina, 291
Remission a very marked character in fevers of children, 3
Remittent fever, 45, 57
Renal degeneration, case of, 200
Retention of urine in tubercular meningitis, 166

Reynolds, Dr., on chorea, 229
,, on bromide of potassium in epilepsy, 370
Rheumatism, connection with chorea, 225
Rheumatism often latent in childhood, 225
Rheumatism after scarlatina, 308
Richardson, Dr., on scarlatina, 287
Rickets, acute and chronic, 98
,, case of, 98
,, definition of, 78
,, frequency of, 78
,, ,, in hospital, 102
,, causes, 80
,, ,, health of father, 80, 82
Rickets, causes, health of mother, 80, 82
Rickets, causes, food, 81, 83
,, arrest of growth in, 88
,, fracture of bones, 88
,, course of symptoms, 89
,, effect on skull, 89
,, murmur in fontanelles, 90
,, dentition retarded, 91
,, effect on spinal column, 91
,, pathology of, 86
,, urine in, 87
,, catarrh and bronchitis, 96
,, modes of termination, 99
,, with tubercle, 119
,, treatment of, 100
,, ,, prophylactic, 100
Rickets, diet, 130
,, treatment of, attention to hygiene, 101
Rickets, treatment of, cod-liver oil, 101
Rickets, treatment of, lowering measures not admissible, 101
Rickets, mechanical appliances in, 101
Rickets, proportionate frequency of, to other diseases, 102
Ringer, Dr., on temperature in scarlatina, 293
Ringer, Dr., on temperature in tubercular disease, 115
Ringworm, see Tinea Tonsurans, 355
Roseola distinguished from scarlatina, 302
Roseola, epidemic, 303

INDEX.

Rotatory movements of epileptic character, 372
Rubeola notha, 303

S.

Salaam convulsions, 371
Saline mixtures, 389
Savory, Mr., on pyæmia, 208
Scabies, 359
Scarlatina, chapter on, 285 ; great fatality, 285 ; attack, 291 ; critical days in, 294 ; *diagnosis*, 301; from diphtheria, herpetic angina, measles, miliaria, roseola, tonsillitis and urticaria, 302 ; *symptoms*, 292 ; eruption, 292 ; throat, 294 ; nasal mucous membrane and tongue, 295 ; stomach, 295 ; temperature, 293 ; urine, 296
Scarlatina, desquamation in, 296 ; sequelæ, 297, 303 ; albuminuria, 304 ; bronchitis, 307 ; cancrum oris, 308 ; diphtheria, 144, 306 ; convulsions, 306 ; dropsy, 304 ; hydropericardium and hydrothorax, 305 ; otorrhœa, 308 ; pneumonia, 306 ; pyæmia, 308 ; rheumatism, 308 ; serous inflammations, 305
Scarlatina, incubation of, 287 ; recurrence in the same person, 291 ; relapses, 291 ; types of, 286 ; anginosa, 297 ; maligna, 298 ; without eruption, 300 ; virus, properties of, 288, 314
Scarlatina in tubercular disease, 109
Sée, M., on chorea, 227
Skin diseases, chapter on, 346
Small-pox co-existing with scarlatina, 308
Skull, shape of, in rickets, 89
Social position, influence on mortality of scarlatina, 290
Soil, effect of, on production of tubercle, 117
Sordes, diagnosis from diphtheria, 144
Spasm of glottis, 378
,, form of convulsion, 378
,, of muscles of respiration, 378
,, ,, treatment of, 379
Spinal cord, relative activity of growth in children, 2

Spleen in rickets, 95
Spontaneous opening of empyema, 71
Squinting after convulsions, 377
,, in tubercular meningitis, 166
,, in cerebral tubercle, 180
,, after diphtheria, 136
Squire, Mr., on distinction between croup and diphtheria, 129
Steam, for throat in scarlatina, 315
Stimulants in typhoid fever, 339
Strabismus, *see* Squinting
Strophulus pruriginosus, 359
,, treatment of, 360
Strumous glands, 113
Strychnia in chorea, 236
Sun, exposure to, cause of meningitis, 167
Surgical operations followed by scarlatina, 289
Syphilitic skin disease, 366
Syringing nostrils in scarlatina, 317

T.

Tabes mesenterica, 110
Tâche cérébrale, 164
,, diagnostic value of, 164
Talipes produced by paralysis, 257
Tape-worm, formula for, 385
Tapping of chest, *see* Thoracentesis
Tapping of head in hydrocephalus, 192
Tarry preparations in eczema, 362
Tavener's belt, use of in pleurisy, 76
Temperature, important in diagnosis, 6 ; modes of ascertaining it, 6
Thoracentesis, 55, 57
Thorax, shapes of, in tuberculosis, 105
Thrombosis in cerebral sinuses, 218
Thudichum, Dr., plan of syringing nares, 317
Thymus in rickets, 96
Tinea decalvans, 357
,, description of, 357
,, contagious, 357
,, treatment, 358
Tinea favosa, *see* Favus
Tinea tonsurans, 355
,, contagious, 355
,, in lower animals, 355
,, treatment of, 356

Tinea tonsurans, treatment by tar and iodine, 356
Tonics, 389
Tracheotomy in diphtheria, 147; directions for, 149
Tracheotomy, mortality of, 147; effects of, 150
Tracheotomy, causes of death, 148
,, use of chloroform, 148
Tubercle, rapid development of, in childhood, 103
Tubercle, distribution in different organs, 103
Tubercle, order of frequency in which different organs are attacked by, 103
Tubercle, of skin, 104
,, cases of, 106, 108, 118, 119
,, in bronchial glands, 106, 114
,, ,, recovery, 107
,, in mesenteric glands, 109, 111
Tubercle in mesenteric glands, case with recovery, 109
Tubercle of lymphatic glands, 113
,, ,, distinguished from other enlargements of glands, 113
Tubercle in omentum, 112
,, of peritoneum, 112
,, diagnosis of, 115
,, ,, use of thermometer, 115
Tubercle, frequency of on post mortem examination, 105
Tubercle, difficulty in detecting by physical signs, 105, 114
Tuberculosis, 103
,, characteristic features of, 104
Tuberculosis, thorax, in, 105
,, acute diagnosis of, from cerebral disease, 116
Tuberculosis, acute diagnosis of from typhoid fever, 116
Tuberculosis, treatment of, 117
,, prophylactic, 117
,, hygienic, 117
,, diet, 117; aperients, 117
,, hypophosphites, 118;
,, pancreatine, 118; cod-liver oil, 118
Tubercular meningitis, *see* Meningitis
Turpentine for purpura, 350

Typhoid fever, chapter on, 325; diagnosis from acute tuberculosis, 616; symptoms, 329; deafness, delirium, 330; epistaxis, 329; eruption, 330; headache, 330; stools, 329; vomiting, 329; sequelæ, bedsores, diphtheria, 332; debility, 331; ulceration of intestines, 331
Typhoid fever, protracted duration of, 336; changes in muscular fibres, 336. Treatment of, 339; of abdominal pain, 340; of bronchitis, 341; cerebral symptoms, 340; of diarrhœa, 340; diet, 336
Typhus fever, 343; mildness in children, 343; case of, 343; gurgling in iliac fossa, 344; mortality as compared with adults, 345; treatment of, 345

U.

Urea not always decreased in diphtheria, 141
Urine in diphtheria, 141
,, in pemphigus, 348
,, rickets, 87
,, in scarlatina, 296, 304
,, suppression of, in scarlatina, 304

V.

Valvular disease without murmur, 227
Vertigo in tubercular meningitis, 168
Vocal fremitus in pleurisy, 48
Vomiting, 164, 177

W.

Wade, Dr., on iodide of potassium in diphtheria, 146
Water-stroke, 169
Whytt's stages of tubercular meningitis, 161
Wilson, Dr. Charles, on croup, 130
West, Dr., on antimony in chorea, 236
West, Dr., on chorea and rheumatism, 225

Wet sheet in scarlatina, 316
Wine in scarlatina, 317
Worms in connection with convulsions, 374
Worms, treatment of, 380

Z.

Zenker on changes of muscles in typhoid fever, 336
Zinc in chorea, 236

THE END.

www.ingramcontent.com/pod-product-compliance
Lightning Source LLC
Chambersburg PA
CBHW050843300426
44111CB00010B/1112